Selected Titles in This Series

22 **Vlastimil Dlab and László Márki, Editors,** Trends in ring theory

21 **S. W. Drury and M. Ram Murty, Editors,** Harmonic analysis and number theory: Papers in honour of Carl S. Herz

20 **J. Borwein, P. Borwein, L. Jörgenson, and R. Corless, Editors,** Organic mathematics

19 **Raymundo Bautista, Roberto Martínez-Villa, and José Antonio de la Peña, Editors,** Representation theory of algebras and related topics

18 **Raymundo Bautista, Roberto Martínez-Villa, and José Antonio de la Peña, Editors,** Representation theory of algebras

17 **V. Kumar Murty, Editor,** Seminar on Fermat's Last Theorem

16 **Bruce N. Allison and Gerald H. Cliff, Editors,** Representations of groups

15 **Karl Dilcher, Editor,** Number theory

14 **Vlastimil Dlab and Helmut Lenzing, Editors,** Representations of algebras

13 **R. A. G. Seely, Editor,** Category theory 1991

12 **A. Nicas and W. F. Shadwick, Editors,** Differential geometry, global analysis, and topology

11 **H. Tachikawa and V. Dlab, Editors,** Representations of finite dimensional algebras

10 **A. Bialynicki-Birula, J. Carrell, P. Russell, and D. Snow, Editors,** Group actions and invariant theory

9 **Joel S. Feldman and Lon M. Rosen, Editors,** Mathematical quantum field theory and related topics

8 **F. V. Atkinson, W. F. Langford, and A. B. Mingarelli, Editors,** Oscillation, bifurcation and chaos

7 **H. Kisilevsky and J. Labute, Editors,** Number theory

6 **J. Carrell, A. V. Geramita, and P. Russell, Editors,** Proceedings of the 1984 Vancouver Conference in algebraic geometry

5 **D. J. Britten, F. W. Lemire, and R. V. Moody, Editors,** Lie algebras and related topics

4 **I. Hambleton and C. R. Riehm, Editors,** Quadratic and Hermitian forms

3 **Zeev Ditzian, Amram Meir, Sherman D. Riemenschneider, and Ambikeshwar Sharma, Editors,** Second Edmonton conference on approximation theory

2 **Richard M. Kane, Stanley O. Kochman, Paul S. Selick, and Vector P. Snaith, Editors,** Current trends in algebraic topology (Parts 1 and 2)

1 **Carl Herz and R. Rigelhof, Editors,** 1980 Seminar on harmonic analysis

TRENDS IN RING THEORY

Canadian Mathematical Society
Société mathématique du Canada

CONFERENCE PROCEEDINGS • Volume 22

TRENDS IN RING THEORY

Proceedings of a Conference
at Miskolc, Hungary
July 15–20, 1996

Vlastimil Dlab
László Márki
Editors

Published by the American Mathematical Society
for the Canadian Mathematical Society
PROVIDENCE, RHODE ISLAND

CMS CONFERENCE PROCEEDINGS
Series Editors
M. Ram Murty and Niky Kamran
Volume 22

Accepted for publication by the Canadian Mathematical Society
Published by the American Mathematical Society

Proceedings of the Conference on Ring Theory, July 15–20, 1996, Miskolc, Hungary.

1991 *Mathematics Subject Classification.* Primary 16–06.

Library of Congress Cataloging-in-Publication Data

Ring Theory Conference (1996 : Miskolc, Hungary)
Trends in ring theory : Ring Theory Conference, July 15–20, 1996, Miskolc, Hungary / Vlastimil Dlab and László Márki, editors.
p. cm. — (Conference proceedings / Canadian Mathematical Society, ISSN 0731-1036 ; v. 22)
Includes bibliographical references.
ISBN 0-8218-0849-4 (alk. paper)
1. Rings (Algebra)—Congresses. I. Dlab, Vlastimil. II. Márki, L. (László) III. Title. IV. Series: Conference proceedings (Canadian Mathematical Society) ; v. 22.
QA247.R56 1996
512′.4—dc21 97-30267
CIP

∞ The paper used in this book is acid-free and falls within the guidelines
established to ensure permanence and durability.
Visit the AMS home page at URL: http://www.ams.org/

10 9 8 7 6 5 4 3 2 1 01 00 99 98

Contents

Preface ix

List of invited lectures xi

List of contributed talks xiii

List of participants xxi

Filter Dimension and its Application
Vladimir V. Bavula 1

Automorphisms of Polynomial, Free and Generic Matrix Algebras
Vesselin Drensky 13

Matrix Problems and Representation-Theory
Peter Gabriel 27

Prime and Primitive Spectra of Multiparameter Quantum Affine Spaces
K. R. Goodearl and **E. S. Letzter** 39

PI-algebras and Nil Algebras of Bounded Index
Alexander Kemer 59

Representations of Finite Dimensional Algebras and Singularity Theory
Helmut Lenzing 71

Zassenhaus Conjectures for Infinite Groups
Zbigniew S. Marciniak and **Sudarshan K. Sehgal** 99

Graded Rings — an Approach via Semigroups of Matrices
Jan Okniński 113

Semiprimitivity of Group Algebras: Past Results and Recent Progress
Donald S. Passman 127

Asymptotics of Codimensions of some P.I. Algebras
Amitai Regev 159

Problems on Group Rings
Klaus W. Roggenkamp 173

Tame Module Categories of Finite Dimensional Algebras
Andrzej Skowroński 187

Direct Limits of Finite Dimensional Algebras and Finite Groups
Aleksandr E. Zalesskii 221

Preface

The aim of the Ring Theory Conference which took place at the University of Miskolc during July 15–20, 1996 was, on the one hand, to reflect some of the contemporary trends in the subject and, on the other hand, to make it possible for a large number of Eastern European algebraists to meet their colleagues from other parts of the world. We feel that both these goals were successfully accomplished.

The Conference, formally organized as a satellite conference of the Second European Congress of Mathematics in Budapest, met with a great success. The number of participants exceeded any expectations: the meeting was attended by 198 colleagues from all over the world. Among them, there was a very large group coming from Eastern Europe (beside 20 from Hungary, there were 21 participants from Russia and 39 from other countries of the former Soviet block in Europe), many people – especially many students – from Southern Europe (20 from Spain), and also a strong group from overseas (e.g. 18 from the United States).

The Scientific Committee consisting of I. Ágoston (Budapest), P. N. Ánh (Budapest), A. A. Bovdi (Debrecen), V. Dlab (Ottawa), A. Giambruno (Palermo), L. Márki (Budapest), G. Révész (Miskolc), C. Robson (Leeds), L. Rowen (Ramat Gan), J. Szigeti (Miskolc), R. Wiegandt (Budapest) prepared a programme including 19 invited addresses, the list of which appears below.

The present volume contains papers based on invited talks given at the Conference. Some of them are of a survey nature, others present new unpublished results. We wish to express our gratitude to the speakers who submitted their contributions to this volume, as well as to all referees for their generous assistance.

In addition to the invited lectures, there were 124 contributed talks, also listed below, some of which will appear in a special issue of the Journal of Pure and Applied Algebra.

We wish to thank the University of Miskolc for hosting the Conference, and especially Jenő Szigeti for the local arrangements which made our stay in Miskolc so pleasant. Thanks are due to the following organizations for financial support: the International Mathematical Union, the Hungarian Soros Foundation, the International Science Foundation, the National Council for Technical Development, the European Mathematical Society, the Hungarian Academy of Sciences, and the National Foundation for Scientific Research. It was their generous assistance which enabled us to host such a large number of colleagues from less affluent countries. We wish to extend our thanks also to István Ágoston for his assistance, advice and efficiency, without which the Conference could not have run so smoothly.

Vlastimil Dlab and László Márki
Ottawa–Budapest, June 1997

List of invited lectures

Bavula, Vladimir
Krull, Gelfand-Kirillov and filter dimensions of simple affine algebras

Crawley-Boevey, William
Noncommutative deformations of Kleinian singularities

Drensky, Vesselin Stoyanov
Automorphisms of generic matrix algebras and polynomial algebras

Formanek, Edward
Varieties of group representations

Gabriel, Peter
Matrix problems and representation theory

Goodearl, Kenneth R.
Prime spectra of quantum algebras

Kemer, Alexander R.
On the radical of relatively free algebra

Lenzing, Helmut
Representations of finite dimensional algebras and singularity theory

Okniński, Jan
Graded rings – an approach via semigroups of matrices

Passman, Donald S.
Semiprimitivity of group algebras: past results and recent progress

Razmyslov, Yuri P.
The structure theory of algebras with Capelli identities

Regev, Amitai
Asymptotic methods in P.I. algebras

Roggenkamp, Klaus W.
Isomorphisms and automorphisms of group rings

Sehgal, Sudarshan
Zassenhaus conjecture and infinite groups

Skowroński, Andrzej
Tame module categories of finite dimensional algebras

Small, Lance
Affine rings and their representations

Stafford, J. Toby
Non-commutative projective curves

Zalesskii, Aleksandr E.
Direct limits of finite groups and finite dimensional algebras

Zel'manov, Efim I.
On combinatorial ring theory

List of contributed talks

Albu, Toma
Dual Krull dimension and duality

Ánh, Phạm Ngọc
Selfdual modules and Dedekind rings

Ara, Pere
Extensions of exchange rings

Artamonov, Vyacheslav A.
Finitely generated modules over quantum polynomials

Artemovych, O.D.
Differentially trivial and rigid rings of finite rank

Bahturin, Yuri
Identities of graded algebras

Bekkert, Viktor
Schurian vector space categories of polynomial growth

Bokut', Leonid A.
Groebner-Shirshov bases for Lie algebras and (quantum) universal enveloping algebras

Booth, Geoffrey L.
Normal radicals and Γ*-rings*

Bovdi, A.A.
On the central units of a modular group algebra

Bovdi, Viktor
Symmetric units in group rings

Brešar, Matej
Functional identities

Calderón Martin, Antonio Jesus
Dual pair techniques in H^**-theories*

Carvalho, Paula Alexandra A. B.
Prime links in some skew-Laurent rings

Choi, Kyu-Hyuck
Representations and characters of symmetric groups

Clark, John
On polyform and projective $\sum$*-extending modules*

Coelho, Flavio Ulhoa
On the number of modules of infinite projective dimension

Dăscălescu, Sorin
Infinite Hopf–Galois extensions and Morita contexts

Dashdorj, Tserendorj
On minimal ideals of right alternative rings

Del Rio Mateos, Angel
Approximating rings with local units

Diaz, Maria Alicia Avino
On the endomorphism ring of a finite abelian p-group

Dokuchaev, Michael
On finite subgroups of units in integral group rings

Domokos, Mátyás
PI-algebras and invariant theory

Dubrovin, Nikolai I.
Matrix formal series over a left ordered group

Dung, Nguyen Viet
CS-modules with indecomposable decompositions

Erdei, László
Unitary units in modular group algebras of groups of order 16

Eslami, Esfandiar
Essential nilpotent ideals

Facchini, Alberto
Krull-Schmidt fails for serial modules

Farnsteiner, Rolf
Representations of restricted Lie algebras

Feldvoss, Jörg
Residual nilpotency of Hopf algebras

Ferrero, Miguel
Prime ideals in polynomial rings in several indeterminates

Friger, Michael
Analogue for Higman's Theorem for commutative torsion-free rings

Futorny, Vjacheslav
Stratified Lie algebra modules and corresponding module categories

Gateva-Ivanova, Tatiana
Regularity of skew polynomial rings with binomial relations

Geiß, Christof
On $\mathbf{ZA}_{\infty}^{\infty}$ *components for string algebras*

Gentle, Ronald S.
Some comments on TPA, torsion, purity and adjoints

Giambruno, Antonio
Group-graded algebras with polynomial identity

Gómez Pardo, José Luis
CS-rings with finite essential socle

Gómez Torrecillas, José
Locally finite injective modules

Gonçalves, Jairo
Construction of free subgroups in the group of units of modular group algebras

Greferath, Marcus
Projective geometry over rings

Grzeszczuk, Piotr
Invariants of skew derivations

Gudivok, Peter Mikhailovich
Representations of twisted group rings of finite groups over the ring of p*-adic integers*

Guil Asensio, Pedro
Essential embeddings of modules and the FGF problem

Hadas, Ofer
Indices of nillity and absolute nillity in matrix rings

Horváth, Erzsébet
(p, q)*-groups and their applications*

Ivanov, George
A generalization of self-injective von Neumann regular rings

Iwanaga, Yasuo
Duality over Auslander–Gorenstein rings

Jain, S.K.
On maximal quotient rings

Jespers, Eric
Unit groups of integral group rings and abelian centralizers

Jirásko, Josef
On a generalization of QI-rings

Kashu, Alexei I.
Torsion theories and composition of dualities in a Morita-context

Kimmerle, Wolfgang
Automorphisms of general Iwahori-Hecke algebras

Király, Bertalan
The residual Lie-nilpotence of the augmentation ideal

Kirichenko, Vladimir
Semi-perfect semi-distributive rings

Klein, A. A.
On central and noncentral zero divisors

Koryukin, Anatolii
Color superalgebras and Hopf algebras

Koshlukov, Plamen E.
Weak polynomial identities

Krause, Günter
Relative FBN rings and the second layer condition

Kurdics, János
Engel properties of group algebras

Liu, Shaoxue
On tensor products of algebras

Maltsev, Yuri N.
Varieties of rings and commutativity theorems

Marciniak, Zbigniew
Constructing free subgroups of group ring units

Marcus, Andrei
Graded equivalences and their invariants

Marin, Leandro
Categories of modules for associative rings

Markov, Viktor T.
Krull and Gabriel dimension of modules over PI-rings and algebras

Martindale, 3rd, Wallace S.
n-Jordan maps onto prime rings

Martinez Lopez, Consuelo
Jordan algebras of GK dimension 1

Mathieu, Martin
Representation theorems for some additive mappings on semiprime rings

Mazorchuk, Vladimir
Weight modules over generalized Witt algebras

Mazurek, Ryszard
Examples of chain domains

McConnell, John C.
Wedderburn's proof of Wedderburn's 1914 *theorem*

McDougall, Robert
On pseudocomplements and complements in the lattice of all radical classes

Meltzer, Hagen
Exceptional modules, exceptional sequences and tilting complexes for canonical algebras

Mihovski, Stoyl Vassilev
Automorphisms of crossed products of up-groups

Mikhalev, Alexander A.
Primitive elements of free algebras

Mikhalev, Alexander V.
Isomorphisms and anti-isomorphisms of endomorphism rings of modules

Molnár, Lajos
Algebraic differences among group rings of compact infinite groups: the nonexistence of a surjective ring homomorphism from $L^p(G)$ onto $L^q(G')$

Musson, Ian M.
The enveloping algebra of the Lie superalgebra $\mathrm{osp}(1,2r)$

Olivier, Werner A.
Regularities in rings with involution

Paison, Olga
On finite separability in associative ring varieties

Pardo, Enrique
Metric completions of ordered groups and K_0 of exchange rings

Petechuk, Vasil
Sylow p-subgroups of the general linear group over fields

Petrogradsky, Victor M.
Growth in polynilpotent Lie algebras

Puczyłowski, Edmund
Rings which are sums of two subrings

Raphael, Robert
On the epimorphic hull of a ring of continuous functions

Rashkova, Tsetska Grigorova
$*$-Identities of minimal degree in matrix algebras of low order

Riley, David M.
Algebras whose group of units satisfy an identity

Rizvi, Tariq
Countably σ-CS rings with finite uniform dimension

Roomeldi, Raul
On simple and prime $(-1,1)$ superalgebras

Rosenbaum, Kurt
Involutions in group algebras of metacyclic groups

Rowen, Louis Halle
Weakly Azumaya algebras

Salwa, Arkadiusz
Representing idempotents as a sum of two nilpotents — an approach via matrices over division rings

Sands, A.D.
Radicals coinciding with the von Neumann regular radical on Artinian rings

Santa-Clara, Catarina
Direct sums of extending modules

Sapouzhak, Taras
Evolution operator of p-adic harmonic oscillator

Schmidt, Stefan E.
Affine geometry over rings

Seidel, Alexander
A note on parameter-conditions in quantum-polynomial-algebras and an application

Siles-Molina, Mercedes
Orders in associative pairs

Snashall, Nicole
On Hochschild cohomology of preprojective algebras of type A_n

Sommerhäuser, Yorck
Hopf algebras with triangular decomposition

Ştefănescu, Mirela
On some near-rings on groups

Stefanov, Dimitar Georgiev
Automorphisms of free nilpotent associative algebras

Stolin, Alekhandre
Frobenius algebras and YBE

Sviridova, Irina Yu.
Varieties and algebraic algebras of bounded degree

Szakács, Attila
On the φ*-unitary subgroup of the group of units in a finite commutative group algebra*

Szigeti, Jenő
Identities for $n \times n$ *matrices over the Grassmann algebra*

Tasić, Vladimir
Jennings' theorem and Chen's iterated line integrals

Tchebotar, M. A.
On Lie isomorphisms in prime rings with involution

Tsigantshev, Dimitar Georgiev
New polynomial identities for 2×2 *generic matrices in characteristic* 2

Ufnarovski, Victor
Possible and impossible in Gröbner basis calculations

Vámos, Péter
Rings whose projectives have finite representation type

Van den Berg, John
When multiplication of topologizing filters is commutative?

Van Wyk, Leon
On P. Fuchs' characterization of matrix rings

Varadarajan, Kalathoor
Quasi-projective modules with strongly pi-regular endomorphism rings

Watters, John F.
V-rings and regular rings

Wilson, Mark
Bell's criterion for simple Lie superalgebras

Wisbauer, Robert
Spectral torsion theories in module categories

Xi, Changchang
Twisted doubles of algebras

Yakovlev, Anatolii
Homological definability of modules over group rings

Yousif, Mohamed F.
Continuity, perfect dualities and Nakayama permutations

Zaicev, M.V.
On regular varieties of linear algebras

Zand, Hossien
On homogenity of radicals of graded rings

Zolotykh, Andrej A.
Varieties of associative algebras generated by ε-commutative polynomials

List of participants

1. **Ágoston, István,** Mathematical Institute, Hungarian Academy of Sciences, P.O.Box 127, H-1364 Budapest, Hungary
E-mail: agoston@math-inst.hu

2. **Albu, Toma,** Department of Mathematics, University of Bucharest, P.O.Box 2–255, R–78215 Bucuresti, Romania
E-mail: talbu@roimar.imar.ro

3. **Ánh, Phạm Ngọc,** Mathematical Institute, Hungarian Academy of Sciences, P.O.Box 127, H–1364 Budapest, Hungary
E-mail: anh@math-inst.hu

4. **Antunes Simoes, Maria Elisa,** Departamento de Matematica, Faculdade de Ciencias da Universidade de Lisboa, Rua Ernesto de Vasconcelos, bloco C1 – piso 3, P–1700 Lisboa, Portugal
E-mail: mesimoes@cc.fc.ul.pt

5. **Ara, Pere,** Departament de Matemàtiques, Edifici CC, Universitat Autònoma de Barcelona, E–08193 Bellaterra, Spain
E-mail: para@mat.uab.es

6. **Arroyo, Maria Jose,** Departamento de Matematicas, UAM–Iztapalapa, Iztapalapa, CP 09340, Mexico
E-mail: mja@xanum.uam.mx

7. **Artamonov, Vyacheslav A.,** Moscow State University, Department of Mech–math., 119899 Moscow, Russia
E-mail: artamon@nw.math.msu.su

8. **Artemovych, Orest D.,** Department of Mechanics and Mathematics, Lviv State University I. Franko, Universytetska 1, 290602 Lviv, Ukraine
E-mail: mzar@litech.lviv.ua

9. **Baccella, Giuseppe,** Dipartimento di Matematica pura ed Applicata, Università di l'Aquila, I–67100 L'Aquila, Italy
E-mail: baccella@axscaq.aquila.infn.it

10. **Bahturin, Yuri,** Moscow State University, Department of Mech–math., 119899 Moscow, Russia
E-mail: bahturin@nw.math.msu.su

11. **Bartolone, Claudio G.,** Dipartimento di Matematica, Università di Palermo, Via Archirafi 34, I–90123 Palermo, Italy
E-mail: bartolone@ipamat.math.unipa.it

12. **Bavula, Vladimir,** Department of Mathematics, Kiev University, Vladimirskaya 64, 252617 Kiev, Ukraine
E-mail: kinr4%rigkbp@mhsgw.grs.de

13. **Beidar, Konstantin I.,** Department of Mathematics, National Cheng Kung University, Tainan, 70101, Taiwan, ROC
E-mail: t14270@sparc2.cc.ncku.edu.tw

14. **Bekkert, Viktor,** Department of Mathematics, Kiev University, Vladimirskaya 64, 252617 Kiev, Ukraine
E-mail: bekert@uni-alg.kiev.ua

15. **Benanti, Francesca,** Dipartimento di Matematica, Università di Palermo, Via Archirafi 39, I–90123 Palermo, Italy
E-mail: fbenanti@ipamat.math.unipa.it

16. **Bican, Ladislav,** Department of Algebra, Charles University, Sokolovská 83, CZ–18600 Praha, Czech Republic
E-mail: bican@karlin.mff.cuni.cz

17. **Bokut', Leonid A.,** Institute of Mathematics, Siberian Branch of the Russian Academy of Sciences, Universitetskii pr. 4, 630090 Novosibirsk, Russia
E-mail: bokut@math.nsk.su

18. **Booth, Geoffrey L.,** Department of Mathematics, University of Transkei, Private Bag X1, Unitra, Umtata, Eastern cape 5100, South Africa
E-mail: booth@getafix.utr.ac.za

19. **Bovdi, A.A. (Béla),** Department of Mathematics, Kossuth Lajos University, Pf. 12, H–4010 Debrecen, Hungary
E-mail: bodibela@tigris.klte.hu

20. **Bovdi, Viktor,** Department of Mathematics, Bessenyei College, Sóstói u. 31/b, H–4400 Nyíregyháza, Hungary
E-mail: vbovdi@math.klte.hu

21. **Brešar, Matej,** Department of Mathematics, University of Maribor, Pf. Koroška 160, SLO–2000 Maribor, Slovenia
E-mail: bresar@uni-mb.si

22. **Calderón Martin, Antonio Jesus,** Departamento de Algebra, Geometria y Topologia, Universidad de Málaga, Campus de Teatinos, Apdo. 59., E–29080 Málaga, Spain

23. **Carvalho, Paula Alexandra A. B.,** Department of Mathematics, University of Glasgow, University Gardens, Glasgow G12 8QW, United Kingdom
E-mail: pc@maths.gla.ac.uk

24. **Cedo, Ferran,** Departement de Matemàtiques, Edifici CC, Universitat Autònoma de Barcelona, E–08193 Bellaterra, Spain
E-mail: cedo@mat.uab.es

25. **Choi, Kyu–Hyuck,** Department of Mathematics, Won Kwang University, Iksan, Seoul, 570–749, South Korea
E-mail: khchoi@wonnms.wonkwang.ac.kr

26. **Clark, John,** Department of Mathematics and Statistics, University of Otago, Dunedin, New Zealand
E-mail: jclark@maths.otago.ac.nz

27. **Coelho, Flavio Ulhoa,** Departamento de Matematica – IME, Universidade de Sao Paulo, CP 66281 (ag Cidade de Sao Paulo), Sao Paulo, SP CEP 05389–970, Brasil
E-mail: fucoelho@ime.usp.br

28. **Crawley–Boevey, William,** Department of Pure Mathematics, University of Leeds, Leeds LS2 9JT, United Kingdom
E-mail: w.crawley-boevey@leeds.ac.uk

29. **Dăscălescu, Sorin,** University of Bucharest, Facultatea de Matematica, Str. Academiei 14, R–70109 Bucuresti, Romania
E-mail: sdascal@math.math.unibuc.ro

30. **Dashdorj, Tserendorj,** Department of Algebra, University of Mongolia, Ulaanbaatar 46/534, Mongolia
E-mail: numelect@magicnet.mn

31. **Del Rio Mateos, Angel,** Departamento de Matematicas, Universidad de Murcia, E–30100 Murcia, Spain
E-mail: adelrio@fcu.um.es

32. **Diaz, Maria Alicia Avino,** Instituto de Matemáticas, UNAM, Mexico D.F., 04510, Mexico
E-mail: alicia@hardy.fciencias.unam.mx

33. **Dimitrova, Jetchka Mihailova,** Department of Mathematics, University "Prof. Assen Zlatarov", Prof. Yakimov str. 1, BG–8010 Burgas, Bulgaria
E-mail: jmich@btu.bg

34. **Di Vincenzo, Onofrio Mario,** Dipartimento di Matematica, Università della Basilicata, Via N. Sauro 85, I–85100 Potenza, Italy
E-mail: divincenzo@unibas.it

35. **Dlab, Vlastimil,** Dept. of Mathematics and Statistics, Carleton University, Ottawa, Ontario K1S 5B6, Canada
E-mail: vdlab@math.carleton.ca

36. **Dogruoz, Semra,** University of Glasgow, Department of Mathematics, Glasgow, Scotland G12 8QW, United Kingdom
E-mail: sd@euclid.maths.gla.ac.uk

37. **Dokuchaev, Michael,** Institute de Matematica e Estatística, Universidade de Sao Paulo, C.P.66281, Sao Paulo, CEP 05389–970, Brasil
E-mail: dokucha@ime.usp.br

38. **Domokos, Mátyás,** Mathematical Institute, Hungarian Academy of Sciences, P.O.Box 127, H–1364 Budapest, Hungary
E-mail: domokos@math-inst.hu

39. **Draper Fontanals, Cristina,** c/Cister 5. 2^oA, E–29015 Málaga, Spain
E-mail: crisdf@ccuma.sci.uma.es

40. **Drensky, Vesselin Stoyanov,** Section of Algebra, Institute of Mathematics and Informatics, Bulgarian Academy of Sciences, Akad. G. Bonchev Str., Block 8, BG–1113 Sofia, Bulgaria
E-mail: drensky@bgearn.acad.bg

41. **Dubrovin, Nikolai I.,** Vladimir State Polytechnical University, Department of Mathematics, Gorki str., Vladimir, 600026, Russia
E-mail: dubr%rtf@vpti.vladimir.su

42. **Dung, Nguyen Viet,** Department of Algebra and Number Theory, Institute of Mathematics, P.O.Box 631, Bo Ho, Hanoi, Vietnam
E-mail: nvd@thevinh.ac.vn

43. **Enochs, Edgar,** Department of Mathematics, University of Kentucky, Lexington, Kentucky KY 40506–0027, USA
E-mail: enochs@ms.uky.edu

44. **Erdei, László,** Department of Mathematics, Lajos Kossuth University, P.O.Box 12, H–4010 Debrecen, Hungary
E-mail: erdeil@math.klte.hu

45. **Eslami, Esfandiar,** Department of Mathematics, University of Kerman, Kerman, Iran
E-mail: eslami@math.kermanu.ac.ir

46. **Facchini, Alberto,** Università di Udine, Dipartimento di Matematica e Informatica, Via della Scienze 206, I–33100 Udine, Italy
E-mail: facchini@dimi.uniud.it

47. **Farnsteiner, Rolf,** Department of Mathematics, University of Wisconsin, Milwaukee, Wisconsin 53201–413, USA
E-mail: rolf@csd.uwm.edu

48. **Feldvoss, Jörg,** Mathematisches Seminar, Universität Hamburg, Bundesstr. 55, D–20146 Hamburg, Germany
E-mail: feldvoss@math.uni-hamburg.de

49. **Ferrero, Miguel,** Instituto de Matemática, UFRGS, Avda Bento Goncalvez 9500, Porto Alegre, 91509–900, Brasil
E-mail: ferrero@if.ufrgs.br

50. **Finogenov, Anton,** Institute of Mechanics and Mathematics, ul. S. Kovalevskoi 16, Ekaterinburg, 620219, Russia
E-mail: fin@top.imm.intec.ru

51. **Formanek, Edward,** Department of Mathematics, Penn. State University, State College, Pennsylvania 16801, USA
E-mail: formanek@math.psu.edu

52. **Friger, Michael,** Department of Mathematics and Computer Science, Ben Gurion University of the Negev, P.O.Box 653, Beer–Sheva, 84105, Israel
E-mail: friger@indigo.cs.bgu.ac.il

53. **Futorny, Vjacheslav,** Department of Mathematics, Kiev University, Kiev, 252617, Ukraine
E-mail: futorny@uni-alg.kiev.ua

54. **Gabriel, Peter,** Matematisches Institut, Universität Zürich, Winterthurerstr. 190, CH–8057 Zürich, Switzerland
E-mail: gabriel@math.unizh.ch

55. **Galvao, M. Luisa,** R. Ernesto Vasconcelos, Faculdade de Ciencias, Departamento de Matematica, Bloco C1. 3 piso, P–1700 Lisboa, Portugal
E-mail: mlgalvao@ptmat.lmc.fc.ul.pt

56. **Gateva–Ivanova, Tatiana,** Institute of Mathematics, Bulgarian Academy of Sciences, Acad. G. Bonchev Str. Bl. 8, BG–1113 Sofia, Bulgaria
E-mail: tatiana@bgearn.acad.bg

57. **Geiß, Christof,** Instituto de Matemáticas, UNAM, Mexico D.F., 04510, Mexico
E-mail: geiss@servidor.unam.mx

58. **Gentle, Ronald S.,** Department of Mathematics, Eastern Washington University, Cheney, Washington 99004, USA
E-mail: rgentle@ewu.edu

59. **Giambruno, Antonio,** Dipartimento di Matematica, Università di Palermo, Via Archirafi 34, I–90123 Palermo, Italy
E-mail: giambruno@ipamat. math.unipa.it

60. **Glavatsky, Sergei T.,** Department of Mechanics and Mathematics, Moscow State University, Moscow 119899, Russia
E-mail: glavatsky@cnit.chem.msu.su

61. **Gómez Lozano, Miguel Angel,** Departamento de Algebra, Geometria y Topologia, Facultad de Ciencias, Universidad de Málaga, Campus de Teatinos, Apartado 59, E–29080 Málaga, Spain

62. **Gómez Pardo, José Luis,** Departamento de Alxebra, Facultad de Matemáticas, Universidad de Santiago, E–15771 Santiago de Compostela, Spain
E-mail: pardo@zmat.usc.es

63. **Gómez Torrecillas, José,** Departamento de Algebra, Facultad de Ciencias, Universidad de Granada, E–18071 Granada, Spain
E-mail: torrecil@ugr.es, jgomez@andaya.ugr.es

64. **Gonçalves, Jairo,** Department of Mathematics, University of Sao Paulo, P.O.Box 66281, Sao Paulo, SP 05389–970, Brazil
E-mail: jzg@ime.usp.br

65. **González, Nieves Rodríguez,** Departamento de Alxebra, Facultad de Matemáticas, Universidad de Santiago, E–15771 Santiago de Compostela, Spain
E-mail: nieves@zmat.usc.es

66. **González Jimenez, Santos,** Department of Mathematics, University of Oviedo, E–33007 Oviedo, Spain
E-mail: santos@pinon.ccu.uniovi.es

67. **Goodearl, Kenneth R.,** Department of Mathematics, University of California, Santa Barbara, California 93106, USA
E-mail: goodearl@math.ucsb.edu

68. **Graeter, Joachim,** Universität Potsdam, Institut für Mathematik, Pf. 601553, D–14415 Potsdam, Germany
E-mail: graeter@rz.uni-potsdam.de

69. **Greferath, Marcus,** FB11, Universität Duisburg, D–47048 Duisburg, Germany
E-mail: greferath@math.uni-duisburg.de

70. **Grzeszczuk, Piotr,** Institute of Mathematics, University of Warsaw, Białystok Division, Akademicka 2, PL–15267 Białystok, Poland
E-mail: piotr@cksr.ac.bialystok.pl

71. **Gudivok, Peter Mikhajlovich,** F. Tihogo St., 15/58, Uzhgorod, 294018, Ukraine
E-mail: pgres@karpaty.uzhgorod.ua

72. **Guil Asensio, Pedro,** Department of Mathematics, University of Murcia, E–30100 Murcia, Spain
E-mail: paguil@fcu.um.es

73. **Hadas, Ofer,** Tel–Aviv University, School of Mathematical Sciences, Ramat–Aviv, Tel–Aviv, 69978, Israel
E-mail: oferh@math.tau.ac.il

74. **Herbera, Dolors,** Departament de Matemàtiques, Universitat Autònoma de Barcelona, E–08193 Barcelona, Spain
E-mail: dolors@mat.uab.es

75. **Horváth, Erzsébet,** Department of Mathematics, Technical University, H–1111 Budapest, Hungary
E-mail: he@vma.bme.hu

76. **Ivanov, George,** School of MPCE, Macquarie University, Sydney, N.S.W. 2109, Australia
E-mail: ivanov@mpce.mq.edu.au

77. **Ivanyos, Gábor,** Computer Institute of the Hungarian Academy of Sciences, Lágymányosi u., H–1111 Budapest, Hungary
E-mail: gabor.ivanyos@sztaki.hu

78. **Iwanaga, Yasuo,** Faculty of Education, Shinshu University, Nishi–Nagano 6, Nagano 380, Japan
E-mail: iwanaga@gipnc.shinshu-u.ac.jp

79. **Jain, Surender Kumar,** Department of Mathematics, Ohio University, Athens, Ohio 45701, USA
E-mail: jain@ace.cs.ohiou.edu

80. **Jespers, Eric,** Department of Mathematics and Statistics, Memorial University of Newfoundland, St John's, Newfoundland A1G 5S7, Canada
E-mail: ejespers@albert.math.mun.ca

81. **Jirásko, Josef,** Department of Mathematics, Faculty of Civil Engineering, Czech Technical University, Thákurova 7, CZ–16629 Praha 6, Czech Republic
E-mail: jirasko@mbox.cesnet.cz

82. **Kadison, Lars,** Institut for Matematikk og Statistikk, NTNU, N–7055 Dragvoll, Norway
E-mail: kadison@lie.matstat.unit.no

83. **Kashu, Alexei I.,** Institute of Mathematics, Academiei str. 5, MD–2028 Kishinev, Moldova
E-mail: 91casuai@math.moldova.su

84. **Kemer, Alexander R.,** Department of Mathematics, Ulyanovsk Branch of Moscow University, Ulyanovsk, 432700, Russia
E-mail: alex@kemer.univ.simbirsk.su

85. **Kępczyk, Marek,** ul. Orzeszkowej 5A/26, PL–15083 Białystok, Poland
E-mail: kep@katmat.pb.bialystok.pl

86. **Kimmerle, Wolfgang,** Mathematisches Institut B, Universität Stuttgart, Pfaffenwaldring 57, D–70550 Stuttgart, Germany
E-mail: kimmerle@mathematik.uni-stuttgart.de

87. **Király, Bertalan,** Dept. of Mathematics, Eszterházy Károly Teacher Training College, H–3300 Eger, Hungary
E-mail: kiraly@gemini.ektf.hu

88. **Kirichenko, Vladimir,** Department of Mathematics, Kiev University, Kiev, 252617, Ukraine
E-mail: vkir@trion.kiev.ua

89. **Kirkman, Ellen E.,** Department of Mathematics, Wake Forest University, P.O.Box 7388, Winston–Salem, North Carolina 27109, USA
E-mail: kirkman@mthcsc.wfu.edu

90. **Klein, Avraham A.,** School of Mathematical Sciences, Tel–Aviv University, Tel–Aviv, 69978, Israel
E-mail: aaklein@math.tau.ac.il

91. **Koryukin, Anatolii,** Institute of Mathematics, Siberian Branch of the Russian Academy of Sciences, Universitetskii pr. 4, 630090 Novosibirsk, Russia

92. **Koshlukov, Plamen E.,** Department of Mathematics, University of Sofia, James Bourchier 5, P.O.Box 48, BG–1164 Sofia, Bulgaria
E-mail: pek@fmi.uni-sofia.bg

93. **Krause, Günter,** Department of Mathematics and Astronomy, University of Manitoba, Winnipeg, Manitoba R3T 2N2, Canada
E-mail: gkrause@cc.umanitoba.ca

94. **Kurdics, János,** Department of Mathematics, Bessenyei College, Sóstói u. 31/b, H–4400 Nyíregyháza, Hungary
E-mail: kurdics@ny1.bgytf.hu

95. **Kuzmanovich, James,** Department of Mathematics, Wake Forest University, P.O.Box 7388, Winston–Salem, North Carolina 27109, USA
E-mail: kuz@mthcsc.wfu.edu

96. **Lakatos, Piroska,** Institute of Mathematics and Informatics, Lajos Kossuth University, P.O.Box 12, H–4010 Debrecen, Hungary
E-mail: lapi@math.klte.hu

97. **Lakos, Gyula,** Eötvös University, Múzeum krt. 6–8, H–1088 Budapest, Hungary
E-mail: lakos@cs.elte.hu

98. **Latyshev, Viktor N.,** Department of Mechanics and Mathematics, Moscow State University, Moscow, 119899, Russia
E-mail: artamon@nw.math.msu.su

99. **Leavitt, William G.,** 1600 Ridgeway Rd., Lincoln, Nebraska 68506, USA
E-mail: wleavitt@unl.edu

100. **Lenzing, Helmut,** Fachbereich Mathematik–Informatik, Universität–GH Paderborn, D–33095 Paderborn, Germany
E-mail: helmut@taylor.uni-paderborn.de

101. **Lévai, Levente,** Department of Algebra and Number Theory, Eötvös University, Múzeum krt. 6–8, H–1088 Budapest, Hungary
E-mail: levai@cs.elte.hu

102. **Liu, Shaoxue,** Department of Mathematics, Beijing Normal University, Beijing 100875, P. R. China
E-mail: liusx@bnu.ihep.ac.cn or liusx@ns.bnu.edu.cn

103. **Lomp, Christian,** Department of Mathematics, University of Glasgow, Glasgow, Scotland G12 8QW, United Kingdom
E-mail: cl@maths.gla.ac.uk

104. **Maltsev, Yuri N.,** Department of Mathematics, Altai State University, Dimitrova 66, 656099 Barnaul, Russia
E-mail: vup@altgu.mezon.altai.su

105. **Marciniak, Zbigniew,** Institute of Mathematics, Warsaw University, ul. Banacha 2, PL–02997 Warsaw, Poland
E-mail: zbimar@mimuw.edu.pl

106. **Marcus, Andrei,** Faculty of Mathematics and Computer Science, Babes–Bolyai University, str. Mihail Kogalniceanu 1, RO–3400 Cluj–Napoca, Romania
E-mail: marcus@math.ubbcluj.ro

107. **Marin, Leandro,** Department of Mathematics, University of Murcia, E–30071 Murcia, Spain
E-mail: leandro@fcu.um.es

108. **Márki, László,** Mathematical Institute, Hungarian Academy of Sciences, P.O.Box 127, H–1364 Budapest, Hungary
E-mail: marki@math-inst.hu

109. **Marko, František,** Department of Mathematics, Carleton University, Ottawa, Ontario K1S 5B6, Canada
E-mail: fmarko@math.carleton.ca

110. **Markov, Viktor T.,** CNIT, Moscow State University, Vorobjevy Gory, Moscow 119899, Russia
E-mail: markov@cnit.chem.msu.su

111. **Martindale, 3rd, Wallace S.,** Department of Mathematics, University of Massachusetts, Amherst, Massachusetts 01003, USA
E-mail: martndle@math.umass.edu

112. **Martinez Lopez, Consuelo,** Department of Mathematics, University of Oviedo, E–33007 Oviedo, Spain
E-mail: chelo@pinon.ccu.uniovi.es

113. **Masson, Gisli,** Department of Mathematics, Stockholm University, S–10691 Stockholm, Sweden
E-mail: gisli@matematik.su.se

114. **Mathieu, Martin,** Centre de Recerca Matematica, Universitat Autónoma de Barcelona, Apartat 50, E–08193 Bellaterra, Spain
E-mail: mm@uni-tuebingen.de

115. **Mazorchuk, Vladimir,** Department of Mathematics, Kiev University, Vladimrskaya 64, Kiev 252033, Ukraine
E-mail: mazorchuk@uni-alg.kiev.ua

116. **Mazurek, Ryszard,** Institute of Mathematics, University of Warsaw, Białystok Division, Akademicka 2, PL–15267 Białystok, Poland
E-mail: mazurek@cksr.ac.bialystok.pl

117. **McConnell, John C.,** School of Mathematics, University of Leeds, Leeds LS2 9TJ, United Kingdom
E-mail: pmt6jcm@leeds.ac.uk

118. **McDougall, Robert,** Department of Mathematics and Computing, Central Queensland University, Rockhampton, Queensland 4702, Australia
E-mail: r.mcdougall@cqu.edu.au

119. **Meltzer, Hagen,** Fakultät für Mathematik, TU Chemnitz–Zwickau, D–09107 Chemnitz, Germany
E-mail: meltzer@mathematik.tu–chemnitz.de

120. **Merklen, Héctor,** Instituto de Matemaática e Estatística, Universidade de Sao Paulo, C.P. 66281, CEP 05389–970, Sao Paulo, SP, Brazil
E-mail: merklen@ime.usp.br

121. **Miers, C. Robert,** Department of Mathematics and Statistics, University of Victoria, Victoria, British Columbia V8W 3P2, Canada
E-mail: crmiers@castle.uvic.ca

122. **Mihovski, Stoyl Vassilev,** Department of Mathematics, University of Plovdiv "Paisy Hilendarski", Tsar Assen 24, BG–4000 Plovdiv, Bulgaria
E-mail: mihovski@ulcc.uni-plovdiv.bg

123. **Mikhalev, Alexander A.,** Department of Mechanics and Mathematics, Moscow State University, Moscow, 119899, Russia
E-mail: aamikh@cnit.math.msu.su

124. **Mikhalev, Alexander V.,** Department of Mechanics and Mathematics, Moscow State University, Moscow, 119899, Russia
E-mail: mikhalev@cnit.math.msu.su

125. **Molnár, Lajos,** Institute of Mathematics and Informatics, Lajos Kossuth University, P.O.Box 12, H–4010 Debrecen, Hungary
E-mail: molnar@math.klte.hu

126. **Muñoz Albarran, Salvador,** c/Victoria 89 3°A, E–29012 Málaga, Spain

127. **Musson, Ian M.,** Department of Mathematical Sciences, University of Wisconsin–Milwaukee, P.O.Box 413, Milwaukee, Wisconsin 53201–0413, USA
E-mail: musson@csd.uwm.edu

128. **Nogueira, M. Teresa,** Faculdade de Ciências, Universidade de Lisboa, Departament de Matematica, Rua Ernesto de Vasconcelos, Bloco C1, piso 3, P–1700 Lisboa, Portugal
E-mail: mtnoguei@cc.fc.ul.pt

129. **Okniński, Jan,** Institute of Mathematics, Warsaw University, ul. Banacha 2, PL–02097 Warsaw, Poland
E-mail: okninski@mimuw.edu.pl

130. **Olivier, Werner A.,** Department of Mathematics, University of Port Elizabeth, P.O.Box 1600, Port Elisabeth, 6000, South Africa
E-mail: maawao@upe.ac.za

131. **Paison, Olga,** Math. and Mechanics Department, Ural State University, pr. Lenina 51, Ekaterinburg, 620083, Russia
E-mail: Olga.Paison@usu.ru

132. **Pardo, Enrique,** Departament de Matemàtiques, Facultat de Ciències, Edifici Cc, Universitat Autònoma de Barcelona, E–08193 Bellaterra, Spain
E-mail: epardo@mat.uab.es

133. **Passman, Donald S.,** Mathematics Department, University of Wisconsin–Madison, 480 Lincoln Drive, Madison, Wisconsin 53706, USA
E-mail: passman@math.wisc.edu

134. **Patay, Zoltán,** Department of Mathematics, Budapest Technical University, Műegyetem rakpart 3–5, H–1111 Budapest, Hungary
E-mail: patay@vma.bme.hu

135. **Perera, Francesc,** Departament de Matemàtiques, Facultat de Ciències, Edifici Cc, Universitat Autònoma de Barcelona, E–08193 Bellaterra, Spain
E-mail: perera@mat.uab.es

136. **Petechuk, Vasil,** 5 Sechenov St., Uzhgorod, 294014, Ukraine
E-mail: root@gd.uzhgorod.ua

137. **Petrogradsky, Victor M.,** Branch of Moscow State University, Mathematics Department, Lev Tolstoy 42, Ulyanovsk, 432700, Russia
E-mail: wmp@nmf.univ.simbirsk.su

138. **Piacentini Cattaneo, Giulia Maria,** Dipartimento di Matematica, Università di Roma "Tor Vergata", Via della Ricerca Scientifica, I–00133 Roma, Italy
E-mail: piacentini@mat.utovrm.it

139. **Pospíchal, Tomáš,** Department of Mathetamics and Statistics, Carleton University, Ottawa, Ontario K1S 5B6, Canada
E-mail: tpospich@math.carleton.ca

140. **Puczyłowski, Edmund,** Institute of Mathematics, University of Warsaw, Banacha 2, PL–02097 Warsaw, Poland
E-mail: edmundp@mimuw.edu.pl

141. **Ramón, Antoine Riolobos,** Facultat de Ciences, Edifici C. Departament de Matemàtiques, Universitat Autònoma de Barcelona, E–08193 Bellaterra, Spain
E-mail: ramon@mat.uab.es

142. **Raphael, Robert,** Mathematics Department, Concordia University, Loyola Campus, Montreal, Québec H4B 1R6, Canada
E-mail: raphael@ancor.concordia.ca

143. **Rashkova, Tsetska Grigorova,** Department of Algebra and Geometry, Angel Kanchev University of Rousse, ul. Studentska 8, BG–7017 Rousse, Bulgaria
E-mail: tcetcka@ami.ru.acad.bg

144. **Razmyslov, Yuri P.,** Department of Mechanics and Mathematics, Moscow State University, Moscow, 119899, Russia

145. **Regev, Amitai,** Department of Mathematics, Penn. State University, State College, Pennsylvania 16801, USA
E-mail: regev@poincare.math.psu.edu

146. **Reppa, Zoltán,** Department of Algebra and Number Theory, Eötvös University, Múzeum krt. 6–8, H–1088 Budapest, Hungary
E-mail: reppa@cs.elte.hu

147. **Riley, David M.,** Department of Mathematics, University of Alabama, Tuscaloosa, Alabama 35587–0350, USA
E-mail: driley@gp.as.ua.edu

148. **Ríos, Jose,** IMATE, UNAM Ciudad Universitaria, México DF, CP 04510, Mexico
E-mail: jrios@gauss.matem.unam.mx

149. **Rizvi, Tariq,** Department of Mathematics, Ohio State University — Lima, 4240 Campus Drive, Lima, Ohio 45804, USA
E-mail: rizvi@math.ohio-state.edu

150. **Roggenkamp, Klaus W.,** Mathematisches Institut B, Universität Stuttgart, Pfaffenwaldring 57, D–70550 Stuttgart, Germany
E-mail: kwr@phoebus.mathematik.uni-stuttgart.de

151. **Roomeldi, Raul,** Institute of Pure Mathematics, University of Tartu, EE–2400 Tartu, Estonia
E-mail: roomeldi@math.ut.ee

152. **Rosenbaum, Kurt,** Institut für Mathematik, PH Erfurt Mühlhausen, Nordhäuser Str. 63, D–99089 Erfurt, Germany
E-mail: rosenbaum@imath1.mathnat.ph-erfurt.de

153. **Rowen, Louis Halle,** Department of Mathematics, Bar–Ilan University, Ramat–Gan 52900, Israel
E-mail: rowen@bimacs.cs.biu.ac.il

154. **Salwa, Arkadiusz,** Institute of Mathematics, Warsaw University, Banacha 2, PL–02097 Warsaw, Poland
E-mail: asalwa@mimuw.edu.pl

155. **Sands, Arthur D.,** Mathematics Department, Dundee University, Dundee, Scotland DD1 4HN, United Kingdom
E-mail: adsands@mcs.dund.ac.uk

156. **Santa–Clara, Catarina,** Department of Mathematics, University of Glasgow, University Gardens, Glasgow, Scotland G12 8QW, United Kingdom
E-mail: cg@maths.gla.ac.uk

157. **Sapouzhak, Taras,** Mebelschikiv 7, Apt. 7, Uzhgorod, 294000, Ukraine

158. **Schmidt, Stefan E.,** FB Mathematik, Universität Mainz, D–55099 Mainz, Germany
E-mail: schmidt@mat.mathematik.uni-mainz.de

159. **Sehgal, Sudarshan,** Department of Mathematical Sciences, University of Alberta, Edmonton, Alberta T6G 2G1, Canada
E-mail: s.sehgal@ualberta.ca

160. **Seidel, Alexander,** Wolkerstr. 8, D–41462 Neuß, Germany
E-mail: seidelax@mail.rz.uni-duesseldorf.de

161. **Siles Molina, Gema,** Departamento de Algebra–Geometria y Topologia, Facultad de Ciencias, Universidad de Málaga, Campus de Teatinos, Apartado 59, E–29080 Málaga, Spain
E-mail: msilesm@ccuma.sci.uma.es

162. **Siles–Molina, Mercedes,** Departamento de Algebra–Geometria y Topologia, Facultad de Ciencias, Universidad de Málaga, Campus de Teatinos, Apartado 59, E–29080 Málaga, Spain
E-mail: msilesm@ccuma.sci.uma.es

163. **Skowroński, Andrzej,** Faculty of Mathematics and Informatics, Nicholas Copernicus University, Chopina 12/18, PL–87100 Toruń, Poland
E-mail: skowron@mat.uni.torun.pl

164. **Small, Lance,** Department of Mathematics, University of California San Diego, La Jolla, California 92093, USA
E-mail: lwsmall@ucsd.edu

165. **Snashall, Nicole,** Department of Mathematics and Computer Science, University of Leicester, University Road, Leicester, England LE1 7RH, United Kingdom
E-mail: njss@mcs.le.ac.uk

166. **Sommerhäuser, Yorck,** Universität München, Mathematisches Institut, Theresienstr. 39, D–80333 München, Germany
E-mail: sommerh@rz.mathematik.uni–muenchen.de

167. **Stafford, J. Toby,** Department of Mathematics, University of Michigan, Ann Arbor, Michigan 48104, USA
E-mail: jts@math.lsa.umich.edu

168. **Ştefănescu, Mirela,** Department of Mathematics, Ovidius University, Bd. Mamaia 124, R–8700 Constanţa, Romania
E-mail: mirelast@univ.ovidius.ro

169. **Stefanov, Dimitar Georgiev,** Section of Algebra, Institute of Mathematics and Informatics, Bulgarian Academy of Sciences, Acad. G. Bonchev Str. Block 8, BG–1113 Sofia, Bulgaria
E-mail: drensky@bgearn.acad.bg

170. **Stolin, Alekhandre,** Department of Mathematics, Royal Institute of Technology, S–10044 Stockholm, Sweden
E-mail: astolin@math.kth.se

171. **Sviridova, Irina Yu.,** Kirov Str. 28, Apt. 137, Ulyanovsk, 432048, Russia
E-mail: alex@kemer.univ.simbirsk.su

172. **Szakács, Attila,** Department of Mathematics, Kőrösi Csoma Sándor College, H–5600 Békéscsaba, Hungary
E-mail: h7219sza@ella.hu

173. **Szeidl, Ádám,** Department of Algebra, Eötvös University, Múzeum krt. 6–8, H–1088 Budapest, Hungary
E-mail: szeidl@cs.elte.hu

174. **Szigeti, Jenő,** Department of Mathematics, University of Miskolc, H–3515 Miskolc, Hungary
E-mail: matszj@gold.uni-miskolc.hu

175. **Takane, Martha,** Instituto de Matemáticas, UNAM, Mexico D.F., 04510, Mexico
E-mail: takane@servidor.unam.mx

176. **Tasić, Vladimir,** Department of Mathematics and Statistics, University of New Brunswick, Fredericton, New Brunswick E3B 5E3, Canada
E-mail: vlad@conway.math.unb.ca

177. **Tchebotar, Mikhail A.,** Department of Mechanics and Mathematics, Moscow State University, Moscow 119899, Russia

178. **Tortorici, Gaetano,** Dipartimento di Matematica, Università di Palermo, via Archirafi 34, I–90123 Palermo, Italy
E-mail: gaetano@ipamat.math.unipa.it

179. **Tsigantshev, Dimitar Georgiev,** Institute of Mathematics and Informatics, Bulgarian Academy of Sciences, Acad. G. Bonchev Str. Block 8, BG–1113 Sofia, Bulgaria
E-mail: drensky@bgearn.acad.bg

180. **Ufnarovski, Victor,** Department of Mathematics, Lund University, Lund, Sweden
E-mail: ufn@maths.lth.se

181. **Vaccaro, Maria Alessandra,** Dipartimento di Matematica, Università di Palermo, Via Archirafi 34, I–90123 Palermo, Italy
E-mail: vaccaro@ipamat.math.unipa.it

182. **Valenti, Angela,** Dipartimento di Matematica, Università di Palermo, Via Archirafi 34, I–90123 Palermo, Italy
E-mail: a.valenti@ipamat.math.unipa.it

183. **Vámos, Péter,** Department of Mathematics, University of Exeter, North Park Rd., Exeter, England EX4 4QE, United Kingdom
E-mail: vamos@maths.exeter.ac.uk

184. **Van den Berg, John,** Department of Mathematics, University of Natal, Private Bag X01, Scottsville, Pietermaritzburg, 3209, South Africa
E-mail: vandenberg@math.unp.ac.za

185. **Van Wyk, Leon,** Department of Mathematics, University of Stellenbosch, Private Bag X1, Matieland, 7602, South Africa
E-mail: lvw@maties.sun.ac.za

186. **Varadarajan, Kalathoor,** Department of Mathematics, University of Calgary, 2500 University Drive, Calgary, Alberta T2N 1N4, Canada
E-mail: varadara@math.ucalgary.ca

187. **Watters, John F.,** Department of Mathematics and Computer Science, University of Leicester, Leicester, England LE1 7RH, United Kingdom
E-mail: jfw@le.ac.uk

188. **Wiegandt, Richárd,** Mathematical Institute, Hungarian Academy of Sciences, P.O.Box 127, H–1364 Budapest, Hungary
E-mail: wiegandt@math-inst.hu

189. **Wilson, Mark,** Department of Mathematics, University of Auckland, Private Bag, Auckland, 92019, New Zealand
E-mail: wilson@math.auckland.ac.nz

190. **Wisbauer, Robert,** Mathematisches Institut, Universitẗ Düsseldorf, Universitätsstr. 1, D–40225 Düsseldorf, Germany
E-mail: wisbauer@ze9.rz.uni-duesseldorf.de

191. **Xi, Changchang,** Department of Mathematics, Beijing Normal University, Beijing 100875, P. R. China
E-mail: xi@mathematik.uni-bielefeld.de

192. **Yakovlev, Anatolii V.,** Department of Mathematics and Mechanics, St. Petersburg State University, Universitetskaya nab. 7/9, St. Petersburg 199034, Russia
E-mail: yakov@hq.math.lgu.spb.su

193. **Yousif, Mohamed F.,** Department of Mathematics, Ohio State University, Lima, Ohio 45804, USA
E-mail: yousif.1@osu.edu

194. **Zaicev, Mikhail V.,** Department of Algebra, Faculty of Mathematics and Mechanics, Moscow State University, Moscow 119899, Russia
E-mail: zaicev@nw.math.msu.su

195. **Zalesskii, Aleksandr E.,** School of Mathematics, University of East Anglia, Norwich, England NR4 7TJ, United Kingdom
E-mail: a.zalesskii@uea.ac.uk

196. **Zand, Hossien,** Faculty of Mathematics and Computing, Open University, Milton Keynes, MK7 6AA, United Kingdom
E-mail: h.zand@open.ac.uk

197. **Zel'manov, Efim I.,** Department of Mathematics, Yale University, New Haven, Connecticut 06520, USA
E-mail: zelmanov@math.yale.edu

198. **Zolotykh, Andrej A.,** Department of Mechanics and Mathematics, Moscow State University, Moscow 119899, Russia
E-mail: zolotykh@cnit.math.msu.su

Canadian Mathematical Society
Conference Proceedings
Volume **22**, 1998

Filter Dimension and its Application

Vladimir V. Bavula

Abstract. The present paper is a review of the author's articles on the filter dimension and the multiplicity for it. An analog of the Bernstein's inequality for simple affine algebras is established and the filter dimension of rings of differential operators with commutative regular coefficients is calculated (it equals 1). A new definition of holonomic module for simple affine algebras is given (which is "more correct" than the classical one). For a large class of simple affine algebras it is shown that each holonomic module has finite length.

1. Motivations, Main Results

The simple affine algebras is a big class of algebras which can be divided into two parts: finite dimensional and infinite dimensional. Concerning the first part we understand the structure of these algebras and there are general techniques how to study them. The situation is opposite for the second part. We do not understand yet the structure of simple affine infinite dimensional algebras and there are no general methods to study them. One reason for this is the large variety of these algebras. There are many types of simple affine infinite dimensional algebras (rings of differential operators, skew Laurent polynomial rings, generalized Weyl algebras, primitive factors of universal enveloping algebras, quantum groups etc.) and different approaches are used to study them.

A few years ago a new dimension of modules and algebras, **the filter dimension** (fil.dim) and an analog of the multiplicity for this dimension, were introduced, in [Bav 1]. It allows one to study the simple affine infinite dimensional algebras as a class.

The aim of this paper is to give a review of articles [Bav 1, 2], where the filter dimension and the multiplicity of it are studied and are applied to simple affine algebras and rings of differential operators with commutative regular coefficients.

AMS Subject Classification (1991). 16P90.
The author was supported by an INTAS grant, research fellow at U.I.A.

In order to give a flavour of this theory let us formulate main results of [Bav 1, 2] and the present paper.

- **(An analog of the Bernstein's inequality for simple affine algebras).** *Let A be a simple affine algebra and M be a nonzero finitely generated A-module. Then*

$$\mathrm{GK}(A) \leq \mathrm{GK}(M)\big(\text{fil.}\dim A + \max\{\text{fil.}\dim A, 1\}\big).$$

If, in addition, A is infinite dimensional, then

$$\text{fil.}\dim A \geq 1/2.$$

- *Let A be a simple affine finitely partitive algebra with $GK(A) < \infty$. Then the Krull dimension (in sense of Gabriel-Rentschler)*

$$\mathrm{K.}\dim(A) \leq \mathrm{GK}(A)\big(1 - 1/(\text{fil.}\dim A + \max\{\text{fil.}\dim A, 1\})\big).$$

Let B be an integral regular domain, affine over a field of characteristic zero and $\mathcal{D}(B)$ be its ring of differential operators. It is well-known that $\mathcal{D}(B)$ is a simple affine Noetherian algebra.

- *The filter dimension*

$$\text{fil.}\dim \mathcal{D}(B) = 1.$$

Some missed definitions can be found in [Bj], [KL], [MR] or [NVO].

2. An Analog of the Bernstein's Inequality for Simple Affine Algebras

Let K be a field, a module means a left one, $\otimes = \otimes_K$. Let A be an affine algebra with generators $a_1, \dots, a_s$. Then A is equipped with a standard finite dimensional filtration F: $A = \cup_{i\geq 0} A_i$, where $A_0 = K$, $A_1 = K + \sum_{i=1}^s Ka_i$, $A_i := A_1^i$, $i \geq 2$. From now on a "product" of sets means the vector space generated by products of elements, for example, $A_j a A_j$ is the vector space generated by xay, where x, $y \in A_j$.

Let M be a finitely generated A-module and M_0 be a finite dimensional generating subspace of M, the module M can be equipped with a filtration $\{M_n = A_n M_0\}$. Let $M_{i,\text{gen}}$ be any generating subspace of the module M from M_i. Define the **return function** $\nu_{F,M_0} : \mathbf{N}_0 \to \mathbf{N}_0 \cup \{\infty\}$ of the module M associated with F and M_0 by the rule

$$\nu_{F,M_0}(i) = \inf\{j \in \mathbf{N}_0 \cup \{\infty\} : A_j M_{i,\text{gen}} \supseteq M_0 \quad \text{for all} \quad M_{i,\text{gen}}\},$$

where $\mathbf{N}_0 = \{0, 1, 2, \dots\}$.

For an affine algebra B any two standard finite dimensional filtrations $F = \{A_i\}$ and $G = \{B_i\}$ are **equivalent**, ($F \sim G$), i.e. there exist natural numbers a, b, c, d such that

$$A_i \subseteq B_{ai+b} \quad \text{and} \quad B_i \subseteq A_{ci+d} \quad \text{for} \quad i \gg 0.$$

If one of the inclusions holds, say the first, we write $F \leq G$.

For each function f from $\mathbf{N}_0$ to $\mathbf{R}'$, where $\mathbf{R}' = \{r \in \mathbf{R},\ r \geq 1\}$ we can associate a number $\gamma(f) \in \mathbf{R}$, a **degree** of f,

$$\gamma(f) = \inf \{r \in \mathbf{R} : f(n) \leq n^r \text{ for sufficiently large } n \gg 0\}.$$

Lemma 2.1. *Let A be an affine algebra equipped with two standard finite dimensional filtrations $F = \{A_i\}$ and $G = \{B_i\}$ and M be a finitely generated A-module. Then for any finite dimensional generating subspaces M_0 and N_0 of M : $\gamma(\nu_{F,M_0}) = \gamma(\nu_{G,N_0})$.*

Proof. The module M has two standard finite dimensional filtrations $\{M_i = A_i M_0\}$ and $\{N_i = B_i N_0\}$. Put $\nu = \nu_{F,M_0}$ and $\mu = \nu_{G,N_0}$ for short.

Suppose that $F = G$. Choose a natural number s such that $M_0 \subseteq N_s$ and $N_0 \subseteq M_s$, so $N_i \subseteq M_{i+s}$ and $M_i \subseteq N_{i+s}$ for each i. Since $M_0 \subseteq A_{\nu(i+s)} N_{i,\mathrm{gen}}$ for each i and $N_0 \subseteq A_s M_0$, we have $N_0 \subseteq A_{\nu(i+s)+s} N_{i,\mathrm{gen}}$, hence, $\mu(i) \leq \nu(i+s)+s$ and finally $\gamma(\mu) \leq \gamma(\nu)$. By symmetry, there is the opposite inequality, so $\gamma(\mu) = \gamma(\nu)$.

Suppose that $M_0 = N_0$. Since A is affine, all standard finite dimensional filtrations are equivalent, so $F \sim G$ and we choose numbers a, b, c, d as above. Then $M_0 \subseteq A_{\nu(ci+d)} N_{i,\mathrm{gen}} \subseteq B_{a\nu(cj+d)+b} N_{i,\mathrm{gen}}$, therefore $\mu(i) \leq a\nu(ci+d)+b$ for each i, hence, $\gamma(\mu) \leq \gamma(\nu)$. By symmetry, we get the opposite inequality, therefore $\gamma(\mu) = \gamma(\nu)$.

Now, $\gamma(\nu_{F,M_0}) = \gamma(\nu_{F,N_0}) = \gamma(\nu_{G,N_0})$. □

Definition. The **filter dimension** $\mathrm{fil.dim}\, M$ of the A-module M is $\gamma(\nu_{F,M_0})$. The **filter dimension** $\mathrm{fil.dim}\, A$ of A is $\mathrm{fil.dim}({}_A A_A)$, i.e. the filter dimension of the A-bimodule A (or the left $A \otimes A^o$-module A, where A^o is the opposite ring to A).

Lemma 2.1 shows that $\mathrm{fil.dim}\, M$ does not depend on the choice of filtration of the algebra A and a generating subspace of M.

If A is simple, define the **return function** $\nu_F : \mathbf{N}_0 \to \mathbf{N}_0 \cup \{\infty\}$ of A associated with F in the following way

$$\nu_F(i) = \inf \{j \in \mathbf{N}_0 \cup \{\infty\} : A_j a A_j \ni 1 \quad \text{for all} \quad a \neq 0 \in A_i\}.$$

Lemma 2.2. *Let A be a simple affine algebra equipped with a standard finite dimensional filtrations $F = \{A_i\}$ Then $\mathrm{fil.dim}\, A = \gamma(\nu_F)$.*

Proof. The algebra A is simple, it can be considered as a simple $A \otimes A^{op}$-module. The tensor product $F \otimes F^{op} = \{C_n\}$ of standard finite dimensional filtrations $F = \{A_i\}$ and $F^{op} = \{A_i^{op}\}$ is a standard finite dimensional filtration of $A \otimes A^{op}$, where $C_n = \sum \{A_i \otimes A_j^{op}, i+j \leq n\}$. Let $\nu_{F \otimes F^{op},1}$ be the return function of the $A \otimes A^{op}$-module A associated with $F \otimes F^{op}$ and 1. It is easy to verify that the following inequalities hold

$$\nu_F(i) \leq \nu_{F \otimes F^{op},1}(i) \leq 2\nu_F(i), \quad \text{for each} \quad i,$$

then $\mathrm{fil.dim}\, A = \gamma(\nu_F)$. □

Theorem 2.3. (An analog of the Bernstein's inequality for simple affine algebras.) *Let A be a simple affine algebra and M be a nonzero finitely generated A-module. Then*

$$\mathrm{GK}(A) \leq \mathrm{GK}(M)\big(\mathrm{fil.dim}\, A + \max\{\mathrm{fil.dim}\, A, 1\}\big). \tag{2.1}$$

Proof. Let $\{A_i\}$ and $\{M_i = A_iM_0\}$ be standard finite dimensional filtrations of the algebra A and the module M respectively, where M_0 is a finite dimensional generating subspace of M. Let ν be the return function of A associated with $\{A_i\}$. For each $a \neq 0 \in A_i$ using the definition of ν we have

$$A_{\nu(i)}aM_{\nu(i)} = A_{\nu(i)}aA_{\nu(i)}M_0 \supseteq 1M_0 = M_0,$$

hence, the linear map

$$A_i \to \mathrm{Hom}\,\big(M_{\nu(i)}, M_{\nu(i)+i}\big), a \to (m \to am),$$

is injective, so $\dim\, A_i \leq \dim\, M_{\nu(i)} \dim\, M_{\nu(i)+i}$. Using elementary properties of the degree γ ([MR] Lemma 1.7, p. 280) and the definition of the Gelfand-Kirillov dimension we get $\mathrm{GK}(A) = \gamma(\dim A_i) \leq \gamma(\dim M_{\nu(i)}) + \gamma(\dim M_{\nu(i)+i}) \leq \gamma(\dim M_i)\gamma(\nu) + \gamma(\dim M_i)\max\{\gamma(\nu), 1\} = \mathrm{GK}(M)(\mathrm{fil.dim}\, A + \max\{\mathrm{fil.dim}\, A, 1\})$. □

Theorem 2.4. *Let A be an affine algebra, then*

$$\mathrm{GK}(M) \leq \mathrm{GK}(A)\ \mathrm{fil.dim}\, M,$$

for any simple A-module M.

Proof. Let ν be the return function of M associated with a standard finite dimensional filtration $F = \{A_i\}$ of A and $m \neq 0 \in M$. Let $\pi : M \to Km \simeq K$ be a projection map. Then for any $i \geq 0$ we get an injective linear map

$$M_i \to \mathrm{Hom}(A_{\nu(i)}, K), \quad u \to (a \to \pi(au)),$$

hence, $\dim\, M_i \leq \dim\, A_{\nu(i)}$ and $\mathrm{GK}(M) \leq \mathrm{GK}(A)\ \mathrm{fil.dim}\, M$. □

Corollary 2.5. *Let A be a simple affine infinite dimensional algebra. Then*

$$\mathrm{fil.dim}\, A \geq 1/2.$$

Proof. Since A is affine, resp. infinite dimensional, $\mathrm{GK}(A \otimes A^{op}) = 2\,\mathrm{GK}(A)$, resp. $\mathrm{GK}(A) > 0$. Applying Theorem 2.4 to a simple $A \otimes A^{op}$-module $M = A$ we finish the proof. □

3. Krull, Gelfand-Kirillov and Filter Dimension of Simple Affine Algebras

An algebra S is **(left) finitely partitive** ([Jos 1], [MR]) if, given any finitely generated S-module M, there is an integer $n > 0$ such that for every chain

$$M = M_0 \supset M_1 \supset \ldots \supset M_m$$

with $\mathrm{GK}(M_i/M_{i+1}) = \mathrm{GK}(M)$, one has $m \leq n$. As claim McConnell and Robson in ([MR], p. 298): "yet no examples are known which fail to have this property", i.e. being finitely partitive.

Theorem 3.1. *Let A be a simple affine finitely partitive algebra with $GK(A) < \infty$. Then the Krull dimension*

$$\mathrm{K.dim}(A) \leq \mathrm{GK}(A)\,\bigl(1 - 1/\bigl(\mathrm{fil.dim}\, A + \max\{\mathrm{fil.dim}\, A, 1\}\bigr)\bigr) \tag{3.1}$$

Proof. By ([Jos 2], 3.1.6 or [MR], 8.3.18): let a, $b \in \mathbf{N}$ and suppose, for all finitely generated A-modules M with $\mathrm{K.dim}(M) = a$, that $\mathrm{GK}(M) \geq a + b$; then

$$\mathrm{GK}(A) \geq \mathrm{K.dim}(A) + b \tag{3.2}$$

Applying this result to the family of finitely generated A-modules of finite length, $a = 0$, by Theorem 2.3 we can put $b = \mathrm{GK}(A)/(\mathrm{fil.dim}\, A + \max\{\mathrm{fil.dim}\, A, 1\})$, so (3.1) follows from (3.2). □

Corollary 3.2. *Let A be as above and* $\mathrm{fil.dim}\, A < \infty$. *Then the Krull dimension of A is finite, moreover, it is less than the Gelfand-Kirillov dimension of A* $(\mathrm{K.dim}(A) < \mathrm{GK}(A))$. □

Let B be an integral regular domain, affine over a field of characteristic zero and $\mathcal{D}(B)$ be its ring of differential operators. It is well-known that $\mathcal{D}(B)$ is a simple affine Noetherian algebra ([MR], ch. 15). In Section 4 we prove the following result.

Theorem 3.3. *The filter dimension*

$$\mathrm{fil.dim}\,\mathcal{D}(B) = 1.$$

□

As an application we compute the Krull dimension of $\mathcal{D}(B)$.

Corollary 3.4. ([MR], ch. 15)

$$\mathrm{K.dim}\ \mathcal{D}(B) = \mathrm{GK}(\mathcal{D}(B))/2 = \mathrm{K.dim}(B).$$

Proof. The second equality is clear $(\mathrm{GK}(\mathcal{D}(B)) = 2\,\mathrm{GK}(B) = 2\,\mathrm{K.dim}(B))$. It follows from Theorems 3.1, 3.3 that $\mathrm{K.dim}\ \mathcal{D}(B) \leq \mathrm{GK}(\mathcal{D}(B))\,/\,2$. The map $I \to \mathcal{D}(B)$ from the set of left ideals of B to the one of $\mathcal{D}(B)$ is injective, thus $\mathrm{K.dim}(B) \leq \mathrm{K.dim}\ \mathcal{D}(B)$. □

Corollary 3.5. ([MR], 15.4.3) *Let M be a nonzero finitely generated $\mathcal{D}(B)$-module. Then*

$$\mathrm{GK}(M) \geq \mathrm{GK}(\mathcal{D}(B))/2 = \mathrm{K.dim}(B).$$

Proof. It follows from Theorems 2.3, 3.3. □

4. Filter Dimension of Rings of Differential Operators with Regular Commutative Coefficients (Proof of Theorem 3.3)

Let K be a field of characteristic 0 and A be an affine commutative K-algebra which is a regular domain. The set of all K-derivations of A is denoted by $\mathrm{Der}_K\, A$. The K-subalgebra $\Delta = \Delta(A)$ of $\mathrm{End}_K\, A$ generated by A and $\mathrm{Der}_K\, A$ is called the

derivation ring of A. The ring $\Delta(A)$ (A is regular) coincides with the ring of differential operators $\mathcal{D}(A)$ ([MR], 15.5.6) and is a simple finitely partitive (for GK dimension) K-affine algebra ([MR], 15.3.8, 15.1.21). We refer the reader to [MR], chapter 15, for basic definitions.

Let $\{A_i\}$ and $\{\Delta(A)_i\}$ be standard finite dimensional filtrations on A and $\Delta(A)$ respectively such that $A_i \subseteq \Delta(A)_i$ for all $i \geq 0$. Then the enveloping algebra $\Delta^e(A) := \Delta(A) \otimes \Delta(A)^o$ can be equipped with a standard finite dimensional filtration $\{\Delta^e(A)_i\}$ which is the tensor product of filtrations $\{\Delta(A)_i\}$ and $\{\Delta(A)_i{}^o\}$.

Then $A \simeq \Delta(A)/\Delta(A) \operatorname{Der}_K A$ is a simple left $\Delta(A)$-module ([MR], 15.3.8], and $\operatorname{GK}(\Delta(A)) = 2\operatorname{GK}(A)$ ([MR], 15.3.2). Applying Theorem 2.3 in case of the ring $\Delta(A)$ and $M = A$ we see that

$$\operatorname{fil.dim} \Delta(A) \geq 1.$$

It remains to prove the opposite inequality. For this we recall some properties of $\Delta(A)$ (see [MR], ch. 15, for details).

Given $0 \neq c \in A$, denote by A_c the localization of A with respect of the powers of c, then $\Delta(A_c) \simeq \Delta(A)_c$ and the map $\Delta(A) \to \Delta(A)_c$, $d \to d/1$, is injective ([MR], 5.1.25). There is a finite subset $\{c_1, \dots, c_t\}$ of A such that the ring $\prod_{i=1}^t \Delta(A_{c_i})$ is left and right faithfully flat over $\Delta(A)$,

$$\sum_{i=1}^{t} Ac_i = A \quad \text{(see the proof of 15.2.13, [MR]).} \tag{4.1}$$

For each $c = c_i$, $\operatorname{Der}_K(A_c)$ is a free A_c-module on $\partial_j = \partial/\partial x_j$, $j = 1, \dots, n$ for some $x_1, \dots, x_n \in A$ ([MR], 15.2.13). Note that the choice of the x_i depends on the choice of c_j.

Fix $c = c_i$. We aim to prove the following statement:

- *there exist natural numbers a, b, α, and β such that for any $d \in \Delta(A)_m$ there exists*

$$w \in \Delta^e(A)_{am+b}: \qquad wd = c^{\alpha m + \beta}. \tag{$*$}$$

Suppose that we are done. It can be easily seen that a, b, α, and β may be chosen in such a way that $(*)$ holds for all i. It follows from (4.1) that

$$\sum_{i=1}^{t} f_i c_i = 1 \quad \text{for some} \quad f_i \in A.$$

Choose $\nu \in \mathbf{N}$: all $f_i c_i \in \Delta(A)_\nu$ and set $N(m) = \alpha m + \beta$, then

$$1 = \left(\sum f_i c_i\right)^{tN(m)} = \sum g_i c_i^{N(m)} = \sum g_i w_i d = wd,$$

where w_i from $(*)$, i.e. $w_i \in \Delta^e(A)_{am+b}$, $w_i d = c_i^{N(m)}$. So, $w = \sum g_i w_i \in \Delta^e(A)_{\nu t N(m) + am + b}$ and $\operatorname{fil.dim} \Delta(A) \leq 1$.

Fix $c = c_i$. By ([MR], 15.1.24) $\operatorname{Der}_K(A_c) \simeq \operatorname{Der}_K(A)_c$ and $\operatorname{Der}_K A$ can be seen as a finitely generated A-submodule of $\operatorname{Der}_K(A_c)$ ([MR], 15.1.7). Choose $k \in \mathbf{N}$ such that

$$\delta_i := c^k \partial_i \in \operatorname{Der}_K A \quad \text{for all} \quad i = 1, \dots, n.$$

Let M be an A-submodule of $\mathrm{Der}_K A$ generated by all δ_i. Since $M_c = \mathrm{Der}_K(A)_c$ and $\mathrm{Der}_K A$ is a finitely generated A-module, there exists $l \in \mathbf{N}$ such that $c^l \,\mathrm{Der}_K A \subseteq M$. Then

$$c^l \Delta(A)_1 \subseteq A_p + A_p\delta_1 + \cdots + A_p\delta_n \quad \text{for some} \quad p \in \mathbf{N}.$$

Easy induction arguments show that

$$c^{(2m-1)l}\Delta(A)_m \subseteq \sum_{|\alpha|\leq m} A_{mp+(m-1)(lq+\nu)}\, \delta^{\alpha}, \tag{4.2}$$

where we fix natural numbers q and ν with $c \in A_q$ and $\delta_i(A_1) \subseteq A_\nu$ for all $i = 1, \ldots, n$; $\delta^\alpha = \delta_1^{\alpha_1} \ldots \delta_n^{\alpha_n}$, $\alpha = (\alpha_1, \ldots, \alpha_n)$, $|\alpha| = |\alpha_1| + \cdots + |\alpha_n|$.

Let $0 \neq d \in \Delta(A)_m$. Then

$$0 \neq d_1 := c^{(2m-1)}d \in \sum_{|\alpha|\leq m} A_{mp+(m-1)(lq+\nu)}\, \delta^{\alpha}.$$

Applying $ad\, x_i := x_i \otimes 1 - 1 \otimes x_i \in \Delta^e(A)$ $(i = 1, \ldots, n)$ to d_1 in an appropriate way no more than m times and using $ad\, x_i\, \delta_j = -\delta_{ij}c^k \in A_{kq}$ (δ_{ij} is the delta-function) we get

$$0 \neq d_2 \in A_{mp+(m-1)(lq+\nu)+mkq} = A_z. \tag{4.3}$$

Let R be a subalgebra of A generated by all x_i. By ([MR], 15.1.12) $R = K[x_1, \ldots, x_n]$ is a polynomial ring with its field of fractions $Q = K(x_1, \ldots, x_n)$. Since $\mathrm{K.dim}\, R = n = \mathrm{K.dim}\, A$, A is a finitely generated R-module. Thus the field of fractions L of A equals to $L = QA$ and it is a finite field extension over Q, say, of dimension $s = \dim_Q L$. Choose a Q-basis $e_1 = 1, \ldots, e_s$ of L all elements of which belong to A. Set $N := \sum_{i=1}^s Re_i \subseteq A$. Since $QN = L = QA$ and the R-module A is finitely generated, there exists a nonzero element $\alpha \in R$ with $\alpha A \subseteq N$. We may suppose that N is an R-subalgebra ($e_ie_j \in N$ for all i, j). For consider the elements $1, \alpha e_2, \ldots, \alpha e_n$ ($\alpha e_i\, \alpha e_j \in \alpha^2 A \subseteq \alpha N \subseteq \sum_{i=1}^s R\alpha e_i$). The polynomial ring $R = \cup_{i=0}^\infty R_i$ is filtered by the degree of indeterminates. Then $N = \cup_{i=0}^\infty N_i$, where $N_i = \sum R_ie_j$. Pick $r \in \mathbf{N}$ such that $e_ie_j \in N_r$ for all i, j, then

$$N_iN_j \subseteq N_{i+j+r} \quad \text{for all} \quad i, j.$$

Choose $\sigma \in \mathbf{N}$ such that $\alpha A_1 \subseteq N_\sigma$, then

$$\alpha^i A_i \subseteq (\alpha A_1)^i \subseteq N_\sigma^i \subseteq N_{\sigma i + r(i-1)} \subseteq N_{(\sigma+r)i-r} \text{ for all } i.$$

Multiplying d_2 by α^z on the left in (4.3) we have an element

$$0 \neq d_3 := \alpha^z d_2 \in N_{am+b} \quad \text{for some} \quad a, b \in \mathbf{N}.$$

Using the Q-algebra monomorphism

$$L \to \mathrm{Hom}_Q(L, L),\ \nu \to (x \to \nu x),$$

we can identify the ring L with its image in $\mathrm{Hom}_Q(L, L)$. Fixing the Q-basis $e_1, \dots, e_s$ of L the algebra $\mathrm{Hom}_Q(L, L)$ can be identified with the matrix ring $M_s(Q)$, i.e. a linear map with its matrix (with respect to the basis $e_1, \dots, e_s$).

Let $u = (u_{ij})$ be the matrix of the linear map $v \to d_3 v$. The d_3 belongs to the R-subalgebra N, so all $u_{ij} \in R_{am+b+r}$ and the determinant $|u| \in R_{s(am+b+r)}$ is nonzero. The inverse matrix $u^{-1} = |u|^{-1}(w_{ij})$ corresponds to the linear map $x \to d_3^{-1}x$. It is clear that all $w_{ij} \in R_{(s-1)(am+b+r)}$. Choose $\tau \in \mathbf{N}$ such that all $e_i \in A_\tau$ and all $x_j \in A_\tau$. Then

$$|u|d_3^{-1} = |u|d_3^{-1}1 = |u||u|^{-1}(w_{ij})(1, 0, \dots, 0)^T \in \sum R_{(s-1)(am+b+r)}\, e_i \subseteq A_{\tau(s-1)(am+b+r)+\tau}.$$

Multiplying d_3 by $|u|d_3^{-1}$ we get

$$0 \neq d_4 := |u|d_3^{-1}d_3 = |u| \in R_{s(am+b+r)}.$$

We can easily choose $\partial^\alpha = \partial_1^{\alpha_1} \dots \partial_n^{\alpha_n}$, $|\alpha| \leq m$, such that

$$0 \neq d_5 := \partial^\alpha(d_4) \in K.$$

It is clear that

$$w := c^{2km} \prod_{i=1}^{k} (c^{-k} ad\, \delta_i)^{\alpha_i} \in \Delta^e(A)$$

and finally

$$d_5^{-1}wd_4 = d_5^{-1}c^{2km} \prod_{i=1}^{k} (c^{-k} ad\, \delta_i)^{\alpha_i}(d_4) = d_5^{-1}c^{2km}\partial^\alpha(d_4) = c^{2km}.$$

Now $(*)$ follows.

In fact we prove the following corollary.

Corollary 4.1. *There exist natural numbers a and b such that for any $d \in \Delta(A)_m$ there exists $w \in \Delta^e(A)_{am+b}$ satisfying $wd = 1$.*

5. An Analog of Multiplicity for the Filter Dimension, Holonomic Modules over Simple Affine Algebras

Let f be a function from $\mathbf{N}_0$ to $\mathbf{R}'$, the **leading coefficient** of f is a non-zero limit (if it exists)

$$\mathrm{lc}\,(f) = \lim f(i)/i^d \neq 0, \quad i \to \infty,$$

where $d = \gamma(f)$. If $d \in \mathbf{N}$, we define the **multiplicity** $e(f)$ of f by

$$e(f) = d!\ \mathrm{lc}\,(f).$$

The factor $d!$ ensures that the multiplicity $e(f)$ is a positive integer in some important cases.

Lemma 5.1. *Let A be an affine algebra equipped with a standard finite dimensional filtration $F = \{A_i\}$ and M be a finitely generated A-module with generating finite dimensional subspaces M_0 and N_0.*

1. *If* $\mathrm{lc}\,(\nu_{F,M_0})$ *exists, then does* $\mathrm{lc}\,(\nu_{F,N_0})$*, moreover,* $\mathrm{lc}\,(\nu_{F,M_0}) = \mathrm{lc}\,(\nu_{F,N_0})$.
2. *If* $\mathrm{lc}\,(\dim A_i M_0)$ *exists, then does* $\mathrm{lc}\,(\dim A_i N_0)$ *and* $\mathrm{lc}\,(\dim A_i M_0) = \mathrm{lc}\,(\dim A_i N_0)$.

Proof. 1. The module M has two filtrations $\{M_i = A_i M_0\}$ and $\{N_i = A_i N_0\}$. Put $\nu = \nu_{F,M_0}$ and $\mu = \nu_{F,N_0}$ for short. Choose a natural number s such that $M_0 \subseteq N_s$ and $N_0 \subseteq M_s$, so $N_i \subseteq M_{i+s}$ and $M_i \subseteq N_{i+s}$ for $i \geq 0$. Since $M_0 \subseteq A_{\nu(i+s)} N_{i,\mathrm{gen}}$ for each i and $N_0 \subseteq A_s M_0$, we have $N_0 \subseteq A_{\nu(i+s)+s} N_{i,\mathrm{gen}}$, hence, $\mu(i) \leq \nu(i+s)+s$. By symmetry, $\nu(i) \leq \mu(i+s)+s$, hence, $\mathrm{lc}\,(\mu)$ exists and equals to $\mathrm{lc}\,(\nu)$.

2. Since $\dim N_i \leq \dim M_{i+s}$ and $\dim M_i \leq \dim N_{i+s}$ for $i \geq 0$, the statement is clear. □

Lemma 5.1 shows that the leading coefficients of the functions $\dim A_i M_0$ and ν_{F,M_0} do not depend on the choice of the generating subspace M_0. So denote them by $l(M) = l_F(M)$ and $L(M) = L_F(M)$ respectively (if they exist). If $\mathrm{GK}(M)$ (resp. $\mathrm{fil.dim}\, A$) is a natural number, then by $e(M) = e_F(M)$ (resp. $E(M) = E_F(M)$) we denote the multiplicity of the function $\dim A_i M_0$ (resp. ν_{F,M_0}).

We denote by $L(A) = L_F(A)$ the leading coefficient $L_F({}_{A\otimes A^{op}}A)$ of the return function $\nu_{F\otimes F^{op},1}$ of the $A\otimes A^{op}$-module A.

Consider the algebras A having the following properties.

- (S) *A is a simple affine infinite dimensional algebra.*
- (N) *There exists a standard finite dimensional filtration $F = \{A_i\}$ such that the associated graded algebra $\mathrm{gr}\, A := \oplus_{i\geq 0} A_i/A_{i-1}$, $A_{-1} = 0$, is left Noetherian.*
- (D) $\mathrm{GK}(A) < \infty$, $\mathrm{fil.dim}\, A < \infty$, *both $l(A) = l_F(A)$ and $L(A) = L_F(A)$ exist.*
- (H) *For every holonomic A-module M there exists $l(M) = l_F(M)$.*

In many cases we use the weaker form of the condition (D).

- (D') $\mathrm{GK}(A) < \infty$, $d = \mathrm{fil.dim}\, A < \infty$, *there exist $l(A) = l_F(A)$ and a positive number $c > 0$ such that $\tilde{\nu}(i) \leq c i^d$ for $i \gg 0$ where $\tilde{\nu}$ is the return function $\nu_{F\otimes F^{op},1}$ of the $A \otimes A^{op}$-module A.*

It follows from (N) that A is left Noetherian.

Let A_n be the n'th **Weyl algebra**, i.e. the K−algebra with $2n$ generators $X_1, \ldots, X_n$, $\partial_1, \ldots, \partial_n$ and relations

$$\partial_i X_j - X_j \partial_i = \delta_{ij}, \text{ the Kronecker delta, } X_i X_j - X_j X_i = \partial_i \partial_j - \partial_j \partial_i = 0, \quad i \neq j.$$

If $\mathrm{char}\, K = 0$, then A_n is a simple Noetherian algebra.

Lemma 5.2. ([Bav 1])

1. *The Weyl algebra A_n over a field of characteristic zero with the standard finite dimensional filtration $F = \{B_m\}$ associated with $2n$ generators as above satisfies* (S), (N), (D), (H), *moreover, $L_F(A_n) = 1$.*
2. *$L_F(W_n) = 1$, where the A_n-module $W_n = K[X_1, \ldots, X_n] = A_n/A_n(\partial_1, \ldots, \partial_n)$.*

Definition. Let A be a simple affine infinite dimensional algebra. We call an A-module M **holonomic** if the inequality (2.1) becomes the equality, i.e.

$$\mathrm{GK}(A) = \mathrm{GK}(M)(\text{fil.}\dim A + \max\{\text{fil.}\dim A, 1\}).$$

Denote by $\mathrm{hol}(A)$ the family of holonomic A-modules.

Remark. A classical definition of holonomic module is as follows: a module N over an algebra B is holonomic if $\mathrm{GK}(N) = \mathrm{GK}(B/\operatorname{ann} N)/2$. Both definitions of holonomic module coincide for simple affine infinite dimensional algebras which have the filter dimension 1 (e.g. rings of differential operators from Section 4). But there exist simple affine algebras of filter dimension greater than 1 and the sets of holonomic modules are non-empty, [Bav 1], thus the first definition is "more correct" (the idea of holonomicity means that a module has as minimal as possible Gelfand-Kirillov dimension).

Example. 1. The module W_n over the Weyl algebra A_n, char $K = 0$, is holonomic, since $\mathrm{GK}(A_n) = 2n = 2\,\mathrm{GK}(W_n)$, $\text{fil.}\dim A_n = 1$. More generally, let B be an integral regular domain, affine over a field of characteristic zero and $\mathcal{D}(B)$ be its ring of differential operators. It follows from Corollary 3.4 that B is a holonomic $\mathcal{D}(B)$-module ($\mathrm{K.}\dim(B) = \mathrm{GK}(B)$).

2. Let A be a central simple affine infinite dimensional algebra. Then $A \otimes A^{op}$ is a simple affine algebra of filter dimension greater than or equal one and the $A \otimes A^{op}$-module A is holonomic, ([Bav 1], Corollary 1.7), if $\text{fil.}\dim A \otimes A^{op} = 1$.

Theorem 5.3. *Assume that an algebra A satisfies the hypotheses* (S), (H), (D) *resp.* (D') *for some standard finite dimensional filtration $F = \{A_i\}$ of A. Then for every holonomic A-module M:*

$$l(A) \leq l^2(M)(L(A)L'(A))^{\mathrm{GK}(A)/(\text{fil.dim}\, A + \max\{\text{fil.dim}\, A, 1\})},$$

where

$$L'(A) = \begin{cases} L(A), & \text{if } \text{fil.}\dim A > 1; \\ L(A) + 1, & \text{if } \text{fil.}\dim A = 1; \\ 1, & \text{if } \text{fil.}\dim A < 1; \end{cases}$$

resp.

$$l(A) \leq l^2(M)\big(c(c+1)\big)^{\mathrm{GK}(A)/(\text{fil.dim}\, A + \max\{\text{fil.dim}\, A, 1\})}.$$

Proof. Let M_0 be a generating finite dimensional subspace of M and $\{M_i = A_i M_0\}$ be a standard finite dimensional filtration of M. In the proof of Theorem 1.2 we established the following inequality: $\dim A_i \leq \dim M_{\nu(i)} \dim M_{\nu(i)+i}$ for each $i \geq 0$ where ν is the return function of A associated with F. Since $\nu(i) \leq \tilde{\nu}(i)$ for $i \geq 0$ we conclude that $\dim A_i \leq \dim M_{\tilde{\nu}(i)} \dim M_{\tilde{\nu}(i)+i}$, hence, if (D) holds we have

$$l(A) i^{\mathrm{GK}(A)} + \cdots \leq l^2(M)(L(A)L'(A))^{\mathrm{GK}(M)} i^{\mathrm{GK}(M)(\text{fil.dim}\, A + \max\{\text{fil.dim}\, A, 1\})} + \cdots,$$

where three dots denote terms of smaller order.

If (D') holds we have

$$l(A) i^{\mathrm{GK}(A)} + \cdots \leq l^2(M)(c(c+1))^{\mathrm{GK}(M)} i^{\mathrm{GK}(M)(\text{fil.dim}\, A + \max\{\text{fil.dim}\, A, 1\})} + \cdots.$$

The module M is holonomic, i.e. $\mathrm{GK}(A)=\mathrm{GK}(M)(\text{fil.}\dim A+\max\{\text{fil.}\dim A,1\})$, and now comparing "leading" coefficients in the inequalities above we finish the proof. □

Let A be as in Theorem 3.3. We attach to the algebra A two positive numbers c_A and c'_A in the cases (D) and (D') respectively:

$$c_A^2 = l(A)/(L(A)L'(A))^{\mathrm{GK}(A)/(\text{fil.dim}\, A+\max\{\text{fil.dim}\, A,1\})}$$

and

$$(c'_A)^2 = l(A)/(c(c+1))^{\mathrm{GK}(A)/(\text{fil.dim}\, A+\max\{\text{fil.dim}\, A,1\})}.$$

Corollary 5.4. *Assume that an algebra A satisfies the hypotheses* (S), (N), (H), (D) *or* (D'). *Let $0 \to N \to M \to L \to 0$ be an exact sequence of finitely generated A-modules. Then M is holonomic if and only if N and L are holonomic, moreover, $l(M) = l(N) + l(L)$.*

Proof. The algebra A is left Noetherian, so the module M is finitely generated iff both N and L are so. The proof of Proposition 3.11 ([MR], p. 295) shows that we can choose finite dimensional generating subspaces N_0, M_0, L_0 of the modules N, M, L respectively such that the sequences

$$0 \to N_i = A_i N_0 \to M_i = A_i M_0 \to L_i = A_i L_0 \to 0$$

are exact for all i, hence, $\dim M_i = \dim N_i + \dim L_i$ and results follow. □

Theorem 5.5. *Suppose that* (S), (N), (H), (D) (*resp.* (D')) *hold. Then each holonomic A-module M has finite length which is less or equal to $l(M)/c_A$ (resp. $L(M)/c'_A$).*

Proof. If $M = M_1 \supset M_1 \supset \ldots \supset M_m \supset M_{m+1} = 0$ is a chain of distinct submodules, then by Corollary 5.4 and Theorem 5.3

$$l(M) = \sum_{i=1}^{m} l(M_i/M_{i+1}) \geq mc_A, \quad (\text{resp.} l(M) \geq mc'_A),$$

thus $m \leq l(M)/c_A$ (resp. $m \leq l(M)/c'_A$). □

References

[Bav 1] V.V. Bavula, Filter dimension of algebras and modules, a simplicity criterion of generalized Weyl algebras, *Comm. Algebra* **24** (1996), no. 6, 1971–1992.

[Bav 2] V.V. Bavula, Krull, Gelfand-Kirillov and filter dimensions of simple affine algebras, *J. Algebra* (to appear).

[Bj] J.-E. Björk, *Rings of differential operators*, North-Holland Math. Libr., Amsterdam, 1979.

[Be] J. Bernstein, Modules over the ring of differential operators. A study of fundamental solution to equations with constant coefficients, *Funk. Anal. i Pril.* **5** (1971), no. 2, 1–16.

[Jos 1] A. Joseph, *Dimension en algèbre non-commutative*, Cours de troisième cycle, Univ. de Paris 6, mimeographed notes, 1980.

[Jos 2] A. Joseph, *Applications de la thèorie des anneaux aux algèbres enveloppantes*, Cours de troisième cycle, Univ. de Paris 6, mimeographed notes, 1981.

[KL] G.R. Krause and T.H. Lenagan, *Growth of algebras and Gelfand-Kirillov dimension*, Pitman Adv. Publ. Progr., 1985.

[MR] J.C. McConnell and J.C. Robson, *Non-commutative Noetherian rings*, Wiley, 1987.
[NVO] C. Naštašescu and F. Van Oystaeyen, *Dimensions of Ring Theory*, D. Reidel Publ. Co., Boston, 1987.

V.V. Bavula
Department of Mathematics
Kiev University
Vladimirskaya Str, 64
Kiev 252617
Ukraine
e-mail: kinr4%rigkbp@mhsgw.grs.de

Canadian Mathematical Society
Conference Proceedings
Volume **22**, 1998

Automorphisms of Polynomial, Free and Generic Matrix Algebras

Vesselin Drensky

Abstract. The purpose of this paper is to survey some old and recent results on the automorphisms of polynomial, free and generic matrix algebras as well as on the automorphisms of algebras which are related with the algebra of generic matrices. We also give a list of open problems.

Introduction

Throughout this paper we fix a field K of any characteristic and all algebras and vector spaces are over K. Let R be an associative algebra finitely generated by a set $\{r_1, \dots, r_m\}$. The main problem discussed is to describe the automorphism group $\operatorname{Aut} R$ of the algebra R. One possible approach is the following. Let $K\langle x_1, \dots, x_m\rangle$ be the free associative algebra of rank m. The map

$$x_j \longrightarrow r_j, \quad j = 1, \dots, m,$$

can be uniquely extended to a homomorphism

$$\pi : K\langle x_1, \dots, x_m\rangle \longrightarrow R$$

and

$$R \cong K\langle x_1, \dots, x_m\rangle / \operatorname{Ker} \pi.$$

AMS Subject Classification (1991). Primary 16W20, Secondary 13B25, 14E09, 16R30, 16S10, 17B40, 20G05.

Keywords and Phrases: automorphisms of polynomial, free associative and free Lie algebras, algebras with polynomial identity, generic matrices.

Partially supported by Grant MM404/94 of the Bulgarian Foundation for Scientific Research.

The investigation of $\operatorname{Aut} R$ may be carried out in three steps.

Step 1. Describe the automorphisms of $K\langle x_1, \dots, x_m \rangle$.

Step 2. Find the automorphisms ψ of $K\langle x_1, \dots, x_m \rangle$ which induce automorphisms of R, i.e. such that $\psi(\operatorname{Ker} \pi) = \operatorname{Ker} \pi$.

Step 3. Try to understand the automorphisms of R which cannot be obtained from the first two steps, i.e. cannot be lifted to automorphisms of the free algebra.

A similar approach is also possible for other classes of algebras, e.g. for commutative algebras, Lie algebras, groups, etc. Of course in these cases we should replace the free associative algebra $K\langle x_1, \dots, x_m \rangle$ respectively with the polynomial algebra $K[x_1, \dots, x_m]$, the free Lie algebra of rank m, the free group of rank m, etc.

In our considerations we "avoid" the second step and assume that any automorphism of $K\langle x_1, \dots, x_m \rangle$ induces an automorphism of R. We make even the stronger restriction that the kernel of π is invariant under all endomorphisms of $K\langle x_1, \dots, x_m \rangle$. In other words, every map

$$\phi_0 : r_j \longrightarrow R, \quad j = 1, \dots, m,$$

gives rise to an endomorphism of R. Under this restriction, there exists an algebra S (i.e. R itself) such that $\operatorname{Ker} \pi$ coincides with the ideal of all polynomial identities in m variables for S. Then R is called a relatively free algebra of rank m (in the variety of algebras defined by the polynomial identities for S). Such an approach allows to use powerful techniques typical for the theory of algebras with polynomial identities, i.e. PI-algebras.

The purpose of our paper is to survey some results on the automorphism groups of the polynomial algebra $K[x_1, \dots, x_m]$, the free associative algebra $K\langle x_1, \dots, x_m \rangle$ and the free Lie algebra L_m of rank m as well as of some relatively free algebras. One of the classes of relatively free algebras considered of great importance is that of generic matrix algebras. They have many relations with the theory of the structure of rings, theory of division algebras, commutative and noncommutative invariant theory, etc. (see e.g. Formanek [20]). We concentrate our considerations mainly on the automorphisms of the generic matrix algebras. Especially interesting is the algebra of two generic 2×2 matrices, in particular because every of its automorphisms induces an automorphism of the polynomial algebra in five variables.

We do not claim to present the complete picture of the problems concerning the automorphisms of polynomial and free algebras. For example, we do not discuss at all the famous Jacobian conjecture or the problem of recognizing the automorphisms in the class of all endomorphisms. For a more detailed (but not comprehensive) survey on automorphisms of relatively free algebras including some techniques for investigation see [15].

The paper is organized as follows. Section 1 is devoted to the automorphisms of polynomial algebras. In Section 2 we consider the automorphisms of free associative and free Lie algebras. In Section 3 we survey results on the automorphisms of generic matrix algebras. Finally, in Section 4 we present some recent results obtained jointly with C.K. Gupta. We also give some open problems either old and well known or new and arising from our exposition.

1. Polynomial Algebras

We fix an m-dimensional vector space V_m with basis $x_1, \dots, x_m$, $m \geq 2$, over the base field K and denote by $K[V_m]$, $K\langle V_m \rangle$ and L_m respectively the polynomial algebra, the free associative algebra and the free Lie algebra freely generated by $x_1, \dots, x_m$. Clearly, every map $x_j \to K[V_m]$, $j = 1, \dots, m$, induces an endomorphism of $K[V_m]$, similarly for $K\langle V_m \rangle$ and L_m. In the sequel we shall define the automorphisms on the set of the free generators only. One of the main problems under discussion will be the following.

Problem 1.1. *Find a natural and "nice looking" set of generators of the automorphism group of* $K[V_m]$, $K\langle V_m \rangle$ *and* L_m.

One of the first results in this spirit is for free groups and is due to Nielsen [29].

Theorem 1.2. ([29]) *The automorphism group of the free group* G_m *of rank* $m \geq 2$ *is generated by the following automorphisms* $\sigma, \theta_1, \theta_2$ *defined by*

(i) $\sigma(x_j) = x_{\sigma_0(j)}$, *where* σ_0 *belongs to the symmetric group* S_m *of degree* m;
(ii) $\theta_1(x_1) = x_1^{-1}$, $\theta_1(x_j) = x_j$, $j \neq 1$;
(iii) $\theta_2(x_1) = x_1 x_2$, $\theta_2(x_j) = x_j$, $j \neq 1$.

Traditionally the automorphisms of a finitely generated relatively free group G induced by automorphisms of the free group are called tame and the other automorphisms are wild. The natural translation of the notion of tameness in the language of ring theory is the following.

Definition 1.3. An automorphism of $K[V_m]$ is called tame if it belongs to the subgroup of $\operatorname{Aut} K[V_m]$ generated by the affine automorphisms

$$x_j \to \alpha_j + \sum_{i=1}^{m} \beta_{ij} x_i, \quad \alpha_j, \beta_{ij} \in K, \quad \det(\beta_{ij}) \neq 0,$$

and the triangular (or de Jonquières) automorphisms

$$x_j \to \alpha_j x_j + f_j(x_1, \dots, x_{j-1}), \quad 0 \neq \alpha_j \in K, \quad f_j \in K[V_m],$$

where the polynomials $f_j(x_1, \dots, x_{j-1})$ do not depend on the variables $x_j, \dots, x_m$. The other automorphisms of $K[V_m]$ (if they exist) are called wild. The notion of tame and wild automorphisms is defined in a similar way for $K\langle V_m \rangle$ and L_m and is also extended to the automorphisms of the relatively free homomorphic images R of $K\langle V_m \rangle$ and L_m, assuming that the algebra R is generated by a fixed set $\{r_1, \dots, r_m\}$, $R \cong K\langle V_m \rangle / \operatorname{Ker} \pi$ and $\pi : x_j \to r_j$, $j = 1, \dots, m$.

The classical form of Problem 1.1 is the following.

Problem 1.4. *Are all automorphisms of* $K[V_m]$, $K\langle V_m \rangle$ *and* L_m *tame?*

For polynomial algebras the affirmative answer is known for $m = 2$ only and is given by the following classical result due to Jung [22] for $K = \mathbb{C}$ and Van der Kulk [37] for any field K.

Theorem 1.5. ([22, 37]) *All automorphisms of the polynomial algebra in two variables* $K[V_2] = K[x_1, x_2]$ *are tame.*

There exist several proofs of Theorem 1.5 based on different ideas from algebraic geometry and combinatorics (see e.g. Nagata [27], Cohn [10] and Dicks [12]). The proof of Van der Kulk [37] gives also the description of $\operatorname{Aut} K[x_1, x_2]$ (see e.g. [12] for the proof of the next theorem and for references).

Theorem 1.6. ([37]) *The group* $\operatorname{Aut} K[x_1, x_2]$ *is the amalgamated free product of the group of affine automorphisms and the group of triangular automorphisms over their intersection with the upper triangular affine automorphisms.*

There exist many algorithms to determine whether a homomorphism of $K[x_1, x_2]$ is an automorphism and if it is, to find its decomposition into a product of affine and triangular automorphisms (see e.g. Cheng and Wang [8]). We shall also mention a difficult result of Suzuki [35] based on the decomposition of $\operatorname{Aut} K[x_1, x_2]$.

Theorem 1.7. ([35]) *The automorphisms of finite order of* $\mathbb{C}[x_1, x_2]$ *are conjugate to linear automorphisms.*

For a survey on the following problem see Kraft and Schwarz [23].

Problem 1.8. *Is every automorphism of finite order in* $\operatorname{Aut} \mathbb{C}[V_m]$*,* $m \geq 3$*, conjugate to a linear automorphism?*

For $m > 2$ Problem 1.4 is still open and seems to be very difficult. The only positive result known to us is due to Anick [3]. First we recall the definition of the formal power series topology of $\operatorname{Aut} K[V_m]$. Let ω be the augmentation ideal of $K[V_m]$, i.e. the set of all polynomials without constant terms in $K[V_m]$. Considering the powers ω^n of ω, $n = 1, 2, \ldots$, as a family of neighbourhoods of the zero polynomial, we equip $K[V_m]$ with a topology called the formal power series topology of $K[V_m]$. It is described by a metric d on $K[V_m]$ satisfying $d(f, g) = \exp(-\operatorname{ht}(f - g))$, where $f, g \in K[V_m]$ and $\operatorname{ht}(f - g) = n$ if $f - g \in \omega^n \setminus \omega^{n+1}$ (i.e. $\operatorname{ht}(f - g) = n$ if and only if the minimal degree of the terms of $f - g$ is equal to n). This topology induces topology on $\operatorname{End} K[V_m]$ also called the formal power series topology of $\operatorname{End} K[V_m]$. The continuous endomorphisms of $K[V_m]$ are those which send the variables x_j to polynomials without constant terms. A sequence of endomorphisms $\phi_1, \phi_2, \ldots$ converges to $\phi \in \operatorname{End} K[V_m]$ if $\operatorname{ht}(\phi_k(x_j) - \phi(x_j)) \to \infty$ when $k \to \infty$ for $j = 1, \ldots, m$.

Theorem 1.9. ([3]) *Let* $\operatorname{char} K = 0$*. Then the subgroup of tame automorphisms of* $K[V_m]$*,* $m \geq 2$*, is dense in the group of all automorphisms.*

In other words, the automorphisms of $K[V_m]$ can be approximated by tame automorphisms. For every automorphism ϕ of $K[V_m]$ there exists a sequence of tame automorphisms $\phi_1, \phi_2, \ldots$, such that for any $n \geq 1$ and any $j = 1, \ldots, m$, the polynomials $\phi(x_j)$ and $\phi_n(x_j)$ have the same terms of degree up to n. Of course, these polynomials may have different terms of higher degree.

In 1970 Nagata [27] gave his famous example of an automorphism of $K[x_1, x_2, x_3]$ which fixes the variable x_3 and is not tame considered as an automorphism of the $K[x_3]$-algebra $(K[x_3])[x_1, x_2]$. The automorphism of Nagata can be obtained in the following way. For simplicity of exposition we assume that the base field K is of characteristic 0. Let Δ be a locally nilpotent derivation of $K[V_m]$, i.e. Δ is a linear transformation of the vector space $K[V_m]$ satisfying $\Delta(fg) = \Delta(f)g + f\Delta(g)$, $f, g \in K[V_m]$, and for any $f \in K[V_m]$ there exists a positive integer n such that $\Delta^n(f) = 0$. Then

$$\delta = \exp(\Delta) = \mathrm{id} + \frac{\Delta}{1!} + \frac{\Delta^2}{2!} + \dots$$

is an automorphism of $K[V_m]$. Obviously, δ is well-defined because Δ is locally nilpotent. A derivation ∂ of $K[V_m]$ is (lower) triangular if $\partial x_j \in K[x_{j+1}, \dots, x_m]$, $j = 1, \dots, m$. Clearly, ∂ is locally nilpotent. Let $w \in \operatorname{Ker} \partial$. Then $\Delta = w\partial$ is also a locally nilpotent derivation of $K[V_m]$. Up to notation the automorphism of Nagata is obtained from the triangular derivation ∂ given by

$$\partial x_1 = x_2, \quad \partial x_2 = x_3, \quad \partial x_3 = 0$$

and for $w = x_2^2 - 2x_1x_3 \in \operatorname{Ker} \partial$ and is defined by

$$\begin{aligned} \nu(x_1) &= x_1 + \left(x_2^2 - 2x_1x_3\right)x_2 + \frac{1}{2}\left(x_2^2 - 2x_1x_3\right)^2 x_3 \\ \nu(x_2) &= x_2 + \left(x_2^2 - 2x_1x_3\right)x_3 \\ \nu(x_3) &= x_3. \end{aligned}$$

Problem 1.10. *Is the automorphism of Nagata tame?*

Very probably the automorphism of Nagata is wild and there have been some attempts to find evidence for its wildness. For example, Alev [1] found that ν does not allow some special kind of decompositions into product of affine and triangular automorphisms; Drensky, Gutierrez and Yu [19] did some computing experiments based on Gröbner basis techniques indicating that the tame automorphisms should have some properties which do not hold for the Nagata automorphism. Nevertheless there is a result of Martha Smith [34] showing that although the Nagata automorphism may be wild, it is not "too wild".

Theorem 1.11. ([34]) *Let ∂ be a triangular derivation of $K[V_m]$ and let $w \in \operatorname{Ker} \partial$. Then the automorphism ϕ of $K[V_{m+1}]$ defined by* $\exp(w\partial)$ *on $x_1, \dots, x_m$ and fixing x_{m+1} is tame.*

The automorphisms of $K[V_m]$ which become tame on $K[V_{m+n}]$ for some $n > 0$ if we define them identically on the new variables $x_{m+1}, \dots, x_{m+n}$, are called stably tame. In other words, if a wild automorphism of $K[V_m]$ is stably tame, this means that the only reasons for the wildness is that there is no room to work in the polynomial algebra of the m-dimensional vector space V_m.

From some point of view, all known automorphisms of the polynomial algebra arise from natural constructions and are "nice looking". Now we can make Problem 1.1 precise in the following way.

Problem 1.12. *Is the automorphism group of $K[V_m]$, $m > 2$, generated by:*

(i) *automorphisms* $\exp(\Delta)$, *where $\Delta = w\partial$, $w \in \operatorname{Ker} \partial$ and ∂ is a triangular derivation;*
(ii) *automorphisms* $\exp(\Delta)$, *where Δ is a locally nilpotent derivation;*
(iii) *stably tame automorphisms;*
(iv) *other natural automorphisms?*

The above discussion suggests the following question.

Problem 1.13. *Can every nilpotent derivation of $K[V_m]$ be obtained by starting with a locally nilpotent derivation ∂ of some particular simple sort, multiplying it by a member w from the kernel of ∂ and conjugating by an automorphism of $K[V_m]$?*

Remark 1.14. Up till now, we have assumed that all associative algebras are unitary. This is not very essential for our considerations. Let ω be the augmentation ideal of $K[V_m]$; clearly ω is isomorphic to the nonunitary polynomial algebra. Let ϕ be an endomorphism of $K[V_m]$, $\phi(x_j) = \alpha_j + f_j$, where $\alpha_j \in K$ and $f_j \in \omega$, $j = 1, \dots, m$, and let ψ be the endomorphism of ω defined by $\psi(x_j) = f_j$. It is easy to see that ϕ is an automorphism of $K[V_m]$ if and only if ψ is an automorphism of ω and ϕ and ψ are simultaneously tame or wild. Similar arguments hold for free associative algebras in the next section.

2. Free Associative and Lie Algebras

For the automorphisms of the free associative algebra $K\langle V_m\rangle$, the situation is similar to that for the automorphisms of the polynomial algebra but the study in the noncommutative case seems to be less intensive than in the commutative. Makar-Limanov [26] and Czerniakiewicz [11] independently described the automorphisms of $K\langle x_1, x_2\rangle$.

Theorem 2.1. ([26, 11]) *All automorphisms of $K\langle x_1, x_2\rangle$ are tame.*

The definition of triangular automorphisms of $K[V_2]$ and $K\langle V_2\rangle$ involves polynomials in one variable only (which are automatically in commuting variables). As a consequence of Theorem 2.1 and the decomposition of the automorphism group of $K[V_2]$ from Theorem 1.6 one obtains immediately the following.

Corollary 2.2. *The groups* $\operatorname{Aut} K\langle x_1, x_2\rangle$ *and* $\operatorname{Aut} K[x_1, x_2]$ *are isomorphic.*

As in the commutative case, the problem of whether all automorphisms of the free associative algebra are tame is still open for $m > 2$. One of the main difficulties is that the proof of Theorem 2.1 is based on the proof of Theorem 1.5 and we do not know the structure of $\operatorname{Aut} K[V_m]$ for $m > 2$. For other difficulties see below. For a detailed survey on the topic including proofs of Theorems 2.1 and 2.2 see Cohn [10] and Dicks [13]. In particular, in [10] one can find some examples of automorphisms of $K\langle V_m\rangle$, $m > 2$, which are considered as potential candidates to be wild. About their stable tameness see [34]. Of course, one may ask the free associative analogue of Problem 1.12 or the following problem which arises from Theorem 1.9.

Problem 2.3. *Is the subgroup of the tame automorphisms dense in the group of all automorphisms (with respect to the formal power series topology)?*

The picture is much better in the case of free Lie algebras. Cohn [9] proved the following, now classical, result.

Theorem 2.4. ([9]) *For every integer $m \geq 2$ the automorphisms of the free Lie algebra L_m are tame.*

Remark 2.5. Since the only Lie elements in one variable are its scalar multiples, for $m = 2$ the theorem of Cohn gives that

$$\operatorname{Aut} L_2 \cong GL(V_2) \cong GL_2(K) \cong GL_2.$$

The original proof of Cohn is based on the generalization of the Euclidean algorithm to the noncommutative case and on the technique of free ideal rings. Another proof based on the combinatorics of the bases of the free Lie algebra is due to Kukin [24] (see also [15] for other references). Comparing the final form of the results on the automorphisms of free groups and free Lie algebras with the partial results for polynomial and free associative algebras it seems to us that the main difference between groups and Lie algebras on the one hand and polynomial and free associative algebras on the other lies in the good combinatorics in the group and Lie cases. For example, every subgroup of the free group is free again; a similar statement holds for the subalgebras of the free Lie algebras. There exist algorithms which, for any set of generators of the free group and the free Lie algebra, give a minimal system of free generators and transform this system to the canonical system of free generators $\{x_1, \ldots, x_m\}$. Each step of the algorithms corresponds to a tame automorphism from Theorem 1.2 and Definition 1.3. Unfortunately, these arguments do not work in the associative case.

Remark 2.6. In the introduction we restricted our considerations to relatively free algebras R, i.e. $R \cong K\langle V_m\rangle/I$, where I is the ideal of polynomial identities in m variables for an algebra S. This condition is stronger than the condition $\phi(I) = I$, $\phi \in \operatorname{Aut} K\langle V_m\rangle$. For the case of free groups see Neumann [28]. In the case of algebras, an easy example is the free Lie algebra L_2. By virtue of Remark 2.5, every ideal which is invariant under the action of the general linear group GL_2 is invariant under all automorphisms of L_2 and this class of ideals is much bigger than the class of the ideals of polynomial identities.

The following problem generalizes the problem of the description of the automorphisms of relatively free algebras.

Problem 2.7. *Describe the automorphisms of $K\langle V_m\rangle/I$ and L_m/I where the ideal I is invariant under the automorphisms of $K\langle V_m\rangle$ and L_m. Find naturally arising automorphisms of $K\langle V_m\rangle/I$ and L_m/I which generate dense subgroups of the group of all automorphisms. In particular, describe the ideals I such that all automorphisms of the factor algebra modulo I are tame.*

3. Generic Matrix Algebras

In this section we fix a positive integer $k > 1$ and the polynomial algebra in mk^2 variables

$$\Omega = K\big[\xi_{pq}^{(j)} \mid p, q = 1, \ldots, k, \quad j = 1, \ldots, m\big].$$

The algebra R_{mk} of m generic $k \times k$ matrices $y_1, \ldots, y_m$ is the subalgebra of the $k \times k$ matrix algebra $M_k(\Omega) = \Omega \otimes_K M_k(K)$ with entries from Ω generated by the generic matrices $y_j = (\xi_{pq}^{(j)})$, $j = 1, \ldots, m$. If the base field K is infinite, R_{mk} is isomorphic to $K\langle V_m\rangle/I$, where I is the ideal of polynomial identities in m variables for the $k \times k$ matrix algebra $M_k(K)$. Over a finite field K, the ideal I consists of all polynomial identities of the K-algebra $M_k(E)$, where E is an infinite extension of the field K. In the theory of PI-algebras and some other branches of algebra the generic matrix algebras are among the most attractive objects for investigation (see e.g. Rowen [33] and Formanek [20]). In the early 80's Bergman wrote the seminal paper [5] (which unfortunately never appeared in a journal version) where he constructed a family of wild automorphisms for the algebra R_{2k} of two generic $k \times k$ matrices. Recall that the (Lie) commutators are defined inductively by

$$[u_1, u_2] = u_1(\operatorname{ad} u_2) = u_1u_2 - u_2u_1,$$

$$[u_1, \dots, u_{n-1}, u_n] = [[u_1, \dots, u_{n-1}], u_n], \quad n > 2.$$

Theorem 3.1. ([5]) *Let $f(y_1, y_2)$ be a nonzero element of the algebra R_{2k} of two generic $k \times k$ matrices such that one of the following conditions holds:*

(i) *$f(y_1, y_2)$ is in the centre of R_{2k} and the variable y_2 participates in f in commutators only, i.e. f is a linear combination of expressions of the form*

$$y_1^{n_1}[\dots]y_1^{n_2}[\dots]\dots y_1^{n_{t-1}}[\dots]y_1^{n_t},$$

where $[\dots]$ are commutators in y_1 and y_2;

(ii) *$[f(y_1, y_2), y_1] = 0$ and f is a linear combination of*

$$y_1^{n_1}[y_2, y_1]y_1^{n_2}[y_2, y_1]\dots y_1^{n_{t-1}}[y_2, y_1]y_1^{n_t}.$$

Then the endomorphism β_f of R_{2k} defined by

$$\beta_f(y_1) = y_1$$
$$\beta_f(y_2) = y_2 + f(y_1, y_2)$$

is a wild automorphism of R_{2k}.

The idea of Bergman is the following. There are canonical homomorphisms

$$K\langle x_1, x_2\rangle \longrightarrow R_{2k} \longrightarrow K[x_1, x_2]$$

which give rise to group homomorphisms

$$\operatorname{Aut} K\langle x_1, x_2\rangle \longrightarrow \operatorname{Aut} R_{2k} \longrightarrow \operatorname{Aut} K[x_1, x_2].$$

By Corollary 2.2 the composition of the group homomorphisms is an isomorphism. Hence R_{2k} possesses a wild automorphism if the kernel of $\operatorname{Aut} R_{2k} \longrightarrow \operatorname{Aut} K[x_1, x_2]$ is nontrivial. The conditions on the polynomial $f(y_1, y_2)$ are chosen in such a way as to guarantee that β_f is invertible (with inverse β_{-f}) and β_f is in the desired kernel. The simplest possibility is to choose $f(y_1, y_2)$ to be of minimal degree among the multihomogeneous central polynomials in two variables for $M_k(K)$.

Example 3.2. ([5]) *The endomorphism of R_{22} defined by*

$$\beta_f(y_1) = y_1$$
$$\beta_f(y_2) = y_2 + [y_1, y_2]^2$$

is a wild automorphism of R_{22}.

Proof. By the Cayley–Hamilton theorem, any 2×2 matrix a satisfies

$$a^2 - \operatorname{tr}(a) + \det(a)\operatorname{id} = 0.$$

Since the trace of the matrix $a = [y_1, y_2]$ is equal to zero, we obtain that $[y_1, y_2]^2$ is a scalar matrix, i.e. $[y_1, y_2]^2$ satisfies the assumptions of Theorem 3.1.

Problem 3.3. *Find a natural set of generators of the automorphism group of* R_{mk} *and, if this is not possible, for some dense subgroups of* $\operatorname{Aut} R_{mk}$. *For* $m > 2$, *are all automorphisms of* R_{mk} *tame?*

There are several important algebras related with the generic matrix algebra R_{mk}. We denote by $\bar{C}_{mk}$ the algebra generated by the traces of the elements from R_{mk} and by T_{mk} the generic trace algebra generated by R_{mk} and $\bar{C}_{mk}$. It is known that T_{mk} and $\bar{C}_{mk}$ have nice algebraic properties (see e.g. Van den Bergh [36]). Every automorphism of R_{mk} induces an automorphism of $\bar{C}_{mk}$ and T_{mk}. The result of Bergman has an interesting relation with commutative algebra. The algebra $\bar{C}_{22}$ is the only algebra among $\bar{C}_{mk}$ which is isomorphic to a polynomial algebra,

$$\bar{C}_{22} \cong K\left[\operatorname{tr}(y_1), \operatorname{tr}(y_2), \det(y_1), \det(y_2), \operatorname{tr}(y_1 y_2)\right].$$

Therefore constructing wild automorphisms of R_{22} we obtain automorphisms of the polynomial algebra in five variables which are suspected to be also wild. Unfortunately, Alev and Le Bruyn [2] established that a large class of automorphisms obtained in this way gives rise to tame automorphisms of the polynomial algebras. It is also known (see e.g. Procesi [30]) that over a field of characteristic 0

$$T_{mk} \cong K\left[\operatorname{tr}(y_1), \dots, \operatorname{tr}(y_m)\right] \otimes_K R_{mk}^{(0)},$$

where $R_{mk}^{(0)}$ is the associative algebra generated by m generic $k \times k$ trace-zero matrices. It is known that

$$R_{mk}^{(0)} \cong K\langle V_m \rangle / I_{mk},$$

where I_{mk} is the ideal of the weak polynomial identities in m variables for the $k \times k$ matrix algebra $M_k(K)$. In other words, I_{mk} consists of all polynomials which vanish on the canonical k-dimensional representation of the Lie algebra $\mathrm{sl}_k(K)$. The study of the weak polynomial identities was initiated by Razmyslov [31, 32]. In particular he described the ideal I_{mk} for $k = 2$ [31]. It is known that $K\langle V_m \rangle / I_{m2}$ is a generic Clifford algebra for $m = 2, 3$ (see e.g. [25]), and Alev and Le Bruyn [2] studied also the automorphisms of the polynomial algebras related with generic Clifford algebras. Again, they established that classes of automorphisms of Clifford algebras induce tame automorphisms of polynomial algebras.

Problem 3.4. *Understanding better the automorphism group of the generic matrix algebras* R_{mk}, *what kind of consequences for commutative algebra and algebraic geometry can one obtain? In particular, try to find wild automorphisms for the commutative algebras* $\bar{C}_{km}$.

4. Some New Results

In this section we present some new joint results with C.K. Gupta [17, 18] on the automorphisms of the algebra of two generic 2×2 matrices and the polynomial algebra in five variables. The structure of R_{22} was described by Formanek, Halpin and Li [21], for a background on the properties of the related with R_{22} algebras see Le Bruyn [25]. For simplicity of notation we denote by x and y the two generic 2×2 matrices, by R, $\bar{C}$ and T respectively the algebras R_{22}, $\bar{C}_{22}$ and T_{22} and by C the center of R. It is known that $C = R \cap \bar{C}$.

Proposition 4.1. ([21]) *The centre $\bar{C}$ of T is the polynomial algebra in the commuting variables*

$$\operatorname{tr}(x), \quad \operatorname{tr}(y), \quad \det(x), \quad \det(y), \quad \operatorname{tr}(xy).$$

The vector subspace of C consisting of all polynomials without constant terms is a free $\bar{C}$-module generated by $[x,y]^2$.

In the next theorem we give a family of new automorphisms of T, R and $\bar{C}$.

Theorem 4.2. ([17]) *Let R be the algebra generated by two generic 2×2 matrices x and y over a field K and let T be the trace algebra generated by R and the traces of the matrices of R.*

(i) *For any polynomial $f \in K[t_1,t_2,t_3,t_4]$ let*

$$\tilde{f} = f\Big(\operatorname{tr}(x), \operatorname{tr}(y), [x,y]^2, \big(x\operatorname{tr}(y) - y\operatorname{tr}(x)\big)^2\Big).$$

Then the mapping $\phi_f : \{x,y\} \longrightarrow T$ defined by

$$\phi_f(x) = x + \operatorname{tr}(x)\big(x\operatorname{tr}(y) - y\operatorname{tr}(x)\big)\tilde{f},$$

$$\phi_f(y) = y + \operatorname{tr}(y)\big(x\operatorname{tr}(y) - y\operatorname{tr}(x)\big)\tilde{f}$$

gives rise to an automorphism of T which fixes $\operatorname{tr}(x)$, $\operatorname{tr}(y)$, $[x,y]^2$ and $x\operatorname{tr}(y) - y\operatorname{tr}(x)$.

(ii) *If $f(t_1,t_2,t_3,t_4)$ is divisible by t_3, then ϕ_f induces also an automorphism of R. If $f \neq 0$ then this automorphism is wild.*

(iii) *The mapping ϕ_f induces an automorphism of the centre*

$$\bar{C} = K\big[\operatorname{tr}(x), \operatorname{tr}(y), \det(x), \det(y), \operatorname{tr}(xy)\big]$$

of T which fixes the variables $\operatorname{tr}(x)$, $\operatorname{tr}(y)$ and the elements $[x,y]^2$ and $(x\operatorname{tr}(y) - y\operatorname{tr}(x))^2$.

If $\operatorname{char} K \neq 2$, the algebra $\bar{C}$ has a better set of generators which makes it possible to give explicitly the automorphisms from the previous theorem. It turns out that these automorphisms enjoy many properties of the automorphism of Nagata for $K[V_3]$. Let

$$x_0 = x - \frac{1}{2}\operatorname{tr}(x), \quad y_0 = y - \frac{1}{2}\operatorname{tr}(y)$$

be the generic trace-zero 2×2 matrices. As in [2] we may replace in T the generic matrices x and y by x_0 and y_0 and assume that T is generated by x_0, y_0, $\operatorname{tr}(x)$, $\operatorname{tr}(y)$, $\det(x_0)$, $\det(y_0)$, $\operatorname{tr}(x_0y_0)$. Standard arguments based on the Cayley-Hamilton theorem and the Newton formulae for symmetric functions give that

$$\det(x_0) = \frac{1}{2}\big(\operatorname{tr}^2(x_0) - \operatorname{tr}(x_0^2)\big) = -\frac{1}{2}\operatorname{tr}(x_0^2), \quad \det(y_0) = -\frac{1}{2}\operatorname{tr}(y_0^2),$$

and $\bar{C}$ is generated by

$$p = \operatorname{tr}(x), \quad q = \operatorname{tr}(y), \quad u = \operatorname{tr}(x_0^2), \quad v = \operatorname{tr}(y_0^2), \quad t = \operatorname{tr}(x_0y_0).$$

Easy calculations show that

$$[x,y]^2 = t^2 - uv, \quad \big(x\operatorname{tr}(y) - y\operatorname{tr}(x)\big)^2 = q^2u + p^2v - 2pqt$$

and Theorem 4.2 (iii) has the following form.

Theorem 4.3. ([17]) *Let* $\operatorname{char} K \neq 2$, *let* $f(t_1, t_2, t_3, t_4)$ *be a polynomial in four commuting variables and let*

$$\tilde{f} = f(p, q, t^2 - uv, q^2u + p^2v - 2pqt).$$

Then the endomorphism ψ_f *of the polynomial algebra in five variables* $K[p, q, u, v, t]$ *defined by*

$$\psi_f(p) = p, \quad \psi_f(q) = q,$$

$$\psi_f(u) = u + 2p(qu - pt)\tilde{f} + p^2(q^2u + p^2v - 2pqt)\tilde{f}^2,$$

$$\psi_f(v) = v + 2q(qt - pv)\tilde{f} + q^2(q^2u + p^2v - 2pqt)\tilde{f}^2,$$

$$\psi_f(t) = t + (q^2u - p^2v)\tilde{f} + pq(q^2u + p^2v - 2pqt)\tilde{f}^2.$$

is an automorphism which fixes p, q, $t^2 - uv$ *and* $q^2u + p^2v - 2pqt$.

Remark 4.4. Let $\operatorname{char} K \neq 2$. In the special case when $\tilde{f} = f(p, q)$, let

$$z_1 = q^2u + p^2v - 2pqt$$
$$z_2 = q^2u - p^2v$$
$$z_3 = q^2u\tilde{f} - p^2v\tilde{f} - 2t.$$

Then $\{z_1, z_2, z_3\}$ is a basis of the $K(p, q)$-vector space spanned by u, v, t. Direct calculations show that

$$\psi_f(z_1) = z_1$$
$$\psi_f(z_2) = z_2 + 2pq\tilde{f}z_1$$
$$\psi_f(z_3) = z_3 - 2\tilde{f}z_2.$$

Therefore the automorphism ψ_f induces a tame automorphism of $(K(p, q))[u, v, t]$ because acts triangularly. We do not know if it is tame for $K[p, q, u, v, t]$ or for $(K[p, q])[u, v, t]$.

Problem 4.5. *Are the automorphisms of* $K[p, q, u, v, t]$ *defined in Theorem* 4.3 *tame or stably tame?*

Till the end of the section we assume that the base field K is of characteristic 0 and denote by L the Lie subalgebra of R generated by x and y. It is easy to see that L is isomorphic to the relatively free algebra of rank 2 in the variety of Lie algebras defined by the polynomial identities of the Lie algebra $\mathrm{sl}_2(K)$, i.e. L is the factor algebra of the free Lie algebra L_2 modulo the ideal of polynomial identities in two variables for $\mathrm{sl}_2(K)$. The next result is in the spirit of the result of Anick given in Theorem 1.9.

Theorem 4.6. ([18]) *Let the base field* K *be of characteristic* 0, *let* L *be the Lie subalgebra of* R *generated by* x *and* y *and let* ϕ_{kl} *be the endomorphisms of* L *defined by*

$$\phi_{kl}(x) = x, \quad \phi_{kl}(y) = y + y(\operatorname{ad}^{2k} x)(\operatorname{ad}^{2l+1}[y, x]), \quad k \geq 1, l \geq 0.$$

Then ϕ_{kl} *are automorphisms of* L *and the subgroup* G *of* $\operatorname{Aut} L$ *generated by* $GL_2(K)$ *and all* ϕ_{kl} *is dense in* $\operatorname{Aut} L$ *with respect to the formal power series topology. In* *other words, for every $\psi \in \operatorname{Aut} L$ and every $n \geq 1$ there exists an automorphism ψ_n from G such that the polynomials $\psi(x) - \psi_n(x)$ and $\psi(y) - \psi_n(y)$ have no terms of degree $\leq n$.*

Remark 4.7. The automorphisms ϕ_{kl} are obtained by Bergman's trick in Theorem 3.1 (ii) for constructing wild automorphisms of R. Of course, it is not clear whether $GL_2(K)$ and the ϕ_{kl}'s generate $\operatorname{Aut} L$. For example, in a similar situation Bryant and the author [6] proved that the tame automorphisms of the free metabelian algebra L_m/L_m'', $m > 3$, form a dense subgroup of $\operatorname{Aut} L_m/L_m''$ and Bahturin and Nabiyev [4] established that all inner automorphisms of L_m/L_m'', $m \geq 2$, are wild.

Now we give some idea of the techniques used in the proof of Theorem 4.6. The general linear group $GL_2 = GL_2(K)$ acts canonically as a group of automorphisms of L and L is a completely reducible GL_2-module. The irreducible components of L are described by the author in [14]. On the other hand in the joint papers of the author with C.K. Gupta [16] and Bryant [6] the representation theory of the general linear group was involved to investigation of the automorphisms of relatively free algebras. In our particular case, let

$$I_nA = \{\phi \in \operatorname{Aut} L \mid \phi(x) \equiv x, \phi(y) \equiv y \pmod{L^n}\}, \quad n = 2, 3, \dots .$$

In this way we obtain a descending series of normal subgroups of $\operatorname{Aut} L$ such that $\operatorname{Aut} L = GL_2 \cdot I_2A$. The group GL_2 acts by conjugation on $\operatorname{Aut} L$. It turns out that the factor groups $I_nA/I_{n+1}A$ have a natural structure of GL_2-modules. One of the key observations in [16, 6] is that the GL_2-modules $I_nA/I_{n+1}A$ are isomorphic to submodules of the tensor products $V_2^* \otimes_K L^n/L^{n+1}$, $n = 2, 3, \dots$, where V_2^* is the dual of the GL_2-module V_2 with the canonical action of GL_2 as invertible 2×2 matrices. Using the decomposition of L from [14] we obtain explicitly all highest weight vectors of the irreducible components of the GL_2-modules $V_2^* \otimes_K L^n/L^{n+1}$. These highest weight vectors induce families of endomorphisms of L and the problem is to distinguish which family contains an automorphism from I_nA. The endomorphisms under consideration induce also endomorphisms on the polynomial algebra $K[u, v, t]$ and on the K-vector space $K[u, v, t]x_0 \oplus K[u, v, t]y_0$, where x_0 and y_0 are the generic trace-zero matrices and $u = \operatorname{tr}(x_0^2)$, $v = \operatorname{tr}(y_0^2)$, $t = \operatorname{tr}(x_0y_0)$. Starting the process with an automorphism of L we give rise to automorphisms of these objects. Hence we obtain that some matrices with entries from $K[u, v, t]$ are invertible (for example the Jacobian matrix of the induced automorphism of $K[u, v, t]$). Finally we use a trace criterion for invertibility of a matrix with polynomial entries involved successfully by Bryant, C.K. Gupta, Levin and Mochizuki [7] in the study of automorphisms of relatively free groups. We see that the only highest weight vectors of $V_2^* \otimes_K L^n/L^{n+1}$ which correspond to automorphisms of L are those from the statement of Theorem 4.6.

Problem 4.8. *Is the group* $\operatorname{Aut} L$ *generated by the automorphisms from Theorem* 4.6*?*

We have the feeling that the available information about the automorphisms of the generic matrix algebras and the algebras related with them is not enough to have the complete picture. Maybe it is a good idea to ask the following "general" problem.

Problem 4.9. *Do what you can!*

Acknowledgements

The author is very grateful to G.M. Bergman for the comments on the preliminary version of this paper, for the useful suggestions and for asking Problem 1.13.

References

[1] J. Alev, A note on Nagata's automorphism, in Automorphisms of Affine Spaces (A. van den Essen, ed.), Kluwer Acad. Publ., Dordrecht, 1995, pp. 215–221.

[2] J. Alev and L. Le Bruyn, Automorphisms of generic 2 by 2 matrices, in Perspectives in Ring Theory, (Antwerp, 1987), NATO Adv. Sci. Inst. Ser. C: Math. Phys. Sci. **233**, Kluwer Acad. Publ., Dordrecht, 1988, pp. 69–83.

[3] D.J. Anick, Limits of tame automorphisms of $k[x_1, \ldots, x_N]$, *J. Algebra* **82** (1983), 459–468.

[4] Yu.A. Bahturin and S. Nabiyev, Automorphisms and derivations of abelian extensions of some Lie algebras, *Abh. Math. Sem. Univ. Hamburg* **62** (1992), 43–57.

[5] G.M. Bergman, Wild automorphisms of free P.I. algebras, and some new identities, preprint.

[6] R.M. Bryant and V. Drensky, Dense subgroups of the automorphism groups of free algebras, *Canad. J. Math.* **45** (1993), 1135–1154.

[7] R.M. Bryant, C.K. Gupta, F. Levin and H.Y. Mochizuki, Non-tame automorphisms of free nilpotent groups, *Commun. Algebra* **18** (1990), 3619–3631.

[8] C. Cheng and S. Wang, An algorithm that determines whether a polynomial map is bijective, in Automorphisms of Affine Spaces (A. van den Essen, ed.), Kluwer Acad. Publ., Dordrecht, 1995, pp. 169–176.

[9] P.M. Cohn, Subalgebras of free associative algebras, *Proc. London Math. Soc. (3)* **14** (1964), 618–632.

[10] P.M. Cohn, *Free Rings and Their Relations*, 2nd Edition, London Math. Soc. Monographs No.2, Academic Press, London, 1985.

[11] A.J. Czerniakiewicz, Automorphisms of a free associative algebra of rank 2, I, II, Trans. Amer. Math. Soc. **160** (1971), 393–401; **171** (1972), 309–315.

[12] W. Dicks, Automorphisms of the polynomial ring in two variables, *Publ. Mat. Univ. Autònoma de Barcelona* **27** (1983), 155–162.

[13] W. Dicks, Automorphisms of the free algebra of rank two, in Group Actions on Rings (Susan Montgomery, ed.), Contemp. Math. **43** AMS, Providence R.I., 1985, pp. 63–68.

[14] V. Drensky, Representations of the symmetric group and varieties of linear algebras, *Mat. Sb.* **115** (1981), 98–115 (Russian); translation, *Math. USSR Sb.* **43** (1981), 85–101.

[15] V. Drensky, Endomorphisms and automorphisms of relatively free algebras, *Suppl. Rend. Circolo Mat. di Palermo, Ser. II* **31** (1993), 97–132.

[16] V. Drensky and C.K. Gupta, Automorphisms of free nilpotent algebras, *Canad. J. Math.* **42** (1990), 259–279.

[17] V. Drensky and C.K. Gupta, New automorphisms of generic matrix algebras and polynomial algebras, *J. Algebra* (to appear).

[18] V. Drensky and C.K. Gupta, Lie automorphisms of the algebra of two generic 2×2 matrices, in preparation.

[19] V. Drensky, J. Gutierrez and J.-T. Yu, Gröbner bases and the Nagata automorphism, preprint.

[20] E. Formanek, *The Polynomial Identities and Invariants of $n \times n$ Matrices*, CBMS Regional Conf. Series in Math. **78**, Published for the Confer. Board of the Math. Sci. Washington DC, AMS, Providence RI, 1991.

[21] E. Formanek, P. Halpin and W.-C.W. Li, The Poincaré series of the ring of 2 by 2 generic matrices, *J. Algebra* **69** (1981), 105–112.

[22] H.W.E. Jung, Über ganze birationale Transformationen der Ebene, *J. reine und angew. Math.* **184** (1942), 161–174.

[23] H. Kraft and G. Schwarz, Finite automorphisms of affine N-space, in Automorphisms of Affine Spaces (A. van den Essen, ed.), Kluwer Acad. Publ., Dordrecht, 1995, pp. 55–66.

[24] G.P. Kukin, Primitive elements in free Lie algebras, *Algebra i Logika* **9** (1970), 458–472 (Russian); translation, *Algebra and Logic* **9** (1970), 275–284.

[25] L. Le Bruyn, *Trace rings of generic 2 by 2 matrices*, Memoirs of AMS **66**, No. 363, Providence, R.I., 1987.

[26] L.G. Makar-Limanov, On automorphisms of free algebra with two generators, *Funk. Analiz i ego Prilozh.* **4** (1970), No.3, 107–108. (Russian)

[27] M. Nagata, *On the Automorphism Group of* $k[x, y]$, Lect. in Math., Kyoto Univ., Kinokuniya, Tokyo, 1972.

[28] B.H. Neumann, On characteristic subgroups of free groups, *Math. Z.* **94** (1966), 143–151.

[29] J. Nielsen, Die Isomorphismengruppe der freien Gruppen, *Math. Ann.* **91** (1924), 169–209.

[30] C. Procesi, Computing with 2×2 matrices, *J. Algebra* **87** (1984), 342–359.

[31] Yu.P. Razmyslov, Finite basing of the identities of a matrix algebra of second order over a field of characteristic 0, *Algebra i Logika* **12** (1973), 83–113 (Russian); translation, *Algebra and Logic* **12** (1973), 43–63.

[32] Yu.P. Razmyslov, On a problem of Kaplansky, *Izv. Akad. Nauk SSSR, Ser. Mat.* **37** (1973), 483–501 (Russian); translation, *Math. USSR Izv.* **7** (1973), 479–496.

[33] L.H. Rowen, *Polynomial Identities in Ring Theory*, Academic Press, New York, 1980.

[34] M.K. Smith, Stably tame automorphisms, *J. Pure Appl. Alg.* **58** (1989), 209–212.

[35] M. Suzuki, Propriétés topologiques des polynômes de deux variables complexes, et automorphismes algébriques de l'espace C^2, *J. Math. Soc. Japan* **26** (1974), 241–257.

[36] M. Van den Bergh, Trace rings are Cohen-Macaulay, *J. Amer. Math. Soc.* **2** (1989), 775–799.

[37] W. Van der Kulk, On polynomial rings in two variables, *Nieuw Archief voor Wiskunde (3)* **1** (1953), 33–41.

V. Drensky
Institute of Mathematics and Informatics
Bulgarian Academy of Sciences
Acad. G. Bonchev Str., Block 8
1113 Sofia
Bulgaria
e-mail: drensky@banmatpc.math.acad.bg
drensky@bgearn.acad.bg

Canadian Mathematical Society
Conference Proceedings
Volume **22**, 1998

Matrix Problems and Representation-Theory

Peter Gabriel

Abstract. Our objective is to report on the results obtained by a specialized school around L. A. Nazarova and A. V. Roiter. The publications of the school are not so accessible. General features may be well-spread, but the details are not known even to experts of representation theory of finite-dimensional algebras. Yet the results seem to deserve publicity.

§1. The General Context

Let k be an algebraically closed field and $\mathcal{A}$ an *aggregate* over k, i.e. an additive category $\mathcal{A}$ equipped with a homomorphism of rings $\mu : k \to \operatorname{End} \mathbb{I}_{\mathcal{A}}$ such that the conditions A1 and A2 are satisfied:

A1: $\dim_k \mathcal{A}(x, y) < \infty$, $\forall x, y \in \mathcal{A}$.

A2: Each object of $\mathcal{A}$ is a finite direct sum of objects whose rings of endomorphisms are local.

Here we denote by $\mathcal{A}(x, y)$ the group of morphism from x to y. This group is endowed with the structure of a vector space by means of the map

$$k \times \mathcal{A}(x, y) \to \mathcal{A}(x, y), \quad (\lambda, f) \mapsto \mu(\lambda)(y) \circ f = f \circ \mu(\lambda)(x).$$

Let further M be a *locally finite-dimensional module* over $\mathcal{A}$, i.e. a k-linear functor from $\mathcal{A}$ to the category $\operatorname{mod} k$ of finite-dimensional vector spaces. A *space over* M is a triple (V, f, x) formed by a $V \in \operatorname{mod} k$, an $x \in \mathcal{A}$ and a k-linear map $V \to M(x)$. A *morphism* $(V, f, x) \to (W, g, y)$ of spaces over M is given by a linear map $\ell : V \to W$ and a morphism $\eta \in \mathcal{A}(x, y)$ such that $M(\eta) \circ f = g \circ \ell$. With the obvious composition, the spaces over M thus give rise to a new aggregate, which we denote by $M\mathcal{A}$.

The problem is to classify the (indecomposable) spaces over M up to isomorphism [10].

AMS Subject Classification (1991). 16G60, 16G20.

§2. Example 1

Let S be a finite poset and $\mathcal{A}$ the aggregate formed by the finite sequences $s = (s_1, \dots, s_m)$ of points of S ($m \in \mathbb{N}$). A morphism $s = (s_1, \dots, s_m) \to t = (t_1, \dots, t_n)$ is given by a matrix $H \in k^{n \times m}$ such that $H_{ji} = 0$ if $s_i \not\leq t_j$. The composition of $\mathcal{A}$ is induced by matrix multiplication.

We define a module M over $\mathcal{A}$ by setting $M(s) = k^m$ and letting H act on k^m by matrix multiplication. Each space over M is then isomorphic to some (k^d, F, s), where F is identified with a matrix of size $m \times d$. Two such matrices F, $F' \in k^{m\times d}$ give rise to isomorphic spaces over M if F can be converted into F' by performing a sequence of *elementary* transformations of the following types:

a) multiplying a row or a column by a non-zero scalar;
b) adding one column to another;
c) adding row i to row j if $s_i \leq s_j$.

In the case of a finite poset, we know that $M\mathcal{A}$ admits a finite number of indecomposable objects up to isomorphism if and only if S contains no full subposet isomorphic to one of the 5 posets of *Kleiner's first list* (Fig. 1, [1], [2]).

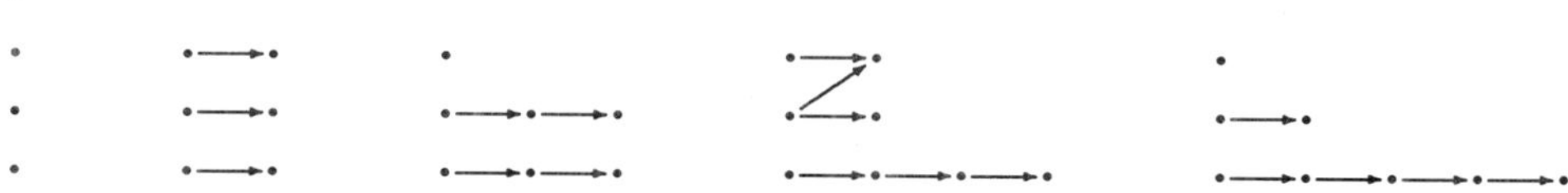

FIGURE 1. Kleiner's first list

The *support* of a space (V, f, s) over M is the full subposet of S formed by the points occurring in the sequence s. If $M\mathcal{A}$ has only finitely many indecomposable objects up to isomorphism, the support of an indecomposable object T is isomorphic to one of the 14 posets of *Kleiner's second list* (Fig. 2, [3]).

§3. Example 2

Let A be a finite-dimensional algebra, $\mathcal{A}(A)$ the aggregate formed by the finite-dimensional projective right A-modules, m a finite-dimensional left A-module and $M(m)$ the module over $\mathcal{A}$ defined by $M(m)(p) := p \otimes_A m$.

It is more or less clear that the aggregate $M\mathcal{A}$ considered in §1 is equivalent to some $M(m)\mathcal{A}(A)$ provided $\mathcal{A}$ admits only finitely many indecomposable objects up to isomorphism. In the particular case of example 1 for instance, we may set $A = \oplus kS^{\geq} := \oplus_{t\geq s} k(t,s)$ and $m = \oplus ks := \oplus_{s\in S} ks$, where $S^{\geq} := \{(t,s) \in S \times S \mid t \geq s\}$. The multiplication law of $kS^{\geq}$ is such that $(t,s)(r,q) = 0$ if $s \neq r$ and $(t,s)(s,q) = (t,q)$. The action of $kS^{\geq}$ on m is such that $(t,s)r = 0$ if $s \neq r$ and $(t,s)s = t$.

§4. Example 3

Let Λ be a finite-dimensional algebra over k, I a minimal two-sided ideal which

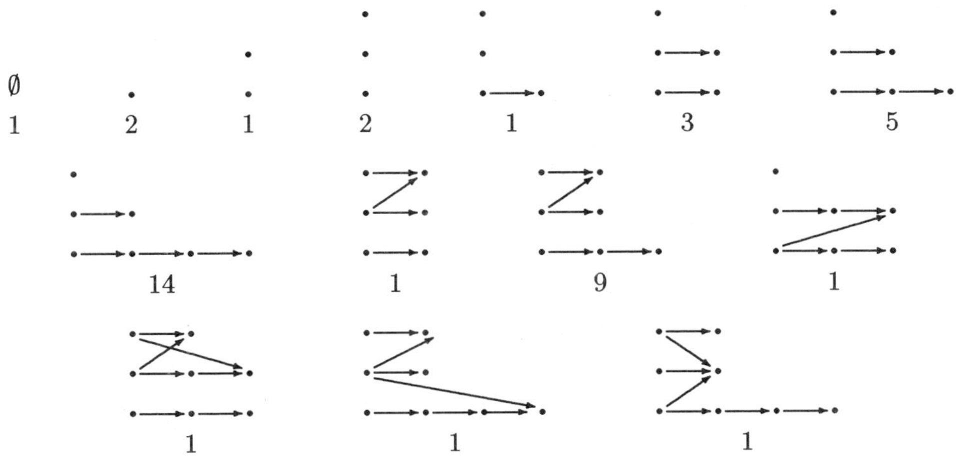

FIGURE 2. Kleiner's second list

The number attached to a poset P of the list is the number of isoclasses of indecomposable triples (V, f, p) with support P.

is contained in the radical of Λ and $\overline{\Lambda} = \Lambda/I$. Our problem is to try and relate $\operatorname{mod}\Lambda$ (the category of finite-dimensional right Λ-modules) to $\mathcal{A} = \operatorname{mod}\overline{\Lambda}$. For this sake we attach to each $m \in \operatorname{mod}\Lambda$ the exact sequence

$$0 \longrightarrow \underline{m} \xrightarrow{\text{incl}} m \xrightarrow{\text{proj}} m/\underline{m} \longrightarrow 0, \qquad (\ddagger)$$

where $\underline{m}$ denotes the largest submodule n of m such that $nI = 0$.

As a simple Λ-bimodule, I it is isomorphic to some $t^\top \otimes_k s$, where $s, t \in \operatorname{mod}\Lambda$ are simple and $t^\top$ denotes the dual of the k-vector space t. The equation $m(\operatorname{Ann}t)I = 0$, where $\operatorname{Ann}t$ denotes the annihilator of t in Λ, then shows that $m \operatorname{Ann}t \subset \underline{m}$, hence that $m/\underline{m} \xrightarrow{\sim} \operatorname{Hom}_\Lambda(t, m/\underline{m}) \otimes_k t$ is a semi-simple Λ-module of type t. Accordingly, $\operatorname{Ext}^1_\Lambda(m/\underline{m}, \underline{m})$ is identified with $\operatorname{Hom}_k(V, \operatorname{Ext}^1_\Lambda(t, \underline{m}))$ where $V = \operatorname{Hom}_\Lambda(t, m/\underline{m})$, and $(\ddagger)$ induces a linear map $\varepsilon_m \in \operatorname{Hom}_k(V, \operatorname{Ext}^1_\Lambda(t, \underline{m}))$.

Denote by $\epsilon \in \Lambda$ an idempotent which lies in all maximal ideals of Λ except $\operatorname{Ann}s$ and is mapped to the unit element of $\Lambda/\operatorname{Ann}s$. Then we have $I\Lambda\epsilon = I\epsilon = I$, hence $Ic \xrightarrow{\sim} t^\top$ if c is the projective cover of $s^\top$. This implies by duality that the injective hull h of s satisfies $h/\underline{h} \xrightarrow{\sim} t$. The resulting exact sequence

$$0 \to \underline{h} \longrightarrow h \longrightarrow t \to 0$$

induces long exact sequences

$$\operatorname{Hom}_\Lambda(\underline{h}, \underline{m}) \to \operatorname{Ext}^1_\Lambda(t, \underline{m}) \to \operatorname{Ext}^1_\Lambda(h, \underline{m}) \to \operatorname{Ext}^1_\Lambda(\underline{h}, \underline{m})$$

and composed maps

$$\eta_m : V \xrightarrow{\varepsilon_m} \operatorname{Ext}^1_\Lambda(t, \underline{m}) \xrightarrow{\text{proj}} M(\underline{m}),$$

where M is a module over $\mathcal{A} = \operatorname{mod}\overline{\Lambda}$ such that

$$M(\underline{m}) = \operatorname{Coker}\big(\operatorname{Hom}_\Lambda(\underline{h}, \underline{m}) \to \operatorname{Ext}^1_\Lambda(t, \underline{m})\big) \xrightarrow{\sim} \operatorname{Ker}\big(\operatorname{Ext}^1_\Lambda(h, \underline{m}) \to \operatorname{Ext}^1_\Lambda(\underline{h}, \underline{m})\big).$$

Theorem. ([12]) *The functor* $\operatorname{mod}\Lambda \to M\mathcal{A}$, $m \mapsto (V, \eta_m, \underline{m})$ *defined above induces a bijection* {*isoclasses of indecomposable* Λ*-modules not isomorphic to* $\underline{h}$} $\xrightarrow{\sim}$ {*isoclasses of indecomposable spaces over* M}.

The theorem allows an investigation of $\operatorname{mod}\Lambda$ by induction on $\dim\Lambda$. The induction step consists of a matrix problem.

§5. Tame-Wild Dichotomy

a) Modules of Wild Space-Type

Let $\mathcal{W}$ denote the aggregate whose objects are the pairs $v = (v(1), v(2)) \in k^{m\times m} \times k^{m\times m}$ of quadratic matrices of the same size. A morphism from $v \in k^{m\times m} \times k^{m\times m}$ to $w \in k^{n\times n} \times k^{n\times n}$ is given by a matrix $z \in k^{n\times m}$ such that $w(1)z = zv(1)$ and $w(2)z = zv(2)$.

With the notations of §1, we attach the k-linear functor

$$[f, g, h] : \mathcal{W} \to M\mathcal{A},\ v \mapsto (V^m,\ \mathbb{I}_m \otimes f + v(1) \otimes g + v(2) \otimes h,\ x^m)$$

to each triple $(f, g, h) \in \operatorname{Hom}_k(V, M(x))^3$ (For each $u \in k^{m\times m}$ and each $\ell \in \operatorname{Hom}_k(V, M(x))$, the *Kronecker-product* $u \otimes \ell \in \operatorname{Hom}_k(V^m, M(x^m)) \xrightarrow{\sim} \operatorname{Hom}_k(V, M(x))^{m\times m}$ denotes the matrix with entries $u_{ij}\ell$). The module M is said to be *of wild space-type* if there exists a triple (f, g, h) such that $[f, g, h]$ preserves indecomposability and heteromorphism. Otherwise, M is *of tame space-type*.

b) Modules of Tame Space-Type

In a similar way, we can attach spaces $[f, g](\mu, n) = (V^n,\ \mathbb{I}_n \otimes f + \mathrm{S}(\mu, n) \otimes g,\ x^n)$ over M to all $f, g \in \operatorname{Hom}_k(V, M(x))$ and all $(\mu, n) \in k \times (\mathbb{N} \setminus \{0\})$. Here $\mathbb{I}_n \otimes f + \mathrm{S}(\mu, n) \otimes g$ denotes the matrix $a \in \operatorname{Hom}_k(V, M(x))^{n\times n}$ such that $a_{ii} = f + \mu g$, $a_{i,i+1} = g$ and $a_{ij} = 0$ if $j - i \notin \{0, 1\}$.

Theorem. ([4], [8], [12], [15]) *Suppose that* $\mathcal{A}$ *has finitely many indecomposable objects up to isomorphism. Then* M *is tame if and only if there exists a family* $((f_i, g_i, E_i))_{i\in I}$ *formed by maps* $f_i\ g_i \in \operatorname{Hom}_k(k^{n_i}, M(x_i))$ *with* $n_i \in \mathbb{N} \setminus \{0\}$ *and by finite subsets* E_i *of* k *such that the following conditions are satisfied:*

a) $[f_i, g_i](\mu, m) \in M\mathcal{A}$ *is indecomposable for all* $i \in I$, $\mu \in k \setminus E_i$ *and* $m \in \mathbb{N} \setminus \{0\}$.
b) $[f_i, g_i](\mu, m)$ *is not isomorphic to* $[f_j, g_j](\nu, n)$ *if* $\mu \notin E_i$, $\nu \notin E_j$ *and* $(i, \mu, m) \neq (j, \nu, n)$.
c) *For each* $m \in \mathbb{N}$, *there are at most finitely many isoclasses of indecomposable spaces* $(V, f, x) \in M\mathcal{A}$ *with* $m = \dim V$ *which are not isomorphic to some* $[f_i, g_i](\mu, m)$.
d) *For each* $n \in \mathbb{N} \setminus \{0\}$, *the number of indices* $i \in I$ *such that* $n_i = n$ *is* $\leq \mathrm{e}^{3^n d^2 n^{2n}}$ *with* $d = \sum_x \dim M(x)$, *where* x *runs through representatives of the indecomposable objects of* $\mathcal{A}$.

§6. Tame and Wild Algebras

Using §4, we can apply the results of §5 to a finite-dimensional algebra Λ: For this sake, we consider the tensor algebra $\otimes\Lambda = k \oplus \Lambda \oplus (\Lambda \otimes_k \Lambda) \oplus (\Lambda \otimes_k \Lambda \otimes_k \Lambda) \oplus \cdots$ and the projection $\varpi : \otimes\Lambda \to \Lambda$ which is induced by the multiplication of Λ. For each $d \in \mathbb{N}$, $\mathrm{Hom}_k(\Lambda, k^{d\times d})$ is then identified with the space of $\otimes\Lambda$-module structures on k^d. The general linear group $\mathrm{GL}_d k$ acts on $\mathrm{Hom}_k(\Lambda, k^{d\times d})$ by conjugation, and two points of $\mathrm{Hom}_k(\Lambda, k^{d\times d})$ are isomorphic as $\otimes\Lambda$-modules if and only if they lie in the same orbit. Furthermore, the maps $\Lambda \to k^{d\times d}$ which factor through ϖ are identified with the Λ-module structures on k^d. They form a $\mathrm{GL}_d(k)$-stable subvariety $\mathrm{mod}_d(\Lambda)$ of $\mathrm{Hom}_k(\Lambda, k^{d\times d})$.

a) Wild Algebras

Each triple $(f,g,h) \in \mathrm{Hom}_k(\Lambda, k^{n\times n})$ provides an exact functor $< f,g,h >$: $\mathcal{W} \to \mathrm{mod}\,\otimes\Lambda$ which maps $v \in k^{m\times m} \times k^{m\times m}$ to $k^m \otimes_k k^n$ equipped with the endomorphisms

$$\mathbb{I}_m \otimes f(\lambda) + v(1) \otimes g(\lambda) + v(2) \otimes h(\lambda), \quad \text{where} \quad \lambda \in \Lambda \subset \otimes\Lambda.$$

Λ is called *wild* if, for some (f,g,h), $< f,g,h >$ factors through $\mathrm{mod}\,\Lambda$ and preserves indecomposability and heteromorphism. Otherwise, Λ is called *tame*.

Clearly, if $< f,g,h >$ factors through $\mathrm{mod}\,\Lambda$, the affine subspace $f + k\,g + k\,h$ of $\mathrm{Hom}_k(\Lambda, k^{d\times d})$ must be contained in $\mathrm{mod}_d(\Lambda)$. Conversely, if the last condition is satisfied, the restriction of $< f,g,h >$ to the full subcategory $\mathcal{W}_c$ of $\mathcal{W}$ formed by the pairs v such that $v(1)v(2) = v(2)v(1)$ factors through $\mathrm{mod}\,\Lambda$. In fact, it is well-known that Λ is wild if and only if $< f,g,h > |\mathcal{W}_c$ preserves indecomposability and heteromorphism for some (f,g,h).

b) Tame Algebras

In a similar way, we can attach Λ-modules $<f,g>(\mu,m)$ to all $(\mu,m) \in k \times (\mathbb{N} \setminus \{0\})$ and all $f,g \in \mathrm{Hom}_k(\Lambda, k^{d\times d})$ such that $f + k\,g \subset \mathrm{mod}_d(\Lambda)$. The underlying space of $< f,g > (\mu,m)$ is $k^{m\times d}$, the multiplication by $\lambda \in \Lambda$ is the map $x \mapsto xf(\lambda) + \mathrm{S}(\mu,m)xg(\lambda)$, where $\mathrm{S}(\mu,m)_{ii} = \mu$, $\mathrm{S}(i,i+1) = 1$ and $\mathrm{S}(\mu,m)_{ij} = 0$ if $j - i \notin \{0,1\}$.

Theorem. ([4], [8], [12], [16]) *The finite-dimensional algebra Λ is tame if and only if there exists a family $((f_i,g_i,E_i))_{i\in I}$ formed by maps $f_i, g_i \in \mathrm{Hom}_k(\Lambda, k^{n_i\times n_i})$ with $n_i \in \mathbb{N} \setminus \{0\}$ and by finite subsets E_i of k such that the following conditions are satisfied:*

a) $f_i + k\,g_i \subset \mathrm{mod}_{n_i}(\Lambda)$ *for all* $i \in I$.
b) $< f_i, g_i > (\mu,m) \in \mathrm{mod}\,\Lambda$ *is indecomposable for all* $i \in I, \mu \in k \setminus E_i$ *and* $m \in \mathbb{N} \setminus \{0\}$.
c) $< f_i\, g_i > (\mu,m)$ *is not isomorphic to* $< f_j\, g_j > (\nu,n)$ *if* $\mu \notin E_i, \nu \notin E_j$ *and* $(i,\mu,m) \neq (j,\nu,n)$.
d) *For each* $m \in \mathbb{N}$, *there are at most finitely many isoclasses of indecomposable* Λ*-modules with dimension* m *which are not isomorphic to some* $< f_i\, g_i > (\mu,m)$.

e) *For each* $n \in \mathbb{N} \setminus \{0\}$, *the number of indices* $i \in I$ *such that* $n_i = n$ *is* $\leq rne^{22(3n^2)^n}$, *where* r *is the dimension of the radical of* Λ *(over* k*).*

The bound $\leq rne^{22(3n^2)^n}$ given in e) seems to be quite rough. Experimental results suggest that an exponential function of n should suffice.

§7. Modules of Finite Space-Type

The classification of general tame algebras lies beyond the range of the techniques at our disposal. For tame algebras of polynomial growth, we refer to the report of A. Skowronski in the present proceedings. A satisfactory understanding has been attained only in the case of algebras having a finite number of indecomposable modules (up to isomorphism) [6], [10]. Our objective in this paper is to present the results obtained by the school of L. A. Nazarova and A. V. Roiter for modules *of finite space-type*, i.e. which admit only finitely many indecomposable spaces over themselves.

Theorem. ([13]) *For each n there are, up to 'equivalence', only finitely many pairs* $(\mathcal{A}, M)$ *where the* $\mathcal{A}$*-module* M *is faithful, of dimension n and of finite space-type.*

(The *dimension* of M is the sum of the dimensions of the vector spaces $M(x)$, where x runs through representatives of the indecomposable objects of $\mathcal{A}$)

The first step of the proof is to show that, if M is of finite space-type and $x \in \mathcal{A}$ indecomposable, $(\mathcal{A}(x,x), M(x))$ must be isomorphic to one of the following pairs:

$$(k,k)\ ,\quad \left(\begin{bmatrix} * & 0 \\ \circ & * \end{bmatrix}, k^2\right)\ ,\quad \left(\begin{bmatrix} * & 0 & 0 \\ \circ & * & 0 \\ \bullet & \circ & * \end{bmatrix}, k^3\right)\ ,\quad \left(\begin{bmatrix} * & 0 & 0 \\ \circ & * & 0 \\ \triangleleft & \bullet & * \end{bmatrix}, k^3\right)\ ,$$

For instance, the algebra $\mathcal{A}(x,x)$ of the third pair is formed by the matrices

$$\begin{bmatrix} \lambda & 0 & 0 \\ \mu & \lambda & 0 \\ \nu & \mu & \lambda \end{bmatrix}$$

with $\lambda, \mu, \nu \in k$.

The first step shows in particular that $\dim_k M(x) \leq 3$ for each indecomposable $x \in \mathcal{A}$. In fact, Roiter and Sergeichuk show that the faithful modules of finite space-type belong to some wider class of modules which admit a purely combinatorial description. In the general case, the description is not quite smooth, and the characterization of the modules of finite space-type within the wider combinatorial class is not yet completed. We therefore restrict our report to the case where $\dim_k M(x) \leq 2$ for each indecomposable $x \in \mathcal{A}$. In this particular case, the wider combinatorial class is provided by 'dyadic sets'.

§8. Dyadic Sets

A *dyadic set* is a finite set S equipped with an order relation $S^{\leq} \subset S \times S$ and an equivalence relation $S^{\sim} \subset S^{\leq} \times S^{\leq}$ on $S^{\leq}$. These data are subjected to the following conditions, where $x \leq y \sim x' \leq y'$ means that $x \leq y$, $x' \leq y'$ and $(x,y) \sim (x',y')$:

a) $x \leq y \leq z$ and $x \leq z \sim x' \leq z'$ imply the existence of a unique y' such that $x \leq y \sim x' \leq y'$ and $y \leq z \sim y' \leq z'$.
b) The equivalence classes of $S^{\leq}$ have cardinality ≤ 2.
c) $x \leq y \sim x \leq y'$ and $x \leq y \sim x' \leq y$ imply $y = y'$ and $x = x'$ respectively.

In the sequel, a dyadic set S is always equipped with the equivalence relation formed by the pairs $(x,y) \in S \times S$ such that $x \leq x \sim y \leq y$. The objects of the aggregate $\mathcal{A}$ associated with S are the sequences $c = (c_1, \dots c_m)$ of equivalence classes of S. We set $M(c) = \oplus_{i=1}^{i=m} M(c_i) = \oplus_i \oplus_{s \in c_i} ks$ and $\mathcal{A}(c,d) = \oplus_{i=1,j=1}^{i=m,j=n} \mathcal{A}(c_i, d_j)$, where $d = (d_1, \dots, d_n)$ denotes a second object and $\mathcal{A}(c_i, d_j)$ is identified with a subspace of $\mathrm{Hom}_k(M(c_i), M(d_j))$ defined as follows: each equivalence class $e = \overline{(x,y)}$ of $S^{\leq}$ with $x \in c_i$ and $y \in d_j$ provides a basis vector f_e of $\mathcal{A}(c_i, d_j)$ which maps a point $s \in c_i \subset M(c_i)$ to t if $s \leq t \sim x \leq y$ and to 0 if there is no t such that $s \leq t \sim x \leq y$. The action of $\mathcal{A}$ on M is determined by the inclusions $\mathcal{A}(c_i, d_j) \subset \mathrm{Hom}_k(M(c_i), M(d_j))$.

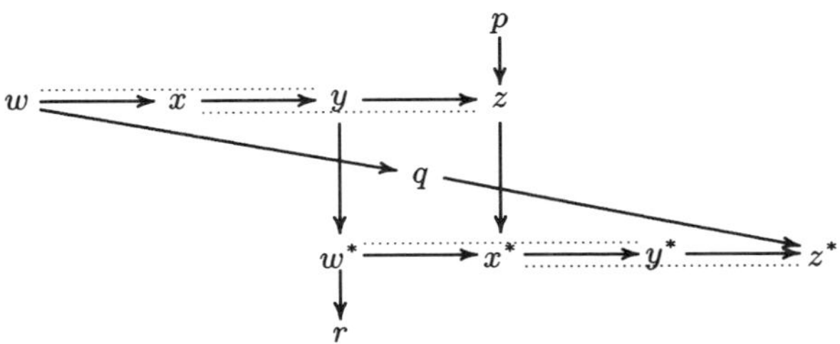

FIGURE 3

In this example, the dyadic set is $S = \{p, r, w, x, y, z, w^*, x^*, y^*, z^*\}$. The order relation is described by the arrows; thus we have $p \leq y^*$ and $x \not\leq p$. Furthermore, w is equivalent to w^*, x to x^*, y to y^*, z to z^*, whereas $\overline{p} = \{p\}$ and $\overline{r} = \{r\}$. The dotted line from w to y, for instance, signalizes that $w \leq y \sim w^* \leq y^*$, hence that $w \leq x \sim w^* \leq x^*$ and $x \leq y \sim x^* \leq y^*$ [18].

On the matrix-level the problem set by $(\mathcal{A}, M)$ is the following: A space over M is determined by a matrix $F \in k^{m \times n}$, an involution $\iota : \{1, \dots, m\} \to \{1, \dots, m\}$ and a map $\mu : \{1, \dots, m\} \to S$ such that $\iota(i) = i$ if $\overline{\mu(i)} = \{\mu(i)\}$ and $\mu(\iota(i)) = \mu(i)^\bullet$ if $\overline{\mu(i)} = \{\mu(i), \mu(i)^\bullet\}$ consists of two points. The triples (F, μ, ι) and (F', μ, ι) with $F' \in k^{m \times n}$ provide isomorphic spaces over M if and only if F can be converted into F' by a sequence of *elementary* transformations of the following types:

a) multiplication of a row or a column by a non-zero scalar.
b) addition of one column to another.
c) addition of row i to row j if $\mu(i) \leq \mu(j)$ and if the equivalence class of $(\mu(i), \mu(j))$ has cardinality 1.
d) simultaneous addition of column i to column j and of column $\iota(i)$ to column $\iota(j)$ if $\iota(i) \neq i$ and $\mu(i) \leq \mu(j) \sim \mu(\iota(i)) \leq \mu(\iota(j))$.

§9. Zigzags of Dyadic Sets

A dyadic set S is said to be *of finite space-type* if so is the module M attached to S in §8. A characterization of dyadic sets of finite space-type, based on a list similar to Kleiner's first list, is worked out in [9]. But the new list is much longer, the directions of use are more involved and no analogue of Kleiner's second list is known. Therefore L. A. Nazarova and A. V. Roiter searched for a new approach reducing dyadic sets of finite space-type to posets.

A *zigzag* of S is a sequence $z = (z_0, z_1, \dots, z_\ell)$, $\ell \in \mathbb{N}$, where $z_i \in S$ is a *doublet* (i.e. $\overline{z_i} = \{z_i, z_i^\bullet\}$) has cardinality 2) for $i < \ell$, a *singleton* (i.e. $\overline{z_i} = \{z_i\}$) if $i = \ell = 0$, a singleton or one of the symbols 0,1 if $i = \ell > 0$. We set $y \leq z$, where $y = (y_0, \dots, y_h)$ is another zigzag, if there exists a number $j \leq h, \ell$ such that $y_i \leq z_i \sim y_i^\bullet \leq z_i^\bullet$ for $i < j$ and that one of the three following cases occurs:

a) $y_j = 0$ or $y_j < z_{j-1}^\bullet$;

b) $z_j = 1$ or $y_{j-1}^\bullet < z_j$;

c) $(y_j, z_j) \in S^{\leq}$ and $\overline{(y_j, z_j)}$ has cardinality 1. For the relation thus defined, the zigzags form a poset $\mathrm{C}(S)$.

Theorem. ([9], [14]) *A dyadic set S is of finite space-type if and only if the poset of zigzags $\mathrm{C}(S)$ is finite and of finite space-type.*

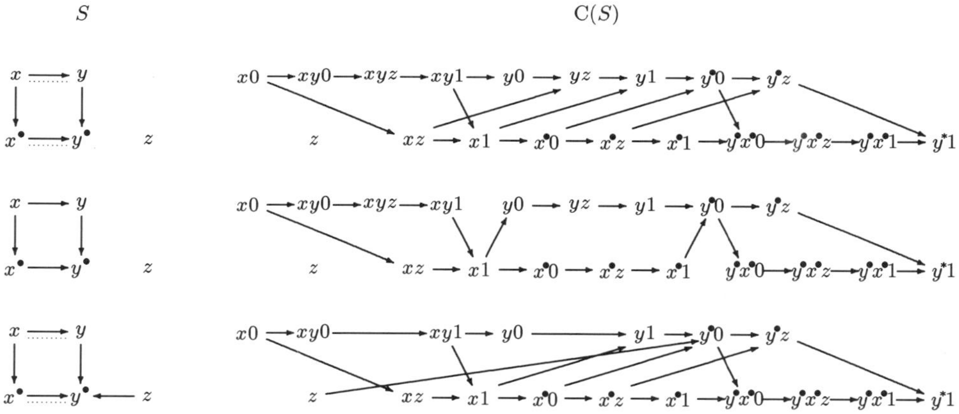

FIGURE 4

The first two dyadic sets are not of finite space-type because the poset formed by the points $z, y0, yz, y1, x^\bullet 0, xz^\bullet z, x^\bullet 1$ of $\mathrm{C}(S)$ is isomorphic to the third poset of Kleiner's first list. In the third case, $\mathrm{C}(S)$ is of finite space-type, hence so is S.

§10. Normal Spaces over a Dyadic Set and Sputniks

The preceding theorem of Nazarova-Roiter provides a theoretical characterization of dyadic sets of finite space-type. In practice, the description of $\mathrm{C}(S)$ and the determination of its space-type may be quite strenuous. Still more arduous is

$$E = \begin{bmatrix} \upsilon_1 \cdots\cdots \upsilon_n \\ \phi_1 \cdots\cdots \phi_n \\ \psi_1 \cdots\cdots \psi_n \\ \cdots\cdots\cdots \\ \eta_1 \cdots\cdots \eta_n \\ \zeta_1 \cdots\cdots \zeta_n \end{bmatrix} \qquad \begin{array}{l} \lambda(1) = x_0x_1x_20 \\ \lambda(2) = r \\ \lambda(3) = s_0s_1 \\ \cdots\cdots\cdots \\ \lambda(m-1) = y_0y_1y_2y_3 \\ \lambda(m) = z_0z_11 \end{array}$$

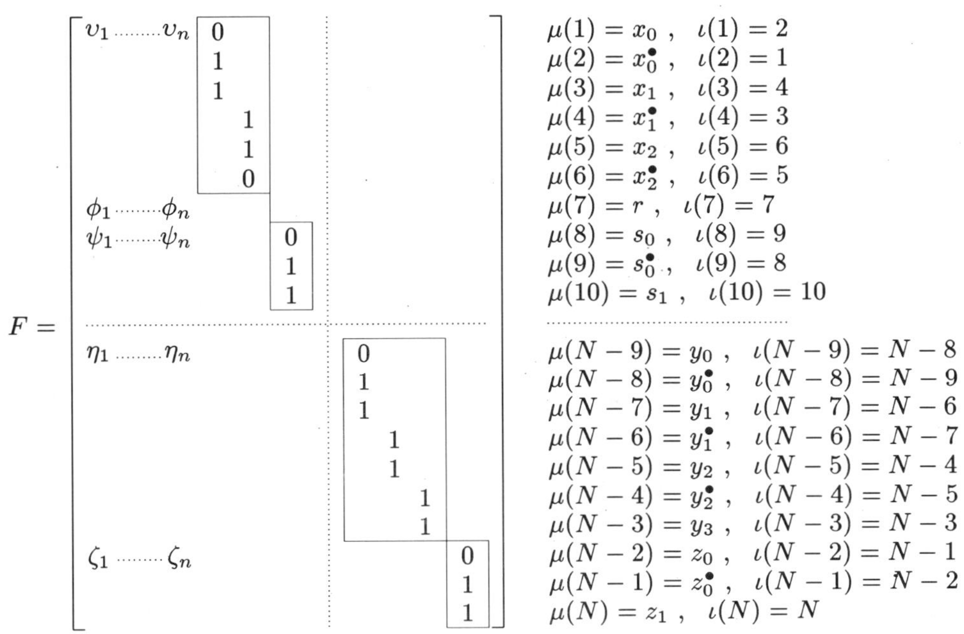

FIGURE 5

N denotes the number of rows of F. With the possible exception of the scalars υ_i, ϕ_i, $\psi_i, \dots, \eta_i$, ζ_i, $(1 < i < n)$, the non marked entries of F are zero.

the description of the indecomposable spaces [17], [18], [19], [20]. On the level of matrices it works as follows:

For each $E \in k^{m\times n}$ and each $\lambda : \{1, \dots, m\} \to \mathrm{C}(S)$ we construct a matrix $F \in k^{p\times q}$, an involution $\iota : \{1, \dots, p\} \to \{1, \dots, p\}$ and a map $\mu : \{1, \dots, p\} \to S$ according to a rule illustrated in Fig. 5. The triple (F, ι, μ) thus obtained satisfies the conditions of §9 and gives birth to a space $\Theta(E, \lambda)$ over (the module M attached to) S. Pairs (E, λ) providing isomorphic spaces 'over $\mathrm{C}(S)$' are mapped to isomorphic spaces $\Theta(E, \lambda)$. Indecomposable spaces isomorphic to some $\Theta(E, \lambda)$ are called *normal*. The remaining indecomposable spaces over S are *'sputniks'* with the following description: Let E_i be a dyadic set of Fig. 6–9 ($i = 1, 2, 3, 4, 5, 6$) and $\varepsilon : E_i \to S$ a map such that

a) $\forall j$, the restriction $\varepsilon | E_{ij}$ is a full embedding of posets;
b) singletons are mapped to singletons, equivalent distinct doublets to equivalent distinct doublets and equivalent inequalities to equivalent inequalities.

The letters at the right edge of each matrix $N \in k^{m\times n}$ escorting E_i are the values of a map $\mu : \{1, \dots , m\} \to E_i$. The spaces over S associated with $(N, \iota, \varepsilon \circ \mu)$, where $\iota : \{1, \dots , m\} \to \{1, \dots , m\}$ is the obvious involution, are the announced sputniks.

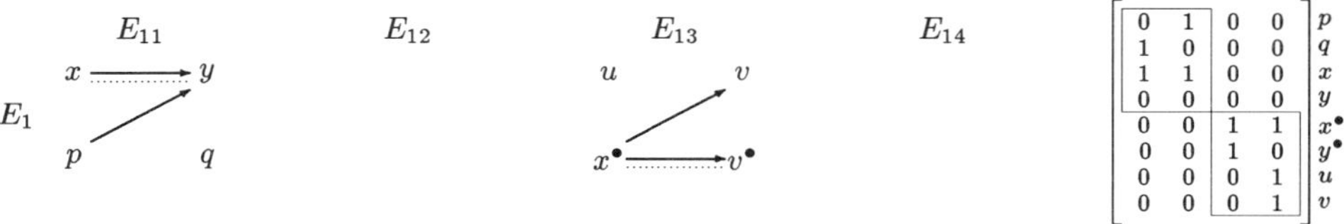

FIGURE 6

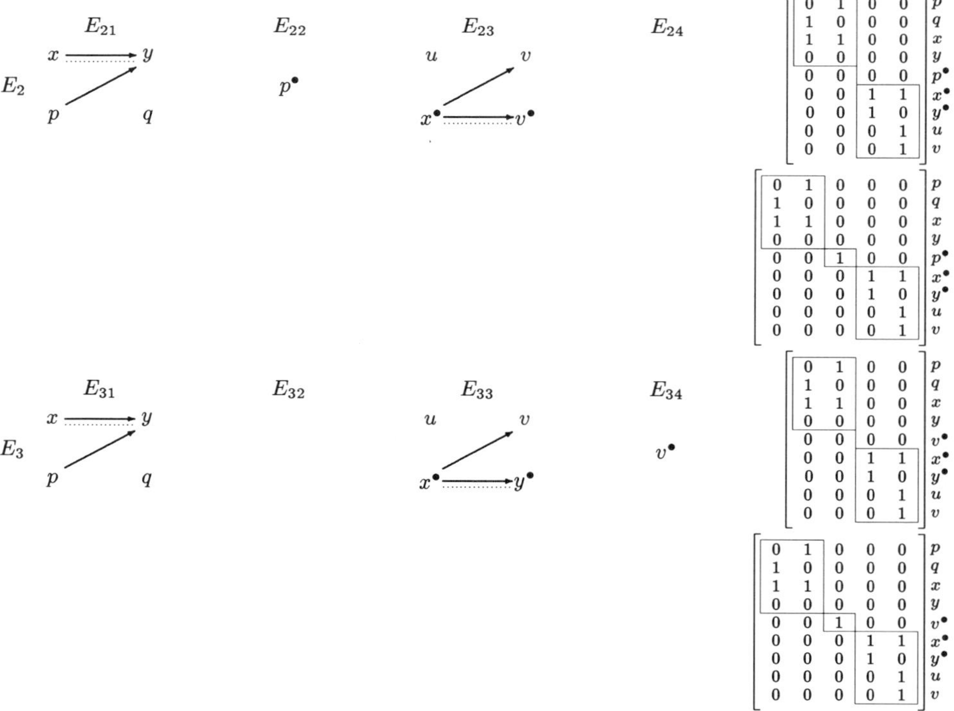

FIGURE 7

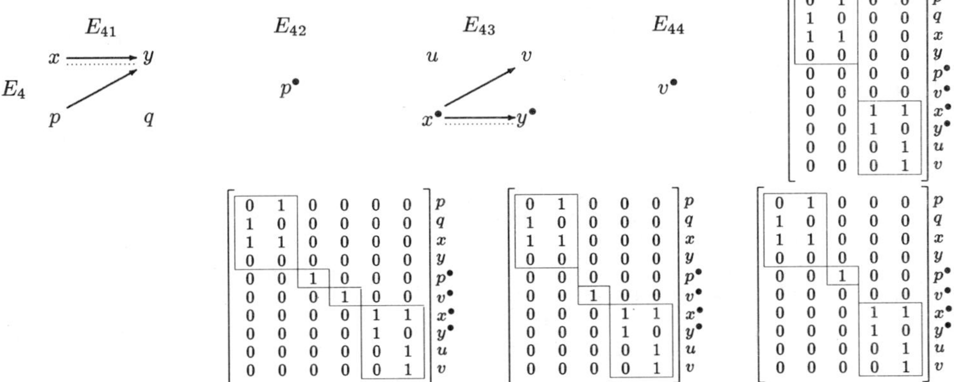

FIGURE 8

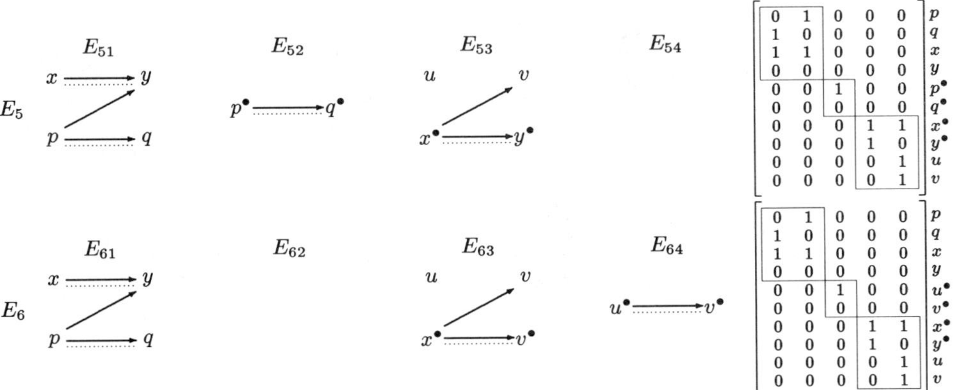

FIGURE 9

References

[1] L.A. Nazarova and A.V. Roiter, Representations of partially ordered sets, in Investigations on Representation Theory, *Zap. Naučn. Sem. LOMI* **28** (1972), 5–31 (Russian); Engl. Transl.: *Sov. Math.* **23** (1975), 585–607.

[2] M.M. Kleiner, Partially ordered sets of finite type, in Investigations on Representation Theory, *Zap. Naučn. Sem. LOMI* **28** (1972), 32–41 (Russian); Engl. Transl.: *Sov. Math.* **23** (1975), 607–615.

[3] M.M. Kleiner, On faithful representations of partially ordered sets of finite type, *Zap. Naučn. Sem. LOMI* **28** (1972), 42–59 (Russian); Engl. Transl.: *Sov. Math.* **23** (1975), 616–628.

[4] Ju.A. Drozd, *Springer Lect. Notes*, 1980 832, pp. 242–258.

[5] L.A. Nazarova and A.V. Roiter, Representations and forms of weakly completed posets, Preprint Acad. of Sc. of Ukraine (1983), 19–54. (Russian)

[6] R. Bautista, P. Gabriel, A.V. Roiter and L. Salmeron, Representation-finite algebras and multiplicative bases, *Inv. math.* **81** (1985), 217–285.

[7] L.A. Nazarova and A.V. Roiter, Representations of bipartite completed posets, *Comm. Math. Helv.* **63** (1988), 498–526.

[8] W.W. Crawley-Boevey, On tame algebras and bocses, *Proc. London Math. Soc.* (1988), 451–483.

[9] L.A. Nazarova and A.V. Roiter, Representations of biinvolutive posets, Preprint Ac. Sc. Ukraine (1991), 31. (Russian)

[10] P. Gabriel and A.V. Roiter, *Representations of finite-dimensional algebras*, Encycl. of Math. Sc. Vol. 73, Algebra VIII, Springer Verlag, 1992, pp. 177.

[11] D. Vossieck, Representation-finite Weakly Completed Posets, phd-thesis, Uni.Zürich (1993), 20p.

[12] P. Gabriel, L.A. Nazarova, A.V. Roiter, V.V. Sergeichuk and D. Vossieck, Tame and wild subspace problems, *Ukr. Math. J.* (1993), 313–352.

[13] A.V. Roiter and V.V. Sergeichuk, A finitely spaced module over an aggregate admits a multiplicative basis, *Ukr. Math. J.* (1994), 567–579.

[14] L.A. Nazarova and A.V. Roiter, Representations of biinvolutive posets II, Preprint Ac. Sc. Ukraine (1994), 72p. (Russian)

[15] Th. Brüstle, Unterraumprobleme, phd-thesis, Uni. Zürich (1995), 51.

[16] Th. Brüstle, *On the growth function of tame algebras*, Comptes-Rendus Ac. Sc., Paris, 1995.

[17] U. Hassler, Representations of Dyadic Sets I, phd-thesis, Uni. Zürich (1996), 62.

[18] Th. Guidon, Reprsentations of Dyadic Sets II, phd-thesis, Uni. Zürich (1996), 47.

[19] Th. Guidon, U. Hassler, L.A. Nazarova and A.V. Roiter, *Two reduction algorithms for dyadic sets*, Comptes-Rendus Ac. Sc., Paris, 1996.

[20] Th. Guidon, U. Hassler, L.A. Nazarova and A.V. Roiter, *Dyadic sets: A dichotomy for indecomposable S-matrices*, Comptes-Rendus Ac. Sc., Paris, 1996.

P. Gabriel
Institut für Mathematik
Universität Zürich-Irchel
Switzerland
e-mail: gabriel@math.unizh.ch

Canadian Mathematical Society
Conference Proceedings
Volume **22**, 1998

Prime and Primitive Spectra of Multiparameter Quantum Affine Spaces

K. R. Goodearl and E. S. Letzter

Abstract. The prime and primitive spectra of the algebra $\mathcal{O}_{\mathbf{q}}(k^n)$, the quantized coordinate ring of affine n-space over a field k, are studied for arbitrary parameter matrices $\mathbf{q}$. In particular, Gelfand-Kirillov dimensions and Goldie ranks of primitive factors are calculated, and the Dixmier–Moeglin equivalence (among primitive, rational, and locally closed prime ideals) is established. There is a natural finite stratification of $\operatorname{spec} \mathcal{O}_{\mathbf{q}}(k^n)$ such that the primitive ideals occurring in a given stratum are precisely the maximal elements of the stratum. In case k is algebraically closed, each stratum of primitive ideals is homeomorphic to a torus obtained as a quotient of a natural torus of automorphisms of $\mathcal{O}_{\mathbf{q}}(k^n)$. These results are based on detailed analyses of noncommutative Laurent polynomial algebras, which are quantized coordinate rings of tori (and are also known as McConnell–Pettit algebras).

As an application, the Dixmier–Moeglin equivalence is established for the algebra $\mathcal{O}_{\mathbf{p},\lambda}(M_n(k))$, the multiparameter quantized coordinate ring of $n \times n$ matrices over k, in the case when λ is a root of unity.

Introduction

Quantum affine spaces — the well-known deformations of commutative polynomial rings — are the most elementary noncommutative algebras occurring among quantized coordinate rings. First of all, the multiparameter quantum affine spaces $\mathcal{O}_{\mathbf{q}}(k^n)$ are iterated skew polynomial rings, over the field k, in which each iteration is twisted only by an automorphism. Furthermore, standard methods show

---

AMS Subject Classification (1991). 16D30, 16D60, 16P40, 16P90, 16S36, 17B37.

The research of both authors was partially supported by National Science Foundation research grants.

that $\mathcal{O}_{\mathbf{q}}(k^n)$ is Auslander regular and Cohen–Macaulay, with global and Gelfand-Kirillov dimension n (cf. [9, 2.5]). Also, the previous results can be used to show that spec $\mathcal{O}_{\mathbf{q}}(k^n)$ is catenary [9, 2.6]. Because of their accessibility, various examples of the algebras $\mathcal{O}_{\mathbf{q}}(k^n)$ have been used as illustrations in many papers on quantum groups. Moreover, such algebras form an essential part of the analysis of quantum groups at roots of unity (see, e.g., [6]). However, to date no full analysis of the general $\mathcal{O}_{\mathbf{q}}(k^n)$ has appeared in the literature. Our purpose in this paper is to provide such a study, insofar as the prime and primitive ideals are concerned.

In our present study, the stratification of spec $\mathcal{O}_{\mathbf{q}}(k^n)$ — noted by De Concini, Kac, and Procesi in the case that the parameters are roots of unity [5, §2; 6, §7], and by Brown and the first author in the case that the parameters generate a torsionfree multiplicative group [4, §4] — is extended to the general case. This general strategy (in a much more subtle form) was employed by Hodges and Levasseur, by Joseph, and by others in the study of quantum function algebras of semisimple Lie groups (cf., e.g., [15; 18] and references therein). One consequence, in this paper, of the above approach is a proof of the Dixmier–Moeglin equivalence for spec $\mathcal{O}_{\mathbf{q}}(k^n)$.

When k is algebraically closed and the parameters are roots of unity, calculations of the Goldie ranks of the (finite codimensional) primitive factors of $\mathcal{O}_{\mathbf{q}}(k^n)$ follow from the prior work of De Concini, Kac, and Procesi [5; 6]. We calculate the Goldie ranks, and the corresponding Gelfand–Kirillov dimensions, of the primitive factors that occur in general.

In any analysis of $\mathcal{O}_{\mathbf{q}}(k^n)$, the natural — and, in retrospect, obvious — first step is to pass to the localizations obtained by factoring out some of the generators of $\mathcal{O}_{\mathbf{q}}(k^n)$ and inverting the remaining ones; this process yields the "stratification" referred to above. The resulting algebras, which in the current context can be conveniently termed "quantum tori", are of the type investigated by McConnell and Pettit in [23]. It is no surprise then that much of our effort is invested in analyzing the prime and primitive spectra of quantum tori. This task is accomplished in Section 1, and the applications to quantum affine spaces are developed in Section 2.

A short final section of this paper contains an application to quantum matrices. It is proved that the Dixmier–Moeglin equivalence holds for $\mathcal{O}_{\mathbf{p},\lambda}(M_n(k))$ in the case that λ is a root of unity. This equivalence is also established in [12], by other means, in the case that λ is not a root of unity.

1. Quantum Tori

We begin with a thorough analysis of the prime and primitive spectra of quantum tori, also known as McConnell–Pettit algebras. Some of our results were obtained in the root of unity case by De Concini, Kac, and Procesi [5, §2; 6, §7], and in the case when the parameters generate a torsionfree multiplicative group by Brown and the first author [4, §4]. It turns out that many of the methods of these papers can be adapted to the general case, where no assumptions are placed upon the parameters.

1.1. Let $\boldsymbol{\lambda} = (\lambda_{ij})$ be a multiplicatively antisymmetric $n \times n$ matrix over the base field k; that is, $\lambda_{ii} = 1$ and $\lambda_{ji} = \lambda_{ij}^{-1}$ for all i, j. The *McConnell–Pettit algebra* $P(\boldsymbol{\lambda})$ is the k-algebra generated by elements $x_1^{\pm 1}, \dots, x_n^{\pm 1}$ subject only to the relations

$x_i x_j = \lambda_{ij} x_j x_i$ for all i, j. These algebras may be viewed as quantizations of the coordinate ring of the torus $(k^\times)^n$.

Set $R = P(\boldsymbol{\lambda})$ and write monomials in R using standard multi-index notation. Thus R has a basis of monomials x^s for $s \in \mathbb{Z}^n$. Define $\sigma\colon \mathbb{Z}^n \times \mathbb{Z}^n \to k^\times$ by

$$\sigma(s,t) = \prod_{i,j=1}^{n} \lambda_{ij}^{s_i t_j}$$

for $s, t \in \mathbb{Z}^n$. Then σ is an *alternating bicharacter*, that is,

$$\sigma(s,s) = 1$$
$$\sigma(t,s) = \sigma(s,t)^{-1}$$
$$\sigma(s, t+t') = \sigma(s,t)\sigma(s,t')$$

for $s, t, t' \in \mathbb{Z}^n$. The last condition says that $\sigma(s,-)$ is a group homomorphism; hence, $\sigma(s,0) = 1$ and $\sigma(s,-t) = \sigma(s,t)^{-1}$ for $s, t \in \mathbb{Z}^n$. The commutation relations in R can be given in the form

$$x^s x^t = \sigma(s,t) x^t x^s$$

for $s, t \in \mathbb{Z}^n$.

Conversely, if α is any alternating bicharacter on $\mathbb{Z}^n$, the k-algebra with a basis of monomials subject to the commutation rules $x^s x^t = \alpha(s,t) x^t x^s$ is a McConnell–Pettit algebra.

Note that R can be written as an iterated skew-Laurent algebra

$$k\big[x_1^{\pm 1}\big]\big[x_2^{\pm 1}; \tau_2\big] \cdots \big[x_n^{\pm 1}; \tau_n\big],$$

for a suitable choice of automorphisms $\tau_2, \dots, \tau_n$. In particular, R is a noetherian domain (see, e.g., [24, 1.4.5]). Since R is a crossed product of k by the polycyclic group $\mathbb{Z}^n$, it satisfies the Nullstellensatz over k [24, 9.4.22]; in particular, since R is also noetherian it is a Jacobson ring.

1.2 Lemma. (Cf. [5, 2.2; 6, 7.1].) *If* $S = \{s \in \mathbb{Z}^n \mid \sigma(s,-) \equiv 1\}$, *then*

$$Z(R) = k\big[x^s \mid s \in S\big].$$

Proof. It is clear from the commutation rule $x^s x^t = \sigma(s,t) x^t x^s$ that a monomial x^s is central in R precisely when $s \in S$. Given $z \in Z(R)$, write $z = \sum_{v \in \mathbb{Z}^n} \alpha_v x^v$ for some $\alpha_v \in k$, all but finitely many of which are zero. For any $t \in \mathbb{Z}^n$,

$$\sum_{v \in \mathbb{Z}^n} \alpha_v x^t x^v = x^t z = z x^t = \sum_{v \in \mathbb{Z}^n} \alpha_v \sigma(v,t) x^t x^v,$$

and since the products $x^t x^v$ are linearly independent, it follows that $\sigma(v,t) = 1$ whenever $\alpha_v \neq 0$. Hence, the support of z is contained in S, and therefore z lies in the k-span of $\{x^s \mid s \in S\}$. □

1.3. The following facts can be easily deduced from the preceding lemma (cf. [5, §2; 6, §7]):

(i) $Z(R)$ is a Laurent polynomial ring in the variables $(x^{b_1})^{\pm 1}, \ldots, (x^{b_r})^{\pm 1}$, where $\{b_1, \ldots, b_r\}$ is a basis for S.

(ii) R is a free $Z(R)$-module with basis $\{x^t \mid t \in T\}$, where T is any transversal for S in $\mathbb{Z}^n$.

(iii) $Z(R)$ is a $Z(R)$-module direct summand of R, since we may choose $0 \in T$.

(iv) R is a finitely generated $Z(R)$-module if and only if the parameters λ_{ij} are all roots of unity. Namely, it follows from (ii) that R is finite over its center if and only if $\mathbb{Z}^n/S$ is finite, and it is immediate that $\mathbb{Z}^n/S$ is finite if and only if some power of x_i is central for each $i = 1, \ldots, n$.

The next proposition may be well known; closely related results can be found, for example, in [3, 5.1], [4, 4.4], [5, §2], [27, Chapter 11, §3], and [9, 2.2]. To prove part (a), we follow an argument outlined in [14, 2.1].

1.4 Proposition.

(a) *If I is an ideal of R, then $I = R(I \cap Z(R))$.*
(b) *If J is an ideal of $Z(R)$, then $J = (RJ) \cap Z(R)$.*

Proof. (a) Choose a transversal T for S in $\mathbb{Z}^n$; then as noted in (1.3.ii), R is a free $Z(R)$-module with basis $\{x^t \mid t \in T\}$. There is an action $\alpha\colon \mathbb{Z}^n \to \operatorname{Aut} R$, where $\alpha(v)$ is the inner automorphism $r \mapsto x^v r (x^v)^{-1}$. Note that each x^t is an eigenvector for this action, with eigenvalue $\sigma(-,t)$. For distinct $t, u \in T$,

$$\sigma(-,t)\sigma(-,u)^{-1} = \sigma(-,t-u) \not\equiv 1$$

because $t - u \notin S$, whence $\sigma(-,t) \neq \sigma(-,u)$. It follows that the eigenspaces for the action α are precisely the cyclic $Z(R)$-modules $Z(R)x^t$ for $t \in T$.

Now let I be an arbitrary ideal of R. Since R is the direct sum of its α-eigenspaces, and since I is invariant under this action, I must be spanned by α-eigenvectors. Hence, I is spanned by the set

$$\{zx^t \mid z \in Z(R), \quad t \in T, \quad zx^t \in I\}.$$

Since the monomials x^t are invertible in R, it follows that I is generated by the elements z occurring in the set above, that is, $I = (I \cap Z(R))R$.

(b) This follows from the fact, noted in (1.3.iii), that $Z(R)$ is a $Z(R)$-module direct summand of R. □

1.5. The set of prime ideals of a ring A will be denoted $\operatorname{spec} A$ and given the Zariski topology; the sets of (left) primitive (respectively, maximal) ideals of A will be denoted $\operatorname{prim} A$ (respectively, $\max A$) and given the Jacobson topologies.

Corollary. *Contraction and extension provide:*

(a) *Mutually inverse isomorphisms between the lattices of ideals of R and $Z(R)$;*
(b) *Mutually inverse homeomorphisms between* $\operatorname{spec} R$ *and* $\operatorname{spec} Z(R)$*;*
(c) *Mutually inverse homeomorphisms between* $\operatorname{prim} R$ *and* $\max Z(R)$. *In particular,* $\operatorname{prim} R = \max R$.

Proof. Part (a) follows immediately from (1.4), and (b) is easily deduced from (a).

To prove (c), first observe from (a) that $\max R$ is homeomorphic to $\max Z(R)$ via contraction and extension. Next, because $Z(R)$ is a Laurent polynomial ring over k (see (1.3.i)), it is an affine commutative k-algebra, and hence a Jacobson ring. Furthermore, if we view R as constructed from $Z(R)$ by successively adjoining the variables $x_1, x_1^{-1}, \dots, x_n, x_n^{-1}$, then R is obtained from $Z(R)$ by an iterated sequence of almost normalizing extensions (see, e.g., [24, 1.6.10]). Hence, R is a constructible $Z(R)$-algebra in the sense of [24, 9.4.12], and so all primitive ideals of R contract to maximal ideals of $Z(R)$ (see, e.g., [24, 9.4.21(i)]). It now follows from (a) that all primitive ideals of R are maximal. □

1.6. We next show that R can be constructed as a skew-Laurent extension of both a simple subalgebra and a subalgebra finite over $Z(R)$. To start, set $S = \{s \in \mathbb{Z}^n \mid \sigma(s,-) \equiv 1\}$ as in (1.2), and let $\overline{S}/S$ be the torsion subgroup of $\mathbb{Z}^n/S$. Then $\overline{S}$ and S are free abelian groups of the same rank. Consequently, there exists a basis $\{s_1, \dots, s_t\}$ for $\overline{S}$ along with positive integers $m_1, \dots, m_t$ such that $\{m_1 s_1, \dots, m_t s_t\}$ is a basis for S; see, for example, [17, II.1.5]. Since $\mathbb{Z}^n/\overline{S}$ is torsionfree and hence free, $\{s_1, \dots, s_t\}$ extends to a basis $\{s_1, \dots, s_n\}$ for $\mathbb{Z}^n$.

Now set $z_i = x^{s_i}$ for $i = 1, \dots, n$. Then R is a McConnell–Pettit algebra $P(\boldsymbol{\mu})$ on the generators $z_1^{\pm 1}, \dots, z_n^{\pm 1}$, where $\mu_{ij} = \sigma(s_i, s_j)$ for all i, j. Since $\{m_1 s_1, \dots, m_t s_t\}$ is a basis for S, it follows from (1.2) that $Z(R) = k[z_1^{\pm m_1}, \dots, z_t^{\pm m_t}]$.

Set $B = k\langle z_1^{\pm 1}, \dots, z_t^{\pm 1}\rangle$ and $C = k\langle z_{t+1}^{\pm 1}, \dots, z_n^{\pm 1}\rangle$; these are McConnell–Pettit algebras and B is finite over the central subalgebra $Z(R)$. We can express R as an iterated skew-Laurent extension of either B or C, since

$$R = B[z_{t+1}^{\pm 1}; \beta_{t+1}] \cdots [z_n^{\pm 1}; \beta_n] = C[z_1^{\pm 1}; \gamma_1] \cdots [z_t^{\pm 1}; \gamma_t]$$

for suitable automorphisms β_i and γ_j.

We claim that C is a simple algebra. By [23, 1.3], it suffices to show that $Z(C) = k$. Now C is spanned by the monomials x^s for s in the subgroup $S' = \mathbb{Z}s_{t+1} + \cdots + \mathbb{Z}s_n$. By (1.2), $Z(C)$ is spanned by monomials x^r for those $r \in S'$ such that $\sigma(r,-) \equiv 1$ on S'. Set $m = m_1 m_2 \cdots m_t$. Then $ms_i \in S$ for $i = 1, \dots, t$, whence $\sigma(mr, s_i) = \sigma(r, s_i)^m = \sigma(ms_i, r)^{-1} = 1$ for $i = 1, \dots, t$ and any $r \in \mathbb{Z}^n$. If $r \in S'$ with $\sigma(r,-) \equiv 1$ on S', we also have $\sigma(mr, s_i) = 1$ for $i = t+1, \dots, n$, whence $\sigma(mr,-) \equiv 1$ and $mr \in S$. As $S \cap S' = 0$, it follows that $r = 0$. This shows that $Z(C) = k$, and therefore C is simple, as claimed.

The Gelfand–Kirillov dimension of an algebra A will be denoted $\operatorname{GKdim}(A)$.

Recall that the prime or primitive spectrum of a noetherian ring is a noetherian topological space, i.e., it enjoys the ascending chain condition on open subsets. The *dimension* of such a space X is the supremum of the lengths ℓ of chains $X_0 \subsetneq X_1 \subsetneq \cdots \subsetneq X_\ell$ of irreducible closed subsets; we shall denote this dimension simply $\dim X$. We shall need the standard fact that $\dim \max A = \operatorname{K.dim} A$ for any commutative affine k-algebra A.

1.7 Lemma. *Let t be the rank of S. Then $\dim \operatorname{prim} R = t$ and $\operatorname{GKdim}(R/P) = n-t$ for all $P \in \operatorname{prim} R$.*

Proof. As in (1.6), $Z(R)$ is a Laurent polynomial ring in the independent indeterminates $z_1^{\pm m_1}, \dots, z_t^{\pm m_t}$. Thus $\dim \max Z(R) = t$, and so $\dim \operatorname{prim} R = t$ by (1.5).

Given $P \in \operatorname{prim} R$, we have $P = MR$ for some $M \in \max Z(R)$. Now $Z(R)/M$ is finite dimensional over k by the Nullstellensatz, and the algebra B is a finite module over $Z(R)$. Hence, B/MB is finite dimensional over k. As noted in (1.6), R is an iterated skew-Laurent extension of B, of the form

$$B[z_{t+1}^{\pm 1}; \beta_{t+1}] \cdots [z_n^{\pm 1}; \beta_n].$$

Since the elements of M commute with the z_i, they are fixed by the automorphisms β_i. Thus MB is invariant under the β_i, and so

$$R/P = R/MR \cong (B/MB)[z_{t+1}^{\pm 1}; \beta_{t+1}] \cdots [z_n^{\pm 1}; \beta_n].$$

Further, $z_{t+1}, \dots, z_{i-1}$ are eigenvectors for β_i, from which we see that each β_i acts locally finite dimensionally on $(B/MB)[z_{t+1}^{\pm 1}; \beta_{t+1}] \cdots [z_{i-1}^{\pm 1}; \beta_{i-1}]$. Therefore we conclude from [19, Proposition 1] that $\operatorname{GKdim}(R/P) = n - t$. □

1.8. The left Goldie rank (uniform rank) of a ring A will be denoted $\text{G.rank}(A)$. By [22, 2.7], Goldie rank is preserved in skew-Laurent extensions: $\text{G.rank}(A[x^{\pm 1}; \tau]) = \text{G.rank}(A)$.

1.9. We calculate the Goldie ranks of primitive factors of R in terms of the following subgroups of $\mathbb{Z}^n$:

$$\begin{aligned} S &= \{ s \in \mathbb{Z}^n \mid \sigma(s, -) \equiv 1 \} \\ \overline{S}/S &= \text{torsion subgroup of } \mathbb{Z}^n/S \\ \widetilde{S} &= \{ s \in \overline{S} \mid \sigma(s, -) \equiv 1 \text{ on } \overline{S} \}. \end{aligned}$$

Note that the quotient groups $\overline{S}/\widetilde{S}$ and $\widetilde{S}/S$ are finitely generated torsion groups, hence finite.

Theorem. *Let P be a primitive ideal of R.*

(a) $\text{G.rank}(R/P) = cd$ *for some positive integers c, d such that c divides $|\widetilde{S}/S|$ and d^2 divides $|\overline{S}/\widetilde{S}|$. In particular,* $\text{G.rank}(R/P)$ *divides* $|\overline{S}/S|$.

(b) *If P has finite codimension, then $\overline{S} = \mathbb{Z}^n$ and the square of* $\text{G.rank}(R/P)$ *divides* $|\mathbb{Z}^n/S|$.

Now assume that k is algebraically closed. Then:

(c) $\text{G.rank}(R/P) = |\overline{S}/\widetilde{S}|^{1/2} |\widetilde{S}/S|$.

(d) (Cf. [5, 2.3; 6, 7.2].) *If P has finite codimension then* $\text{G.rank}(R/P) = |\mathbb{Z}^n/S|^{1/2}$.

Proof. (a) (c) Retain the notation of (1.6). Write $P = MR$ for some $M \in \max Z(R)$. As in the proof of (1.7), B/MB is finite dimensional over k and R/P is isomorphic to an iterated skew-Laurent extension of B/MB, namely

$$R/P \cong (B/MB)[z_{t+1}^{\pm 1}; \beta_{t+1}] \cdots [z_n^{\pm 1}; \beta_n],$$

for suitable automorphisms β_i. (Let us identify R/P with this iterated skew-Laurent ring.) Hence,

$$\text{G.rank}(R/P) = \text{G.rank}(B/MB),$$

by iterated applications of (1.8).

Let N/MB be the radical of the finite dimensional algebra B/MB. Since the β_i restrict to automorphisms of B/MB, they must leave N/MB invariant. Consequently, N/MB induces a nilpotent ideal of R/P. However, R/P is prime, and thus $N/MB = 0$. Therefore B/MB is semisimple. Similarly, it follows from the simplicity of R/P (recall (1.5)) that B/MB is G-simple, where G is the subgroup of $\operatorname{Aut} B/MB$ generated by the restrictions of $\beta_{t+1}, \dots, \beta_n$. Therefore G acts transitively on the maximal ideals of B/MB. Consequently, B/MB is isomorphic to a direct product of copies of a finite dimensional simple k-algebra, say a direct product of c copies of $M_d(\Delta)$, where Δ is a finite dimensional division algebra over k. Thus $\dim_k(B/MB) = cd^2\delta$, where $\delta = \dim_k(\Delta)$, and $\text{G.rank}(R/P) = \text{G.rank}(B/MB) = cd$.

Recall that B is a McConnell–Pettit algebra in the generators $z_1^{\pm1}, \dots, z_t^{\pm1}$ and that $Z(R) = k[z_1^{\pm m_1}, \dots, z_t^{\pm m_t}]$. Consequently, B is a free $Z(R)$-module with rank $m_1 m_2 \cdots m_t = |\overline{S}/S|$. Hence, B/MB has dimension $|\overline{S}/S|$ over the field $F = Z(R)/M$, and thus $\dim_k(B/MB) = f|\overline{S}/S|$, where $f = \dim_k(F)$.

Let Q/MB be a maximal ideal of B/MB; then $B/Q \cong M_d(\Delta)$. Since B is a McConnell–Pettit algebra, (1.5) shows that $Q = M'B$, where $M' = Q \cap Z(B) \in \max Z(B)$. Note that $M \subseteq Q \cap Z(R) \subseteq M'$. Now $B = k\langle x^s \mid s \in \overline{S}\rangle$, because $z_i = x^{s_i}$ and $s_1, \dots, s_t$ generate $\overline{S}$. From (1.2) and the definition of $\widetilde{S}$, we thus have $Z(B) = k[x^s \mid s \in \widetilde{S}]$. Hence, B is a free $Z(B)$-module of rank $|\overline{S}/\widetilde{S}|$. Therefore B/Q has dimension $|\overline{S}/\widetilde{S}|$ over the field $F' = Z(B)/M'$, and consequently $\dim_k(B/Q) = f'|\overline{S}/\widetilde{S}|$, where $f' = \dim_k(F')$.

Observe that F' embeds in the center of B/Q, which is isomorphic to the center of Δ. Hence, $f' \mid \delta$. Since $Z(R) \subseteq Z(B)$ and $M \subseteq M'$, there exists a homomorphism from $F = Z(R)/M$ to $F' = Z(B)/M'$, and thus $f \mid f'$. Now

$$cd^2\delta = \dim_k(B/MB) = f|\overline{S}/S| = f|\overline{S}/\widetilde{S}||\widetilde{S}/S|$$

$$d^2\delta = \dim_k(M_d(\Delta)) = \dim_k(B/Q) = f'|\overline{S}/\widetilde{S}|,$$

from which we obtain $|\widetilde{S}/S| = cf'/f$ and $|\overline{S}/\widetilde{S}| = d^2\delta/f'$. Since $f \mid f'$ and $f' \mid \delta$, this shows that c divides $|\widetilde{S}/S|$ and d^2 divides $|\overline{S}/\widetilde{S}|$. Part (a) is therefore proved.

If k is algebraically closed, then $\delta = f = f' = 1$. In this case, $|\widetilde{S}/S| = c$ and $|\overline{S}/\widetilde{S}| = d^2$, and part (c) follows.

(b) (d) Suppose that P has finite codimension in R. Then $\operatorname{GKdim}(R/P) = 0$ and (1.7) shows that S has rank n. This forces $\mathbb{Z}^n/S$ to be torsion, whence $\overline{S} = \mathbb{Z}^n$. Consequently, $\widetilde{S} = S$, and parts (b) and (d) follow from (a) and (c). □

The computations in (1.7) and (1.9) lead to other expressions and estimates for the GK-dimensions and Goldie ranks of primitive factors of R — see (1.11) and (1.16).

1.10 Lemma. *Let Λ be the multiplicative subgroup of $k^\times$ generated by the λ_{ij}. Then each of the integers $m_1, \dots, m_t$, as defined in (1.9), is the order of an element of the product group Λ^n. In particular, each m_i divides the minimum exponent of the torsion subgroup of Λ, and no m_i is divisible by* char k.

Proof. Fix $i \in \{1, \dots, t\}$. For $m \in \mathbb{Z}$, we have $x^{ms_i} \in Z(R)$ if and only if $\prod_{j=1}^n \lambda_{jl}^{ms_{ij}} = 1$ for all $l = 1, \dots, n$. On the other hand, $x^{ms_i} \in Z(R)$ if and only if $ms_i \in S$, if and only if $m_i \mid m$. Therefore m_i equals the order of the n-tuple

$$\left(\prod_{j=1}^{n} \lambda_{j1}^{s_{ij}}, \dots, \prod_{j=1}^{n} \lambda_{jn}^{s_{ij}}\right)$$

in Λ^n. The remaining statements of the lemma follow directly. □

1.11 Theorem. *Let $P \in \operatorname{prim} R$, and set $d = \operatorname{GKdim}(R/P)$. Let $\Lambda = \langle \lambda_{ij} \rangle \subseteq k^\times$, let r be the torsionfree rank of Λ (that is, the rank of Λ modulo its torsion subgroup), and let e be the minimum exponent of the torsion subgroup of Λ.*

(a) *If $r = 0$ then $d = 0$, while if $r > 0$, then $d \geq \frac{1}{2}(1 + \sqrt{8r+1})$.*

(b) *If k is algebraically closed,* G.rank(R/P) *divides e^{n-d}.*

Proof. (a) If $r = 0$, the λ_{ij} are all roots of unity, and so R is finite over $Z(R)$ by (1.3.iv). In this case, R/P is finite dimensional over the field $Z(R)/(P \cap Z(R))$. By the Nullstellensatz, this field is finite dimensional over k, whence $\dim_k(R/P) = 0$ and thus $d = 0$.

Now assume that $r > 0$, and rewrite R as $P(\boldsymbol{\mu}) = k\langle z_1^{\pm 1}, \dots, z_n^{\pm 1}\rangle$ using the notation of (1.6). Recall from (1.7) that $d = n - t$, where $t = \operatorname{rank} S$. Observe that the multiplicative subgroup of $k^\times$ generated by the μ_{ij}, for $i, j \geq t+1$, can have torsionfree rank at most $d(d-1)/2$. On the other hand, since $z_1^{m_1}, \dots, z_t^{m_t}$ are central in R, we have $\mu_{ij}^{m_i} = 1$ for $i \leq t$ and all j. Thus the subgroup of $k^\times$ generated by all the μ_{ij} is a finite extension of the subgroup generated by those with $i, j \geq t+1$, and so this larger subgroup also has torsionfree rank at most $d(d-1)/2$.

Since each $\mu_{ij} = \sigma(s_i, s_j)$ and the s_i form a basis for $\mathbb{Z}^n$, we have

$$\langle \mu_{ij} \rangle = \langle \sigma(s,t) \mid s, t \in \mathbb{Z}^n \rangle = \Lambda,$$

and therefore $r \leq d(d-1)/2$. In other words, $d^2 - d - 2r \geq 0$. Since $d \geq 0$ and $r > 0$, it follows easily that $d \geq \frac{1}{2}(1 + \sqrt{8r+1})$.

(b) Each $m_i \mid e$ by (1.10), and therefore $m_1 m_2 \cdots m_t \mid e^t$. The choice of basis $\{s_1, \dots, s_t\}$ for $\overline{S}$ made in (1.6) ensures that $|\overline{S}/S| = m_1 m_2 \cdots m_t$. Since G.rank$(R/P)$ divides $|\overline{S}/S|$ by (1.9), it follows that G.rank(R/P) divides e^t. Finally, $t = n - d$ by (1.7), and so part (b) is proved. □

1.12. The estimates given in (1.11) cannot be improved to exact formulas, since neither GKdim(R/P) nor G.rank(R/P) is solely determined by the group Λ. For example, let $q \in k^\times$ and set

$$\boldsymbol{\lambda} = \begin{pmatrix} 1 & q & 1 & 1 \\ q^{-1} & 1 & 1 & 1 \\ 1 & 1 & 1 & q \\ 1 & 1 & q^{-1} & 1 \end{pmatrix} \quad \text{and} \quad \boldsymbol{\lambda}' = \begin{pmatrix} 1 & q & 1 & 1 \\ q^{-1} & 1 & 1 & 1 \\ 1 & 1 & 1 & 1 \\ 1 & 1 & 1 & 1 \end{pmatrix}.$$

Let S and S' denote the groups $\{s \in \mathbb{Z}^4 \mid \sigma(s, -) \equiv 1\}$ corresponding respectively to $P(\boldsymbol{\lambda})$ and $P(\boldsymbol{\lambda}')$.

If q is not a root of unity, then $S = 0$ and $S' = 0^2 \times \mathbb{Z}^2$. Hence, it follows from (1.7) that primitive factors of $P(\boldsymbol{\lambda})$ have GK-dimension 4, while those of $P(\boldsymbol{\lambda}')$ have GK-dimension 2. (Note that 2 is the minimum GK-dimension allowed by (1.11.a) when Λ has torsionfree rank 1.) Now suppose that k is algebraically closed and that q is a primitive ℓ-th root of unity; then $S = (\mathbb{Z}\ell)^4$ and $S' = (\mathbb{Z}\ell)^2 \times \mathbb{Z}^2$. In this case, it follows from (1.9.d) that primitive factors of $P(\boldsymbol{\lambda})$ have Goldie rank ℓ^2, while those of $P(\boldsymbol{\lambda}')$ have Goldie rank ℓ.

1.13. Let Γ denote the torus $(k^\times)^n$. There is a natural action of Γ on R, where $\gamma \in \Gamma$ acts via the unique k-algebra automorphism of R such that $x_i^\gamma = \gamma_i x_i$ for $i = 1, \dots, n$. To express this action of Γ on R in a basis-free form, identify Γ with the character group of $\mathbb{Z}^n$; then $(x^s)^\gamma = \gamma(s)x^s$ for $s \in \mathbb{Z}^n$ and $\gamma \in \Gamma$.

We will also be concerned with the canonically induced actions of Γ on $\operatorname{spec} R$, $\operatorname{prim} R$, $Z(R)$, $\operatorname{spec} Z(R)$, and $\max Z(R)$. Note that the actions on $\operatorname{spec} R$ and $\operatorname{spec} Z(R)$ commute with the bijection between these spectra discussed in (1.5).

Lemma. *If k is infinite, then R is Γ-simple (i.e., the only Γ-invariant ideals of R are 0 and R).*

Proof. Since R is spanned by Γ-eigenvectors, so is any Γ-invariant ideal. Hence, it suffices to show that all nonzero Γ-eigenvectors in R are invertible. Now each monomial x^s is a Γ-eigenvector, and since k is infinite, any two distinct monomials must have different Γ-eigenvalues. It follows that the Γ-eigenspaces in R are the 1-dimensional subspaces kx^s for $s \in \mathbb{Z}^n$. Thus all nonzero Γ-eigenvectors are invertible in R, as desired. □

1.14. (Cf. [5, §2; 6, §7].) Suppose that k is algebraically closed.

(i) Let S be as in (1.2), and choose a basis $\{b_1, \dots, b_r\}$. Then $Z(R)$ is a Laurent polynomial ring in the variables $(x^{b_1})^{\pm 1}, \dots, (x^{b_r})^{\pm 1}$. Since k is algebraically closed, the abelian group $k^\times$ is divisible and hence injective. In particular, Γ maps onto the character group of S by restriction. Consequently, for any r-tuple $(\alpha_1, \dots, \alpha_r) \in (k^\times)^r$, there is some $\gamma \in \Gamma$ such that $\gamma(b_i) = \alpha_i$ for $i = 1, \dots, r$, whence $(x^{b_i})^\gamma = \alpha_i x^{b_i}$ for all i. Thus the restriction of the action of Γ from R to $Z(R)$ maps Γ onto the standard torus of automorphisms of $Z(R) = k[(x^{b_1})^{\pm 1}, \dots, (x^{b_r})^{\pm 1}]$. Hence, Γ acts transitively on $\max Z(R)$.

(ii) As noted above, the action of Γ commutes with contraction, and so it follows from (i) and (1.5) that Γ acts transitively on $\operatorname{prim} R$.

(iii) Fix a primitive ideal P of R, and set $M = P \cap Z(R)$. It follows from (1.5) that M is a maximal ideal of $Z(R)$ and that $\operatorname{Stab}_\Gamma(P) = \operatorname{Stab}_\Gamma(M)$. However, in the notation of (i), M must be generated by $x^{b_1} - \beta_1, \dots, x^{b_r} - \beta_r$ for some $\beta_1, \dots, \beta_r \in k^\times$, and consequently

$$\operatorname{Stab}_\Gamma(M) = \{\gamma \in \Gamma \mid \gamma(b_i) = 1 \text{ for } i = 1, \dots, r\}.$$

Hence, $\operatorname{Stab}_\Gamma(P)$ is both a torus and a (Zariski) closed subgroup of Γ. Moreover, it follows from (i) and (ii) that the Γ-action on $\max Z(R)$ provides an isomorphism of affine varieties between $\Gamma/\operatorname{Stab}_\Gamma(P)$ and $\max Z(R)$ (see, e.g., [16, 12.1]). Consequently, the action of Γ on $\operatorname{prim} R$ produces a homeomorphism of $\Gamma/\operatorname{Stab}_\Gamma(P)$ onto $\operatorname{prim} R$. In particular, the dimension of $\operatorname{prim} R$ as a topological space is equal to the rank of $\Gamma/\operatorname{Stab}_\Gamma(P)$.

There is a group homomorphism $\phi\colon \Gamma \to (k^\times)^r$ given by $\phi(\gamma) = (\gamma(b_1), \dots, \gamma(b_r))$. This homomorphism is surjective by (i), and we have just seen that its kernel is $\operatorname{Stab}_\Gamma(P)$. Therefore $\Gamma/\operatorname{Stab}_\Gamma(P) \cong (k^\times)^r$.

Our first use of (1.14) is the following criterion for R to be an Azumaya algebra. (Cf. [34, Corollary 10] and [35, 3.2] for some general conditions under which strongly graded PI rings are Azumaya.)

1.15 Proposition. *If k is algebraically closed, the following conditions are equivalent:*

(a) *The parameters λ_{ij} are all roots of unity.*
(b) *R has a nonzero finite dimensional representation.*
(c) *R is an Azumaya algebra.*

Proof. That (c) $\Longrightarrow$ (a) $\Longrightarrow$ (b) follows easily from (1.3.iv). The implication (a) $\Longrightarrow$ (c) is shown in [5, 2.3; 6, 7.2] (cf. [2, 2.3]). We conclude by proving that (b) $\Longrightarrow$ (c).

Assume that condition (b) holds. Then R has a finite dimensional irreducible module with finite codimensional annihilator P, and it follows from (1.14.ii) that the primitive spectrum of R is equal to the Γ-orbit of P. However, R is a prime Jacobson ring, and so its Jacobson radical is zero. Hence, R embeds into a direct product of algebras that are all isomorphic to the finite dimensional algebra R/P. We now see that R is a PI domain, and that every primitive factor of R has the same PI degree. Therefore, R is an Azumaya algebra, by the Artin-Procesi Theorem (e.g., [24, 13.7.14]), and so (c) holds. □

1.16 Theorem. *Let P be a primitive ideal of R, and assume that k is algebraically closed.*

(a) *The* GK*-dimension of R/P is equal to the rank of* $\operatorname{Stab}_\Gamma(P)$*, which is equal to its dimension as a variety.*
(b) *The Goldie rank of R/P divides* $|\operatorname{Stab}_\Gamma(P)/\operatorname{Stab}_\Gamma(P)^\circ|$.
(c) (Cf. [5, 2.3; 6, 7.2].) *If P has finite codimension, then* $\operatorname{Stab}_\Gamma(P)$ *is finite and* $\operatorname{G.rank}(R/P) = |\operatorname{Stab}_\Gamma(P)|^{1/2}$.

Proof. (a) Let $m_1, \dots, m_t \in \mathbb{Z}$, $s_1, \dots, s_n \in \mathbb{Z}^n$, and $z_1, \dots, z_n \in R$ be as chosen in (1.6). Also recall, from (1.13), the identification of Γ with the character group of $\mathbb{Z}^n$. After a suitable change of coordinates, we may assume that the n-tuples $\gamma \in \Gamma$ act on R by the rule $z_i^\gamma = \gamma_i z_i$. The stabilizer of the subalgebra B is thus the subgroup

$$T = \{\gamma \in \Gamma \mid \gamma_i = 1 \text{ for } i = 1, \dots, t\},$$

which is a torus of rank $n - t$. Therefore, by (1.7), $\operatorname{GKdim}(R/P)$ equals the rank of T.

Note that $P = MR$ for some maximal ideal M of $Z(R) = k[z_1^{\pm m_1}, \dots, z_t^{\pm m_t}]$. Moreover, $\operatorname{Stab}_\Gamma(M) = \operatorname{Stab}_\Gamma(P)$, as noted in (1.14.iii). Also, since k is algebraically closed, it follows that

$$M = \langle z_1^{m_1} - \alpha_1, \dots, z_t^{m_t} - \alpha_t \rangle$$

for some nonzero scalars $\alpha_1, \dots, \alpha_t \in k^\times$. Consequently,

$$\operatorname{Stab}_\Gamma(P) = \operatorname{Stab}_\Gamma(M) = \{\gamma \in \Gamma \mid \gamma_i^{m_i} = 1 \quad \text{for} \quad i = 1, \dots, t\}.$$

It was shown in (1.10) that none of the integers $m_1, \dots, m_n$ is divisible by $\operatorname{char} k$, and so $\operatorname{Stab}_\Gamma(P)/T$ is isomorphic to $\bigoplus_{i=1}^t \mathbb{Z}/m_i\mathbb{Z}$. Therefore,

$$|\operatorname{Stab}_\Gamma(P)/T| = m_1 m_2 \cdots m_t = |\overline{S}/S|.$$

Since the torus T has finite index in $\operatorname{Stab}_\Gamma(P)$, it must be a maximal torus. Hence, the rank of T is equal to that of $\operatorname{Stab}_\Gamma(P)$, and therefore the GK-dimension of R/P is equal to the rank of $\mathrm{Stab}_\Gamma(P)$. Next, observe that $\mathrm{Stab}_\Gamma(P)$ is the disjoint union of finitely many subvarieties isomorphic to T. However, T is an irreducible variety of dimension $n-t$. Therefore $\mathrm{GKdim}(R/P)$ equals the dimension of the variety $\mathrm{Stab}_\Gamma(P)$.

(b) Because $T = \mathrm{Stab}_\Gamma(P)^\circ$ (see, e.g., [16, 7.3]), it follows from the above that

$$\left| \mathrm{Stab}_\Gamma(P)/\,\mathrm{Stab}_\Gamma(P)^\circ \right| = |\overline{S}/S|.$$

By (1.9.i), the Goldie rank of R/P divides the integer $|\,\mathrm{Stab}_\Gamma(P)/\,\mathrm{Stab}_\Gamma(P)^\circ|$, and (b) follows.

(c) Since P has finite codimension,

$$\mathrm{G.rank}(R/P) = \left| \mathrm{Stab}_\Gamma(P)/\,\mathrm{Stab}_\Gamma(P)^\circ \right|^{1/2}$$

by (1.9.ii). Also, $\mathrm{GKdim}(R/P) = 0$, and so T is trivial. Therefore, $\mathrm{G.rank}(R/P) = |\,\mathrm{Stab}_\Gamma(P)|^{1/2}$. □

1.17. Theorems 1.9, 1.11, and 1.16 all provide information on the GK-dimensions and Goldie ranks of primitive factors of R. Estimates of GK-dimensions and Goldie ranks for the prime factors of R follow readily from known results. For instance, if Q is a prime ideal of R, then it follows from [9, 2.6] and [19, Theorem 2] that Tauvel's Height Formula holds:

$$\mathrm{GKdim}(R/Q) = n - \mathrm{height}(Q).$$

Now suppose that k is algebraically closed. In view of (1.14.ii), all primitive factors of R have the same Goldie rank, say u. Since R is a Jacobson ring, every prime ideal of R is an intersection of primitive ideals. Results of Stafford therefore ensure that the Goldie rank of R/Q divides u; see, for example, [31, 3.9; 24, 4.6.17.iii]. Of course, if R has a finite codimensional primitive ideal P, with $\mathrm{G.rank}(R/P) = u$, then R is an Azumaya algebra, by (1.15), and all of the prime factors of R have PI degree equal to u.

2. Quantum Affine Spaces

We now turn to the algebras $\mathcal{O}_{\mathbf{q}}(k^n)$. The first part of the section shows that the properties of $\operatorname{spec} \mathcal{O}_{\mathbf{q}}(k^n)$, exhibited in [5, 2.4; 6, 7.3] when the parameters q_{ij} are roots of unity, and in [4, §4] when the q_{ij} generate a torsionfree group, actually hold in general. An alternate description of the primitive spectrum of $\mathcal{O}_{\mathbf{q}}(k^n)$ can be found in work of Vancliff, in the case where $q_{ij}q_{jm} = q_{im}$ for all i, j, m [33, §1.3].

2.1. (i) Let $\mathbf{q} = (q_{ij})$ be a multiplicatively antisymmetric $n \times n$ matrix over k, and let $A = \mathcal{O}_{\mathbf{q}}(k^n)$ be the k-algebra generated by elements $x_1, \dots, x_n$ subject only to the relations $x_i x_j = q_{ij} x_j x_i$ for $i, j = 1, \dots, n$; this algebra is the general multiparameter quantum coordinate ring of affine n-space over k. It is clear that A is a constructible k-algebra [24, 9.4.12] and so it satisfies the Nullstellensatz over k [24, 9.4.21]; in particular, A is a Jacobson ring.

(ii) Let Γ denote the torus $(k^\times)^n$, acting on A analogously to (1.6); that is, $\gamma \in \Gamma$ acts via the k-algebra automorphism of A such that $x_i^\gamma = \gamma_i x_i$, for $i = 1, \dots, n$. Let W denote the set of subsets of $\{1, \dots, n\}$, and for $w \in W$ let $\operatorname{spec}_w A$ be the set of those prime ideals in A such that

$$P \cap \{x_1, \dots, x_n\} = \{x_i \mid i \in w\}.$$

Then $\operatorname{spec} A$ is the disjoint union of the sets $\operatorname{spec}_w A$. Likewise, $\operatorname{prim} A$ is the disjoint union of the sets $\operatorname{prim}_w A = \operatorname{prim} A \cap \operatorname{spec}_w A$. Note that the sets $\operatorname{spec}_w A$ and $\operatorname{prim}_w A$ are invariant under the action of Γ. In the case that k is infinite, we shall rewrite this stratification of $\operatorname{spec} A$ in terms of the Γ-prime ideals of A (see (2.11) and (2.12)).

(iii) For each $w \in W$, the ideal $J_w = \langle x_i \mid i \in w \rangle$ is the unique minimum element of $\operatorname{spec}_w A$, and J_w is obviously Γ-invariant. Let E_w denote the multiplicative set in A/J_w generated by the images of those x_j with $j \notin w$. These generators are normal Γ-eigenvectors, and since they are normal, the prime ideals in A/J_w that are disjoint from E_w are precisely the prime ideals that do not contain $x_j + J_w$ for any $j \notin w$. These are exactly the prime ideals of the form P/J_w for $P \in \operatorname{spec}_w A$. Since E_w consists of normal elements, it is an Ore set in A/J_w. Set $A_w = (A/J_w)[E_w^{-1}]$, and let $\phi_w \colon A \to A/J_w \to A_w$ denote the localization map. Then $\operatorname{spec}_w A \approx \operatorname{spec} A_w$ via extension ($P \mapsto \phi_w(P)A_w$) and contraction ($Q \mapsto \phi_w^{-1}(Q)$). Note that there is an induced action of Γ on A_w, and that the extension and contraction maps between $\operatorname{spec}_w A$ and $\operatorname{spec} A_w$ intertwine the Γ-actions.

2.2. Observe that each A_w is a McConnell–Pettit algebra with respect to the generators $(x_i + J_w)^{\pm 1}$, for $i \notin w$, to which the results of the previous section apply. For instance, $\operatorname{prim} A_w = \max A_w$ by (1.5). Let K_w denote the kernel of the induced action of Γ on A_w. In other words,

$$K_w = \{\gamma \in \Gamma \mid \gamma_i = 1 \quad \text{for all} \quad i \notin w\}.$$

Then Γ/K_w is a torus of rank $n - \operatorname{card}(w)$, and its action on A_w coincides with the standard action of the torus $(k^\times)^{n-\operatorname{card}(w)}$ on A_w as in (1.13). In particular, it follows from (1.14.ii) that if k is algebraically closed, Γ acts transitively on $\operatorname{prim} A_w$.

2.3 Theorem. *Let $w \in W$.*

(a) $\operatorname{prim}_w A$ *equals the set of maximal elements of* $\operatorname{spec}_w A$.

(b) *Extension and contraction provide mutually inverse homeomorphisms between the spaces* $\operatorname{prim}_w A$ *and* $\operatorname{prim} A_w$.

(c) *Each prime ideal in the set* $\operatorname{spec}_w A$ *is an intersection of primitive ideals from* $\operatorname{prim}_w A$.

(d) *If k is algebraically closed, Γ acts transitively on* $\operatorname{prim}_w A$.

Proof. As in [4, 4.2], Assumptions 1–5 of [4, 1.1] are clear, and Assumption 6 follows from (1.4). Consequently, parts (a), (b), (c) follow from [4, 1.3, 1.4]. Part (d) follows from part (b) and the fact that Γ acts transitively on $\operatorname{prim} A_w$ (see (2.2)). □

2.4. Part (d) of (2.3) also follows from a general result of Moeglin and Rentschler concerning algebraic groups acting as automorphisms [25, Théorème 2.12]. See (2.12) for further detail.

It follows from (2.3.d), in the algebraically closed case, that the sets $\text{prim}_w A$ are exactly the Γ-orbits in $\text{prim}\, A$. Hence, the decomposition $\text{prim}\, A = \bigcup_{w \in W} \text{prim}_w A$ coincides with the stratification given by DeConcini and Procesi in the case that the q_{ij} are roots of unity [5, 2.4; 6, 7.3].

2.5. Recall that a k-algebra R satisfies the *Dixmier–Moeglin equivalence* provided the following three sets of prime ideals coincide:

(a) $\text{prim}\, R$.

(b) The set of prime ideals that are *locally closed* in $\text{spec}\, R$. (Note that $P \in \text{spec}\, R$ is locally closed exactly when the intersection of those prime ideals properly containing P is strictly larger than P. Locally closed prime ideals are also known as *G-prime* ideals.)

(c) The set of *rational* prime ideals (i.e., $P \in \text{spec}\, R$ such that the center of the Goldie quotient ring $\text{Fract}(R/P)$ is algebraic over k).

If R is Jacobson, then it follows immediately that the locally closed prime ideals are primitive. Moreover, if R satisfies the Nullstellensatz then the primitive ideals are all rational. (By e.g. [7, 4.1.6], for any primitive ideal P of R the center of $\text{Fract}(R/P)$ embeds in the center of the endomorphism ring of a faithful simple (R/P)-module.)

Corollary. *The Dixmier-Moeglin equivalence holds for A.*

Proof. We have "locally closed $\Longrightarrow$ primitive $\Longrightarrow$ rational" since A is a Jacobson ring satisfying the Nullstellensatz.

To prove "primitive $\Longrightarrow$ locally closed", let P be a primitive ideal of A; then $P \in \text{prim}_w A$ for some $w \in W$. By (2.3.a), P is maximal within $\text{spec}_w A$. Let e_w denote the product (in some order) of the elements x_i for $i \in \{1, \dots, n\} \setminus w$. If Q is any prime ideal properly containing P, then $Q \notin \text{spec}_w A$ by the maximality of P. Consequently, $Q \in \text{spec}_z A$ for some $z \supset w$, from which it follows that $e_w \in Q$. Thus the intersection of the prime ideals properly containing P contains e_w. However, since e_w is a product of normal elements not contained in P, it follows that e_w cannot lie in P, and therefore P is locally closed in $\text{spec}\, A$.

Finally, we prove the contrapositive of "rational $\Longrightarrow$ primitive". Thus, let P be a prime ideal of A which is not primitive. Then $P \in \text{spec}_w A$ for some $w \in W$, and by (2.3.b) the extension $\phi_w(P)A_w$ is a non-maximal prime ideal in the localization A_w. Consequently, (1.4) shows that $\phi_w(P)A_w \cap Z(A_w)$ is a non-maximal ideal of $Z(A_w)$, whence $Z(A_w)/(\phi_w(P)A_w \cap Z(A_w))$ is not a field. Hence, this integral domain must be transcendental over k and also naturally embeds in $Z(A_w/\phi_w(P)A_w)$, whence the latter algebra is transcendental over k. Now observe that $A_w/\phi_w(P)A_w$ is isomorphic to the localization of A/P with respect to the image of E_w, and that the image of E_w in A/P consists of regular elements because E_w is disjoint from P. Hence, $\text{Fract}(A/P) \cong \text{Fract}(A_w/\phi_w(P)A_w)$, from which we conclude that $Z(A_w/\phi_w(P)A_w)$ embeds in $Z(\text{Fract}(A/P))$. Therefore $Z(\text{Fract}(A/P))$ is transcendental over k, proving that P is not rational. □

2.6. In order to carry the results of (1.9), (1.11), and (1.16) over to A, we first check that Goldie ranks and GK-dimensions are preserved in the relevant localizations.

Consider $Q \in \operatorname{spec}_w A$ for some $w \in W$, and let $Q_w = \phi_w(Q)A_w$ denote the prime ideal in A_w extended from Q. Now A_w/Q_w is a localization of A/Q with respect to an Ore set of regular normal elements, whence $\operatorname{Fract}(A_w/Q_w) \cong \operatorname{Fract}(A/Q)$, and thus

$$\text{G.rank}(A_w/Q_w) = \text{G.rank}(A/Q).$$

Further, A_w/Q_w is obtained from A/Q by inverting finitely many regular normal elements, namely the elements $x_i + Q$ for $i \notin w$. List these elements as $c_1, \dots, c_t$, and observe that conjugation by any c_j restricts to an automorphism α_j of $(A/Q)\langle c_1^{-1}, \dots, c_{j-1}^{-1}\rangle$ acting locally finite dimensionally (in fact, acting semisimply). Hence, it follows from iterated application of [9, 3.3] that

$$\operatorname{GKdim}(A_w/Q_w) = \operatorname{GKdim}(A/Q).$$

2.7. Define an alternating bicharacter ρ on $\mathbb{Z}^n \times \mathbb{Z}^n$ by the rule

$$\rho(s,t) = \prod_{i,j=1}^{n} q_{ij}^{s_i t_j}.$$

Fix $w \in W$, and set

$$\begin{aligned} L_w &= \{ s \in \mathbb{Z}^n \mid s_i = 0 \quad \text{for all} \quad i \in w \} \\ S_w &= \{ s \in L_w \mid \rho(s,-) \equiv 1 \quad \text{on} \quad L_w \} \\ \overline{S}_w/S_w &= \quad \text{torsion subgroup of} \quad L_w/S_w \\ \widetilde{S}_w &= \{ s \in \overline{S}_w \mid \rho(s,-) \equiv 1 \quad \text{on} \quad \overline{S}_w \}. \end{aligned}$$

Theorem. *Let $P \in \operatorname{prim}_w A$.*

(a) $\dim \operatorname{prim}_w A = \operatorname{rank}(S_w)$.

(b) $\operatorname{GKdim}(A/P) = n - \operatorname{card}(w) - \operatorname{rank}(S_w)$.

(c) $\text{G.rank}(A/P) = cd$ *for some positive integers c, d such that c divides $|\widetilde{S}_w/S_w|$ and d^2 divides $|\overline{S}_w/\widetilde{S}_w|$. In particular,* $\text{G.rank}(A/P)$ *divides* $|\overline{S}_w/S_w|$.

(d) *If P has finite codimension, then $\overline{S}_w = L_w$ and the square of* $\text{G.rank}(A/P)$ *divides* $|L_w/S_w|$.

Now assume that k is algebraically closed. Then:

(e) $\text{G.rank}(A/P) = |\overline{S}_w/\widetilde{S}_w|^{1/2} |\widetilde{S}_w/S_w|$.

(f) (Cf. [5, 2.4; 6, 7.3].) *If P has finite codimension,* $\text{G.rank}(A/P) = |L_w/S_w|^{1/2}$.

Proof. Recall that A_w is a McConnell–Pettit algebra in the generators $(x_i + J_w)^{\pm 1}$ for $i \notin w$. Hence, the restriction of ρ to $L_w \times L_w$ plays the role of the alternating bicharacter σ considered in Section 1. In particular, (1.7) shows that $\dim \operatorname{prim} A_w = \operatorname{rank}(S_w)$ and that

$$\operatorname{GKdim}(A_w/P_w) = \operatorname{rank}(L_w) - \operatorname{rank}(S_w) = n - \operatorname{card}(w) - \operatorname{rank}(S_w)$$

for all $P_w \in \operatorname{prim} A_w$. Thus parts (a) and (b) follow from (2.3.b) and (2.6). Similarly, parts (c)–(f) follow from (1.9) and (2.6). □

2.8 Theorem. *Let $w \in W$, let $\Lambda_w = \langle q_{ij} \mid i,j \notin w \rangle \subseteq k^\times$, let r_w denote the torsionfree rank of Λ_w, and let e_w be the minimum exponent of the torsion subgroup of Λ_w. Let $P \in \operatorname{prim}_w A$, and set $d_w = \operatorname{GKdim}(A/P)$.*

(a) *If $r_w = 0$ then $d_w = 0$, while if $r_w > 0$, then $d_w \geq \frac{1}{2}(1 + \sqrt{8r_w + 1})$.*
(b) *If k is algebraically closed,* G.rank(A/P) *divides* $e_w^{n-\operatorname{card}(w)-d_w}$.

Proof. These results follow from (1.11) and (2.6). □

A special case of (2.8.b) follows from Roseblade's study of group algebras of polycyclic-by-finite groups [30]: When the group $\langle q_{ij} \rangle \subseteq k^\times$ is torsionfree, all prime ideals of A are completely prime (cf. [11, 2.1]).

2.9 Theorem. *Let $w \in W$ and $P \in \operatorname{prim}_w A$, and assume that k is algebraically closed.*

(a) *$\Gamma/\operatorname{Stab}_\Gamma(P)$ is a torus, homeomorphic (with respect to the Zariski topology) to* $\operatorname{prim}_w A$.
(b) *The rank of* $\operatorname{Stab}_\Gamma(P)$ *equals* $\operatorname{GKdim}(A/P) + \operatorname{card}(w)$, *and the order of the quotient group* $\operatorname{Stab}_\Gamma(P)/\operatorname{Stab}_\Gamma(P)^\circ$ *is a multiple of* G.rank(A/P).
(c) *If P has finite codimension in A, then*

$$\text{G.rank}(A/P) = |\operatorname{Stab}_\Gamma(P)/\operatorname{Stab}_\Gamma(P)^\circ|^{1/2}.$$

(d) G.rank(A/Q) *divides* G.rank(A/P) *for all* $Q \in \operatorname{spec}_w A$.

Proof. Let $P_w = \phi_w(P)A_w$ denote the primitive ideal in A_w corresponding to P. Note that the actions of Γ on A and on A_w commute with localization, that is, $\phi_w(a^\gamma) = \phi_w(a)^\gamma$ for $a \in A$ and $\gamma \in \Gamma$. Since $P = \phi_w^{-1}(P_w)$, it follows that $\operatorname{Stab}_\Gamma(P) = \operatorname{Stab}_\Gamma(P_w)$.

(a) From the comments above, we see that

$$\Gamma/\operatorname{Stab}_\Gamma(P) \cong (\Gamma/K_w)/\operatorname{Stab}_{\Gamma/K_w}(P_w).$$

Hence, (1.14.iii) shows that $\Gamma/\operatorname{Stab}_\Gamma(P)$ is a torus homeomorphic to $\operatorname{prim} A_w$. Further, $\operatorname{prim} A_w$ and $\operatorname{prim}_w A$ are homeomorphic by (2.3).

(b) (c) Since $\operatorname{Stab}_\Gamma(P)^\circ$ is the unique maximal torus of $\operatorname{Stab}_\Gamma(P)$, it must contain the torus K_w. Further, the quotient $\operatorname{Stab}_\Gamma(P)^\circ/K_w$ is the maximal torus of $\operatorname{Stab}_\Gamma(P)/K_w$ (maximality follows from the fact that $\operatorname{Stab}_\Gamma(P)^\circ$ has finite index in $\operatorname{Stab}_\Gamma(P)$ [16, 7.3]). According to (1.16), $\operatorname{rank}(\operatorname{Stab}_\Gamma(P)/K_w) = \operatorname{GKdim}(A_w/P_w)$, and $|\operatorname{Stab}_\Gamma(P)/\operatorname{Stab}_\Gamma(P)^\circ|$ is a multiple of G.rank(A_w/P_w); further, if P_w has finite codimension in A_w then

$$\text{G.rank}(A_w/P_w) = |\operatorname{Stab}_\Gamma(P)/\operatorname{Stab}_\Gamma(P)^\circ|^{1/2}.$$

Since $\operatorname{rank}\operatorname{Stab}_\Gamma(P) = \operatorname{rank}(\operatorname{Stab}_\Gamma(P)/K_w) + \operatorname{card}(w)$, parts (b) and (c) now follow from (2.6).

(d) This follows from (1.17) and (2.6). □

2.10. In case k is algebraically closed, (2.8) and (2.9) yield a rough but convenient upper bound for the Goldie ranks of prime factors of A, as follows. Let $\Lambda = \langle q_{ij} \rangle \subseteq k^{\times}$, and let e be the minimum exponent of the torsion subgroup of Λ. For $w \in W$, observe that since Λ_w is a subgroup of Λ, the exponent e_w divides e, whence $e_w^{n-\text{card}(w)-d_w}$ divides e^n. Therefore, it follows from (2.8.b) and (2.9.d) that $\text{G.rank}(A/P) \mid e^n$ for all $P \in \text{spec}\, A$.

It does not seem easy to derive a general upper bound from (2.7), aside from calculating bounds for the prime factors corresponding to each $\text{spec}_w A$. In particular, the maximum value for $\text{G.rank}(A/P)$ need not occur for P in $\text{spec}_{\varnothing} A$. For example, suppose that $n = 3$, that q_{12} is a primitive ℓ-th root of unity for some $\ell > 1$, and that the multiplicative group $\langle q_{13}, q_{23} \rangle$ is free abelian of rank 2. In this case, one easily calculates that $S_{\varnothing} = \{0\}$, and so it follows from (2.7.c) and (2.9.d) that $\text{G.rank}(A/P) = 1$ for all $P \in \text{spec}_{\varnothing} A$. On the other hand, $S_{\{3\}} = \mathbb{Z}\ell \times \mathbb{Z}\ell \times \{0\}$ and $\overline{S}_{\{3\}} = L_{\{3\}} = \mathbb{Z} \times \mathbb{Z} \times \{0\}$, whence $\widetilde{S}_{\{3\}} = S_{\{3\}}$. Consequently, $\text{G.rank}(A/P) = \ell$ for all $P \in \text{prim}_{\{3\}} A$, by (2.7.e).

2.11. Recall that a Γ-*prime ideal* of A is any proper Γ-invariant ideal J such that the product of two Γ-invariant ideals of A is contained in J only if at least one of these ideals itself is contained in J.

Proposition. *Assume that k is infinite.*

(a) *The ideals J_w for $w \in W$ are exactly the Γ-prime ideals of A.*
(b) *For each $w \in W$,*

$$\text{spec}_w A = \left\{ Q \in \text{spec}\, A \;\middle|\; \bigcap_{\gamma \in \Gamma} Q^{\gamma} = J_w \right\}.$$

Proof. (a) Each J_w is prime and Γ-invariant, hence Γ-prime. Conversely, let J be a Γ-prime ideal of A, and set $w = \{i \in \{1, \dots, n\} \mid x_i \in J\}$; then $J_w \subseteq J$. We claim that x_i is regular modulo J for $i \notin w$. Set $L_i = \{a \in A \mid ax_i \in J\}$ and observe that since x_i is a normal Γ-eigenvector and J is Γ-invariant, Ax_i and L_i are Γ-invariant two-sided ideals of A. Since $L_i A x_i \subseteq J$ but $Ax_i \not\subseteq J$, it follows from the Γ-primeness of J that $L_i \subseteq J$. Similarly, the ideal $\{a \in A \mid x_i a \in J\}$ is contained in J, and therefore x_i is regular modulo J, as claimed.

Now the factor ring A/J is E_w-torsionfree, and so $\phi_w(J)A_w$ is an ideal of A_w whose contraction is precisely J (e.g., [13, 9.20, 9.17]). Since J is a proper Γ-invariant ideal of A, the extension $\phi_w(J)A_w$ must be a proper Γ-invariant ideal of A_w. By the Γ-simplicity of A_w (see (1.13)), it follows that $\phi_w(J)A_w = 0$. Therefore $J = \phi_w^{-1}(0) = J_w$.

(b) Let $Q \in \text{spec}_w A$ for some $w \in W$, and set $(Q : \Gamma) = \bigcap_{\gamma \in \Gamma} Q^{\gamma}$. Since Q is prime, it is easily seen that $(Q : \Gamma)$ is Γ-prime, and it is clear that $J_w \subseteq (Q : \Gamma)$. By part (a), $(Q : \Gamma) = J_z$ for some $z \in W$, and obviously $z \supseteq w$. Since $x_i \notin Q$ when $i \notin w$, and since each x_i is a Γ-eigenvector, we see that $z = w$. Therefore we have shown that $(Q : \Gamma) = J_w$ for all $Q \in \text{spec}_w A$.

Finally, any prime ideal Q of A which is not in $\text{spec}_w A$ must be in $\text{spec}_v A$ for some $v \neq w$, whence $(Q : \Gamma) = J_v \neq J_w$. □

2.12. In case k is infinite, (2.11) provides a finite decomposition of $\operatorname{spec} A$ in terms of intersections of Γ-orbits. More precisely, define an equivalence relation $\sim$ on $\operatorname{spec} A$ by the rule

$$P \sim Q \qquad \Longleftrightarrow \qquad \bigcap_{\gamma\in\Gamma} P^\gamma = \bigcap_{\gamma\in\Gamma} Q^\gamma.$$

Then (2.11) shows that the $\sim$-equivalence classes in $\operatorname{spec} A$ are precisely the sets $\operatorname{spec}_w A$ for $w \in W$. Restricting to primitive ideals, we see that the $\sim$-equivalence classes within $\operatorname{prim} A$ are precisely the sets $\operatorname{prim}_w A$.

When k is algebraically closed, each $\operatorname{prim}_w A$ is a single Γ-orbit (2.3.d). Assuming that k has characteristic zero, this conclusion is a special case of a very general result of Moeglin and Rentschler [25, Théorème 2.12] (cf. [28, Theorem 2.1]), who proved that if Γ is a connected affine algebraic group acting rationally by automorphisms on a noetherian k-algebra R (that remains noetherian over all extensions of the base field), then any two rational prime ideals of R whose Γ-orbits have the same intersection must lie in the same Γ-orbit.

3. Application to Quantum Matrices

We now turn to the multiparameter quantum coordinate rings of $n\times n$ matrices studied in [1; 21; 29; 32]; this class of algebras includes the well-known single-parameter quantizations. We will use (2.5) to show that $\mathcal{O}_{\mathbf{p},\lambda}(M_n(k))$ satisfies the Dixmier-Moeglin equivalence when λ is a root of unity. The case where λ is not a root of unity is dealt with in [12] (the two cases appear to require different methods of proof).

3.1. (i) Let n be an integer greater than 1, let $\lambda \in k \setminus \{0, -1\}$, and let $\mathbf{p} = (p_{ij})$ be a multiplicatively antisymmetric $n\times n$ matrix over k. Following [1], let $\mathcal{O}_{\mathbf{p},\lambda}(M_n(k))$ denote the k-algebra with generators $u_{11}, u_{12}, \dots, u_{nn}$ and relations

$$u_{rs}u_{ij} = \begin{cases} p_{ri}p_{js}u_{ij}u_{rs} + (\lambda-1)p_{ri}u_{is}u_{rj} & \text{when } r > i \text{ and } s > j \\ \lambda p_{ri}p_{js}u_{ij}u_{rs} & \text{when } r > i \text{ and } s \le j \\ p_{js}u_{ij}u_{rs} & \text{when } r = i \text{ and } s > j. \end{cases}$$

(ii) Apply the lexicographic ordering to the set of double indices $\{ij \mid 1 \le i, j \le n\}$. As noted in [1, pp. 890–891],

$$\mathcal{O}_{\mathbf{p},\lambda}(M_n(k)) = k[u_{11}][u_{12}; \tau_{12}] \cdots [u_{ij}; \tau_{ij}, \delta_{ij}] \cdots [u_{nn}; \tau_{nn}, \delta_{nn}],$$

where τ_{rs} is the k-algebra automorphism of $k\langle u_{ij} \mid ij < rs\rangle$ determined by

$$\tau_{rs}(u_{ij}) = \begin{cases} p_{ri}p_{js}u_{ij} & \text{when } r \ge i \text{ and } s > j \\ \lambda p_{ri}p_{js}u_{ij} & \text{when } r > i \text{ and } s \le j \end{cases}$$

and δ_{rs} is the k-linear τ_{rs}-derivation of $k\langle u_{ij} \mid ij < rs\rangle$ determined by

$$\delta_{rs}(u_{ij}) = \begin{cases} (\lambda-1)p_{ri}u_{is}u_{rj} & \text{when } r > i \text{ and } s > j \\ 0 & \text{when } r > i \text{ and } s \le j \\ 0 & \text{when } r = i \text{ and } s > j. \end{cases}$$

Moreover, $\delta_{rs}\tau_{rs} = \lambda^{-1}\tau_{rs}\delta_{rs}$; see, for example, [8, §5].

(iii) We now assume that λ is a primitive ℓ-th root of unity, for some integer $\ell > 2$ (not necessarily odd). Since $\delta_{rs}\tau_{rs} = \lambda^{-1}\tau_{rs}\delta_{rs}$, it follows from [10, 3.8] that δ_{rs}^{ℓ} is a τ_{rs}^{ℓ}-derivation on $k\langle u_{ij} \mid ij < rs\rangle$. Note from (ii) that $\delta_{rs}^2(u_{ij}) = 0$ for $ij < rs$. Hence, δ_{rs}^{ℓ} vanishes on a set of generators for the algebra $k\langle u_{ij} \mid ij < rs\rangle$, and since this map is a skew derivation we obtain that $\delta_{rs}^{\ell} = 0$.

It now follows from [10, 3.8] that $u_{rs}^{\ell}a = \tau_{rs}^{\ell}(a)u_{rs}^{\ell}$ for $a \in k\langle u_{ij} \mid ij < rs\rangle$, and hence

$$u_{rs}^{\ell}u_{ij} = p_{ri}^{\ell}p_{js}^{\ell}u_{ij}u_{rs}^{\ell}$$

for $ij < rs$. Viewing $\mathcal{O}_{\mathbf{p},\lambda}(M_n(k))$ as an iterated Ore extension over k in which the variables u_{ij} are adjoined in reverse lexicographical order, it follows similarly that

$$u_{rs}u_{ij}^{\ell} = p_{ri}^{\ell}p_{js}^{\ell}u_{ij}^{\ell}u_{rs}$$

for $ij < rs$.

(iv) Set $\mathbf{p_0} = \left(p_{ij}^{\ell^2}\right)$ and $\mathcal{O}_{\mathbf{p_0}}(M_n(k)) = \mathcal{O}_{\mathbf{p_0},1}(M_n(k))$, and let $v_{11}, v_{12}, \dots, v_{nn}$ denote the canonical generators of $\mathcal{O}_{\mathbf{p_0}}(M_n(k))$. There is now an injective k-algebra homomorphism

$$Fr^*\colon \mathcal{O}_{\mathbf{p_0}}(M_n(k)) \longrightarrow \mathcal{O}_{\mathbf{p},\lambda}(M_n(k)),$$

sending $v_{ij} \mapsto u_{ij}^{\ell}$ (cf. [26, Chapter 7]). Identify $\mathcal{O}_{\mathbf{p_0}}(M_n(k))$ with its image in $\mathcal{O}_{\mathbf{p},\lambda}(M_n(k))$.

It follows from (iii) that $\mathcal{O}_{\mathbf{p},\lambda}(M_n(k))$ is a finite, free, normalizing extension of the subalgebra $\mathcal{O}_{\mathbf{p_0}}(M_n(k))$. Moreover, $\mathcal{O}_{\mathbf{p_0}}(M_n(k)) \cong \mathcal{O}_{\mathbf{q}}(k^{n^2})$ where $\mathbf{q} = (q_{ij,rs})$ is the multiplicatively antisymmetric $n^2 \times n^2$ matrix with entries $q_{ij,rs} = p_{ir}^{\ell^2}p_{sj}^{\ell^2}$.

3.2 Theorem. *The algebra $\mathcal{O}_{\mathbf{p},\lambda}(M_n(k))$ satisfies the Dixmier-Moeglin equivalence when λ is a root of unity.*

Proof. From (2.5) it follows that $\mathcal{O}_{\mathbf{p_0}}(M_n(k))$ satisfies the Dixmier-Moeglin equivalence, and in (3.1.iv) we saw that $\mathcal{O}_{\mathbf{p},\lambda}(M_n(k))$ is a finite free extension of the subalgebra $\mathcal{O}_{\mathbf{p_0}}(M_n(k))$. The theorem therefore follows from [20, 1.5, 2.3] (cf. [36, Corollary 2]). □

3.3 Remark. One constructs the multiparameter quantization $\mathcal{O}_{\mathbf{p},\lambda}(GL_n(k))$ of $GL_n(k)$ by inverting the quantum determinant, a normal but not necessarily central element of $\mathcal{O}_{\mathbf{p},\lambda}(M_n(k))$; see [1, Theorem 3]. That $\mathcal{O}_{\mathbf{p},\lambda}(GL_n(k))$ must also satisfy the Dixmier-Moeglin equivalence is left as an exercise for the reader.

Acknowledgment

We thank David Jordan for pointing out an error in an earlier version of Theorem 1.11.

References

[1] M. Artin, W. Schelter, and J. Tate, Quantum deformations of GL_n, *Communic. Pure Appl. Math.* **44** (1991), 879–895.

[2] M. Awami, M. Van den Bergh, and F. Van Oystaeyen, Note on derivations of graded rings and classification of differential polynomial rings, *Bull. Soc. Math. Belg.* **40** (1988), 175–183.

[3] A. Bell, Localization and ideal theory in Noetherian strongly group-graded rings, *J. Algebra* **105** (1987), 76–115.

[4] K. A. Brown and K. R. Goodearl, Prime spectra of quantum semisimple groups, *Trans. Amer. Math. Soc.* **348** (1996), 2465–2502.

[5] C. De Concini, V. Kac, and C. Procesi, Some remarkable degenerations of quantum groups, *Comm. Math. Phys.* **157** (1993), 405–427.

[6] C. De Concini and C. Procesi, Quantum groups, in D-Modules, Representation Theory, and Quantum Groups (Venezia, June 1992) (G. Zampieri, and A. D'Agnolo, eds.), Lecture Notes in Math. 1565, Springer-Verlag, Berlin, 1993, pp. 31–140.

[7] J. Dixmier, *Enveloping Algebras: The 1996 Printing of the 1977 English Translation*, Graduate Studies in Math. 11, Amer. Math. Soc., Providence, 1996.

[8] K. R. Goodearl, Uniform ranks of prime factors of skew polynomial rings, in Ring Theory, Proceedings of the Biennial Ohio State–Denison Conference, 1992 (S. K. Jain and S. T. Rizvi, eds.), World Scientific, Singapore, 1993, pp. 182–199.

[9] K. R. Goodearl and T. H. Lenagan, Catenarity in quantum algebras, *J. Pure Appl. Algebra* **111** (1996), 123–142.

[10] K. R. Goodearl and E. S. Letzter, Prime ideals in skew and q-skew polynomial rings, *Mem. Amer. Math. Soc. No.* **521** (1994).

[11] K. R. Goodearl and E. S. Letzter, Prime factor algebras of the coordinate ring of quantum matrices, *Proc. Amer. Math. Soc.* **121** (1994), 1017–1025.

[12] K. R. Goodearl and E. S. Letzter, The Dixmier-Moeglin equivalence in quantum matrices and quantized Weyl algebras (to appear).

[13] K. R. Goodearl and R. B. Warfield, Jr., *An Introduction to Noncommutative Noetherian Rings*, London Math. Soc. Student Texts 16, Cambridge Univ. Press, Cambridge, 1989.

[14] T. J. Hodges, Quantum tori and Poisson tori, Unpublished Notes, 1994.

[15] T. J. Hodges and T. H. Levasseur, Primitive ideals of $\mathbb{C}_q[SL(n)]$, *J. Algebra* **168** (1994), 455–468.

[16] J. Humphreys, *Linear Algebraic Groups*, Springer-Verlag, Berlin, 1975.

[17] T. W. Hungerford, *Algebra*, Springer-Verlag, Berlin, 1974.

[18] A. Joseph, *Quantum Groups and Their Primitive Ideals*, Springer, Berlin, 1995.

[19] A. Leroy, J. Matczuk, and J. Okninski, On the Gelfand-Kirillov dimension of normal localizations and twisted polynomial rings, in Perspectives in Ring Theory (Antwerp, 1987), Kluwer, Dordrecht, 1988, pp. 205–214.

[20] E. S. Letzter, Primitive ideals in finite extensions of Noetherian rings, *J. London Math. Soc. (2)* **39** (1989), 427–435.

[21] Yu. Manin, Multiparametric quantum deformation of the general linear supergroup, *Comm. Math. Phys.* **123** (1989), 163–175.

[22] J. Matczuk, Goldie rank of Ore extensions, *Comm. Algebra* (1995), 1455–1471.

[23] J. C. McConnell and J. J. Pettit, Crossed products and multiplicative analogues of Weyl algebras, *J. London Math. Soc. (2)* **38** (1988), 47–55.

[24] J. C. McConnell and J. C. Robson, *Noncommutative Noetherian Rings*, Wiley-Interscience, New York, 1987.

[25] C. Moeglin and R. Rentschler, Orbites d'un groupe algébrique dans l'espace des idéaux rationnels d'une algèbre enveloppante, *Bull. Soc. Math. France* **109** (1981), 403–426.

[26] B. Parshall and J.-p. Wang, Quantum linear groups, *Mem. Amer. Math. Soc. No.* **439** (1991).

[27] D. S. Passman, *The Algebraic Structure of Group Rings*, Wiley-Interscience, New York, 1977.

[28] R. Rentschler, Primitive ideals in enveloping algebras (General case), in Noetherian Rings and their Applications (L. W. Small, ed.), Math. Surveys and Monographs 24, Amer. Math. Soc., Providence, 1987, pp. 37–57.

[29] N. Reshetikhin, Multiparameter quantum groups and twisted quasitriangular Hopf algebras, *Lett. Math. Phys.* **20** (1990), 331–335.

[30] J. E. Roseblade, Prime ideals in group rings of polycyclic groups, *Proc. London Math. Soc. (3)* **36** (1978), 385–447.

[31] J. T. Stafford, The Goldie rank of a module, in Noetherian Rings and their Applications (L. W. Small, ed.), Math. Surveys and Monographs 24, Amer. Math. Soc., Providence, 1987, pp. 1–20.

[32] A. Sudbury, Consistent multiparameter quantisation of $\mathrm{GL}(n)$, *J. Phys. A* **23** (1990), L697–L704.
[33] M. Vancliff, Primitive and Poisson spectra of twists of polynomial rings (to appear).
[34] F. Van Oystaeyen, Azumaya strongly graded rings and ray classes, *J. Algebra* **103** (1986), 228–240.
[35] F. Van Oystaeyen, Graded PI-rings and generalized crossed product Azumaya algebras, *Chinese Ann. Math., Ser. B* **8** (1987), 13–21.
[36] R. B. Warfield, Jr., Bond invariance of G-rings and localization, *Proc. Amer. Math. Soc.* **111** (1991), 13–18.

K. R. Goodearl
Department of Mathematics
University of California
Santa Barbara, CA 93106
USA
e-mail: goodearl@math.ucsb.edu

E. S. Letzter
Department of Mathematics
Texas A&M University
College Station, TX 77843
USA
e-mail: letzter@math.tamu.edu

Canadian Mathematical Society
Conference Proceedings
Volume **22**, 1998

PI-algebras and Nil Algebras of Bounded Index

Alexander Kemer

Abstract. We study the identities of associative algebras over an infinite field of characteristic p. We prove that any PI-algebra satisfies not very deep partial linearization of the identity $x^n = 0$ for sufficiently large n. We apply this result to obtain a few positive results for the following open problem: Is the radical of a relatively-free algebra nil of bounded index? In particular, this problem has a positive solution, if the matrix algebras are finitely based.

We consider associative algebras over an infinite field F of characteristic $p > 0$.

The author has proved in [1] that any PI-algebra over F satisfies the full linearization of the identity $x^n = 0$ for sufficiently large n. It follows from this that for any PI-algebra there exists a nil algebra of bounded index which has exactly the same multilinear identities.

In this paper we show that PI-algebras are much closer to nil algebras of bounded degree, than it was mentioned above. The main result of the paper is the following Theorem

Theorem 1. *For any PI-algebra A there exists a natural number q such that for every sufficiently large n the algebra A satisfies all the partial linearizations of the identity $x^n = 0$, whose degrees with respect to every variable are $\leq \frac{n}{q}$.*

We apply this Theorem to study the radical of a relatively-free algebra of countable rank of an arbitrary variety. By the well-known theorem of S. Amitsur [2] the radical of such an algebra is nil. In the case of characteristic p there are several reasons to pose the following problem which probably has a positive solution:

Problem 1. *Is the radical of a relatively-free algebra (of countable rank) nil of bounded index?*

We formulate this problem having in mind a stronger conjecture:

AMS Subject Classification (1991). 16R10, 16N40.

Problem 2. *Is it true that a proper variety can be generated by some algebraic algebra of bounded index over a field?*

The second main result of the paper is the following theorem:

Theorem 2. *Let F_V be a relatively-free algebra of countable rank of the variety V. For any k an arbitrary T-ideal of F_V, generated by some set of polynomials from* $\mathrm{Rad}\,(F_V)$ *depending on $\leq k$ variables, is nil of bounded index.*

We mention that the statement of the theorem is not true in the case of characteristic 0. Theorem 2 implies immediately

Corollary 1. *If the full matrix algebras over F are finitely based, then the radical of a relatively-free algebra is nil of bounded index.*

Corollary 2. *The radical of a relatively-free algebra of a non-matrix variety is nil of bounded index.*

1. Definitions

In this chapter we give the main concepts of PI-theory. It will be useful for those readers whose mathematical interests are not close to this theory.

Let $F\langle X\rangle$ be the free associative algebra over (an infinite) field F, generated by a countable set X. The elements of X are called non-commutative variables. The elements of the algebra $F\langle X\rangle$ are called polynomials. Consider the element $f \in F\langle X\rangle$, $x_1, \ldots, x_n \in X$. We write $f = f(x_1, \ldots, x_n)$ if f belongs to the subalgebra of $F\langle X\rangle$, generated by a set $\{x_1, \ldots, x_n\}$ and does not belong to the subalgebra, generated by a proper subset of this set.

Let A be an associative algebra over F, $f = f(x_1, \ldots, x_n) \in F\langle X\rangle$, $a_1, \ldots a_n \in A$. Define a map $\pi : X \to A$ putting $\pi(x) = a_i$ if $x = x_i$, $\pi(x) = 0$ if $x \notin \{x_1, \ldots, x_n\}$. By the definition of the algebra $F\langle X\rangle$ there exists a uniquely defined homomorphism $\phi : F\langle X\rangle \to A$ extending the map π. Put $f(a_1, \ldots, a_n) = \phi(f)$. This element is called the result of the substitution $x_i = a_i$ into the polynomial f. We say that the algebra A satisfies the polynomial identity $f(x_1, \ldots, x_n) = 0$, if for arbitrary $a_1, \ldots, a_n \in A$ the equality $f(a_1, \ldots, a_n) = 0$ is satisfied in A. The following ideal of the algebra $F\langle X\rangle$

$$T[A] = \{f \in F\langle X\rangle | f(a_1, \ldots, a_n) = 0, \quad \forall a_1, \ldots, a_n \in A\}$$

is said to be the ideal of identities of A. The algebra A is called a PI-algebra if $T[A] \neq 0$.

An ideal of $F\langle X\rangle$ that is an ideal of identities of some algebra is called a T-ideal. The following internal characterization of T-ideals is well-known: an ideal is a T-ideal if and only if it is completely characteristic, i.e., stable with respect to every endomorphisms of $F\langle X\rangle$. In other words the T-ideals are stable with respect to arbitrary substitutions of elements of $F\langle X\rangle$ for the variables.

The algebras of the form $F\langle X\rangle/\Gamma$, where Γ is a T-ideal, are called the relatively-free algebras corresponding to Γ.

In the study of identities the concept of variety is used. A variety is a class of all the algebras satisfying a given set of identities. It is well-known that the free algebra of countable rank of a given variety is a relatively-free algebra corresponding to the ideal of identities of the variety.

A variety V is called non-matrix if the algebra of matrices of order two over F does not belong to V.

2. Partial Linearizations of Identities

Consider a polynomial $f = f(x_1, \dots, x_m) \in F\langle X\rangle$ depending on the non-commutative variables $x_i \in X$. Take arbitrary variables $x_{i,j} \in X$ ($x_{i,j} \neq x_{k,l}$ for $(i,j) \neq (k,l)$), where $1 \leq i \leq m$, $1 \leq j \leq \deg f$. The homogeneous with respect to every variable $x_{i,j}$ components of the polynomial

$$f\Big(\sum_j x_{1,j}, \dots, \sum_j x_{m,j}\Big)$$

are called the partial linearizations of the polynomial f. Every identity $h = 0$, where the polynomial h is a partial linearization of f is called a partial linearization of the identity $f = 0$. Since the ground field F is infinite, any partial linearization of $f = 0$ follows from this identity. It is easy to see that if an identity $h = 0$ is a partial linearization of an identity $f = 0$, then every partial linearization of the identity $h = 0$ is a partial linearization of the identity $f = 0$.

First of all we prove two main properties of linearizations.

Lemma 1. *Let $h = h(y_1, \dots, y_n, x_2, \dots, x_m)$ be some homogeneous component of the polynomial*

$$f(x_1, \dots, x_m)\big|_{x_1 = y_1 + \dots + y_n},$$

where $y_i \in X$, $y_i \neq x_j$. If $\deg_{y_1} h = \deg_{y_2} h = \dots = \deg_{y_k} h$ ($k \leq n$), then

$$h = \sum_{\sigma \in S(k)} g(y_{\sigma(1)}, \dots, y_{\sigma(k)}, y_{k+1}, \dots, y_n, x_2, \dots x_m)$$

for some polynomial $g(y_1, \dots, y_n, x_2, \dots, x_m)$. ($S(k)$ is the symmetric group of permutations of the set $\{1, \dots, k\}$).

Proof. We can assume that the polynomial f is homogeneous with respect to every variable. Evidently there exists a polynomial

$$u = u(z_1, \dots, z_l, x_2, \dots, x_m)$$

linear with respect to z_i such that

$$f = u(x_1, \dots, x_1, x_2, \dots, x_m).$$

Calculate the polynomial h. Put $n_i = \deg_{y_i} h$. Denote by Λ the set of all partitions $\lambda = (A_1, \dots, A_n)$ of the set $\{1, \dots, l\}$ such that $|A_i| = n_i$ for all $i \leq n$. Take an arbitrary partition $\lambda = (A_1, \dots, A_n) \in \Lambda$ and make the following substitution into the polynomial u: $z_i = y_j$, where j is such that $i \in A_j$, $i = 1, \dots, l$. We denote by u_λ the result of this substitution. Calculating h, we obtain the equality

$$h = \sum_{\lambda \in \Lambda} u_\lambda. \tag{1}$$

Denote by Λ_0 the set of all $\lambda = (A_1, \dots, A_n) \in \Lambda$, satisfying the condition: for every $i < k$ the minimal element of A_i is less than the minimal element of A_{i+1}. Since

$$|A_1| = \cdots = |A_k|,$$

then for every $\lambda = (A_1, \dots, A_n) \in \Lambda$ there exists a uniquely defined $\lambda_0 = (B_1, \dots, B_k, A_{k+1}, \dots, A_n) \in \Lambda_0$ and a uniquely defined permutation $\sigma \in S(k)$ such that $A_i = B_{\sigma(i)}$ for $i \leq k$. It follows from this and (1)

$$h = \sum_{\sigma \in S(k)} g(y_{\sigma(1)}, \dots, y_{\sigma(k)}, y_{k+1}, \dots, y_n, x_2, \dots, x_m),$$

where

$$g = \sum_{\lambda \in \Lambda_0} u_\lambda.$$

The Lemma is proved.

Two sets of identities are said to be equivalent, if any algebra, satisfying one of these sets satisfies the other one.

Lemma 2. *Let $f = f(x_1, \dots, x_m)$ be a polynomial, and let H be the set of all partial linearizations of f, satisfying the following condition: for every $x \in X$ either h does not depend on x or $\deg_x h = p^k$ for some integer $k \geq 0$. Then the identity $f = 0$ is equivalent to the set of all the identities $h = 0$, $h \in H$.*

Proof. Denote by Γ the T-ideal, generated by H. We need to prove $f \in \Gamma$. Assume $f \notin \Gamma$. Then there exists a partial linearization $g = g(x_1, \dots, x_s) \notin \Gamma$ of the polynomial f, satisfying the condition: for every $i = 1, \dots, s$ the identity

$$g\big|_{x_i = y_i + z_i} = g\big|_{x_i = y_i} + g\big|_{x_i = z_i} \tag{2}$$

holds modulo Γ $(y_i, z_i \in X)$. Take an arbitrary element $\alpha \in F$ and substitute $y_i = x_i$, $z_i = \alpha x_i$ into (2). We obtain an identity modulo Γ

$$(1+\alpha)^{n_i} g = (1 + \alpha^{n_i}) g,$$

where $n_i = \deg_{x_i} g$. Since $g \notin \Gamma$, then we have the equality

$$(1+\alpha)^{n_i} = 1 + \alpha^{n_i}$$

for all $\alpha \in F$. It follows that $n_i = p^{k_i}$ for some k_i, because the field F is infinite. It means $g \in H$. We obtain a contradiction. The Lemma is proved.

Now we study the partial linearizations of the identity $x^n = 0$.

Let n, k be natural numbers, $n \geq p^k$. The number n can be written uniquely in the form $n = (m-1)p^k + r$, where m, r are natural numbers, $0 < r \leq p^k$. Consider the partial linearization $h_{n,k}(x_1, \dots, x_m) = 0$ of the identity $x^n = 0$, satisfying the conditions: $\deg_{x_i} h_{n,k} = p^k$ for $i < m$, $\deg_{x_m} h_{n,k} = r$. This linearization is called the *main linearization of order* k. We prove the following extremal properties of the main linearizations:

Lemma 3. *The main linearization $h_{n,k} = 0$ implies all the other partial linearizations, whose degrees with respect to every variable are $\leq p^k$. The main linearization $h_{n,k} = 0$ does not imply a partial linearizations, whose degree is $\geq p^{k+1}$ with respect to some variable.*

Proof. Prove the first part of the Lemma.

Let $h(y_1, \dots, y_s) = 0$ be a partial linearization of the identity $x^n = 0$, $\deg_{y_i} h = n_i$,

$$p^k \geq n_1 \geq n_2 \geq \cdots \geq n_s. \tag{3}$$

Since the partial linearizations of the identity $h = 0$ are partial linearizations of $x^n = 0$, then by Lemma 2 we can assume that $n_i = p^{k_i}$. It follows from this and (3) that n_l divides n_j for $j \leq l$.

Choose the series of integers $j_0, \dots, j_{m-1}$ in the following way. Put $j_0 = 0$. If j_i is chosen, then denote by j_{i+1} the maximal number with the property:

$$\sum_{j_i < j \leq j_{i+1}} n_j \leq p^k.$$

We prove the equalities

$$\sum_{j_i < j \leq j_{i+1}} n_j = p^k, \quad i < m-1, \tag{4}$$

$$\sum_{j_{m-1} < j \leq s} n_j = r. \tag{5}$$

Indeed, assume that for some i equality (4) is not true. Denote by u the difference between the right and left sides of (4). Put $l = j_{i+1} + 1$. Since n_l divides p^k and n_j for all $j \leq l$, then n_l divides u. In particular, we have the inequality $n_l \leq u$ which contradicts the choice of j_{i+1}. The equalities (4) are proven. Hence we obtain

$$\sum_{j_{m-1} < j \leq s} n_j = n - \sum_{0 < j \leq j_{m-1}} n_j = n - (m-1)p^k = r.$$

The equality (5) is also proven.

Let G be the set of all the homogeneous components g of the polynomial $(y_1 + \cdots + y_s)^n$, satisfying the conditions:

$$\sum_{j_i < j \leq j_{i+1}} \deg_{y_j} g = p^k$$

for all $i < m-1$,

$$\sum_{j_{m-1} < j \leq s} \deg_{y_j} g = r.$$

Put $w = \sum_{g \in G} g$. The equalities (4) and (5) imply that $h \in G$. It means that the polynomial h is some homogeneous component of w. It follows from this and from the following evident equality

$$w = h_{n,k}\left(\sum_{0 < j \leq j_1} y_j, \dots, \sum_{j_{m-2} < j \leq j_{m-1}} y_j, \sum_{j_{m-1} < j \leq s} y_j \right),$$

that the identity $h = 0$ is a partial linearization of the identity $h_{n,k} = 0$. In particular, $h = 0$ follows from $h_{n,k} = 0$.

Prove the second part of the Lemma.

Let $g = g(y_1, \dots, y_s)$ be some partial linearization of x^n, $\deg_{y_1} g \geq p^{k+1}$. Assume that the identity $g = 0$ follows from $h_{n,k} = 0$. Since $\deg g = \deg h_{n,k}$, the polynomial g can be written as a linear combination of polynomials of the form $h_{n,k}(a_1, \dots, a_m)$, where $a_1, \dots, a_m$ are linear combinations of the variables y_j. Every polynomial of this form can be written as a linear combination of polynomials of the form $h(y_{i_1}, \dots, y_{i_l})$, where the polynomials h are the partial linearizations of $h_{n,k}, 1 \leq i_j \leq s$. It follows from this, because the polynomial g is non-zero and $\deg_{y_1} g \geq p^{k+1}$, that for some partial linearization $h(x_1, \dots, x_l)$ of the polynomial $h_{n,k}$ and for some $t \leq l$, satisfying the condition

$$\sum_{i=1}^{t} \deg_{x_i} h \geq p^{k+1}, \tag{6}$$

we have the inequality

$$h(x_1, \dots, x_l)\big|_{x_1 = \dots = x_t = y_1} \neq 0$$

in $F\langle X \rangle$. It remains to show that this inequality does not hold in $F\langle X \rangle$.

Put $n_i = \deg_{x_i} h$. By Lemma 2 we can assume $n_i = p^{k_i}$. Since the polynomial h is a partial linearization of $h_{n,k}$, then $k_i \leq k$. For $j \leq k$ denote by m_j the number of such subscripts i, that $i \leq t$ and $k_i = j$. We prove that $m_j \geq p$ for some j. Indeed, if it is not so, then

$$\sum_{i=1}^{t} n_i = \sum_{j=1}^{k} m_j p^j \leq (p-1) \sum_{j=1}^{k} p^j = p^{k+1} - 1.$$

This inequality contradicts condition (6).

Hence the series $n_1, \dots, n_t$ contains at least p equal elements. Without loss of generality we may assume that $n_1 = \dots = n_p$. Then by Lemma 1

$$h = \sum_{\sigma \in S(p)} g\big(x_{\sigma(1)}, \dots, x_{\sigma(p)}, x_{p+1}, \dots, x_l\big)$$

for some polynomial g. Substituting into this equality $x_1 = \dots = x_t = y_1$, we obtain 0, because $p! = 0$ in F. The Lemma is proved.

3. The Proofs of the Main Results

A finite dimensional algebra C is said to be classical over the field F if it can be presented in the form $C = P(C) + \operatorname{Rad}(C)$, where $\operatorname{Rad}(C)$ is the radical of C, the subalgebra $P(C)$ is a direct sum of full matrix algebras over F, $P(C) \cap \operatorname{Rad}(C) = 0$.

To prove Theorem 1 we need one more Lemma.

Lemma 4. *For any proper variety V there exists a natural number d such that every subvariety of V, generated by a finite dimensional algebra, can be generated by some finite dimensional classical algebra C, satisfying the condition:* $\dim P(C) \leq d$.

Proof. By the Theorem from the paper [3], for some n all the algebras of V satisfy the Capelli identity of n-th order:

$$\sum_{\sigma \in S(n)} (-1)^{\sigma} x_{\sigma(1)} y_1 x_{\sigma(2)} y_2 \cdots x_{\sigma(n)} = 0. \tag{7}$$

Let W be a subvariety of V generated by a finite dimensional algebra A. We can assume the algebra A is classical. Assume the algebra A contains an orthogonal set of idempotents $\{e_1, \dots, e_n\}$. Taking the arbitrary elements $a_1, \dots a_{n-1} \in A$, and substituting into (7) $x_j = e_j$, $y_i = e_i a_i e_{i+1}$, we obtain

$$e_1 a_1 e_2 a_2 e_3 \cdots a_{n-1} e_n = 0. \tag{8}$$

Let E be a maximal orthogonal set of idempotents of the subalgebra $P(A)$. Consider the set Δ of all subsets of E which contain less then n elements. For $\delta \in \Delta$ put

$$A_\delta = \mathrm{Rad}\,(A) + \sum_{e,f \in \delta} eP(A)f.$$

Obviously, A_δ is a classical subalgebra of A.

Denote by Γ, Γ_δ the ideals of identities of the algebras A, A_δ, respectively. We prove the equality

$$\Gamma = \bigcap_{\delta \in \Delta} \Gamma_\delta. \tag{9}$$

Indeed, since A_δ is a subalgebra of A, then $\Gamma \subseteq \Gamma_\delta$ for all δ. If the equality (9) is not true, then there exists a polynomial

$$f = f(x_1, \dots, x_m) \in \bigcap_{\delta \in \Delta} \Gamma_\delta, \tag{10}$$

$$f \notin \Gamma, \tag{11}$$

satisfying the condition: for all i and for all variables y, z

$$f\big|_{x_i = y+z} - f\big|_{x_i = y} - f\big|_{x_i = z} \in \Gamma. \tag{12}$$

It can be proven by making the sequential partial linearization of an arbitrary polynomial $g \in \bigcap_{\delta \in \Delta} \Gamma_\delta$, $g \notin \Gamma$. By (11) there exist elements $b_i \in A$ such that

$$f(b_1, \dots, b_m) \neq 0. \tag{13}$$

Every element $a \in A$ can be written in the form $a = c + r$, where $c \in P(A)$, $r \in \mathrm{Rad}\,(A)$, therefore, applying (12), without loss of generality we can assume that $b_1, \dots b_k \in P(A), b_{k+1}, \dots, b_m \in \mathrm{Rad}\,(A)$ for some k. Applying (12) and the Peirce decomposition with respect to the set of idempotents E, we can also assume that for all $i \leq k$

$$b_i = e_{\alpha(i)} a_i e_{\beta(i)}.$$

Put $\delta = \{e_{\alpha(1)}, e_{\beta(1)}, \dots, e_{\alpha(k)}, e_{\beta(k)}\}$. If the set $\delta \notin \Delta$, then δ contains $\geq n$ elements and the inequality (13) is not true, because of (8). If $\delta \in \Delta$, then $b_i \in A_\delta$ for all i and inequality (13) is not true also, because of (10). We obtain a contradiction. The equality (9) is proved.

Consider a set Π whose elements are the zero algebra and all the algebras of the form

$$\bigoplus_{i=1}^{s} M_{n_i}(F),$$

where $M_{n_i}(F)$ is a full matrix algebra of order n_i, $n_1 \geq n_2 \cdots \geq n_s$, $n_1 + \cdots + n_s < n$. It is evident, that if $\delta \in \Delta$, then the semisimple part $P(A_\delta)$ of the algebra A_δ is isomorphic to one of the algebras from Π.

For an arbitrary algebra R denote by $R^\sharp$ the algebra R with unit adjoined.

Take an arbitrary algebra $B \in \Pi$ and consider the free product with adjoined unit $D = B^\sharp *_F F^\sharp\langle x_1, \dots, x_l\rangle$, where $F\langle x_1, \dots, x_l\rangle$ is the free algebra of rank $l = \dim \mathrm{Rad}\,(A)$. We can assume that B and $F\langle x_1, \dots, x_l\rangle$ are subalgebras of D. We denote by T the subalgebra of D, generated by the set $B \cup \{x_1, \dots, x_l\}$. Let I be the ideal of T, generated by $x_1, \dots, x_l$, and U be the verbal ideal of T, generated by all the elements of the form $f(t_1, \dots, t_m)$, where $f \in \Gamma$, $t_i \in T$. Let c be the index of nilpotency of $\mathrm{Rad}\,(A)$. Put $\widetilde{B} = T/(U + I^c)$. Obviously, an algebra A_δ, where $\delta \in \Delta$ is a homomorphic image of $\widetilde{B}$ for some $B \in \Pi$. It follows from this and (9) that

$$\bigcap_{B \in \Pi} T[\widetilde{B}] \subseteq \Gamma,$$

where $T[\widetilde{B}]$ is the ideal of identities of the algebra $\widetilde{B}$. The inverse inclusion also holds because $\Gamma \subseteq T[\widetilde{B}]$. It means, that the variety W can be generated by the algebra

$$C = \bigoplus_{B \in \Pi} \widetilde{B}.$$

Put $d = \sum_{B \in \Pi} \dim B$. This number depends on n which depends on the variety V. The algebra C is classical, because the algebras $\widetilde{B}$ are classical. The dimension of the semisimple part of the algebra C is less then or equal to d. The Lemma is proved.

Now we can prove the first Theorem.

Theorem 1. *For any PI-algebra A there exists a natural number q such that for every sufficiently large n the algebra A satisfies all the partial linearizations of the identity $x^n = 0$, whose degrees with respect to every variable are $\leq \frac{n}{q}$.*

Proof. Let $V = \mathrm{Var}\,(A)$. Denote by Γ the ideal of identities of A. For every natural m consider a subvariety W_m generated by the algebra

$$F\big\langle x_1, \dots, x_m\big\rangle / \big(F\langle x_1, \dots, x_m\rangle \cap \Gamma\big),$$

where $F\langle x_1, \dots, x_m\rangle$ is the free algebra, generated by $x_1, \dots, x_m \in X$. By the main Theorem from the paper [4] the subvariety W_m is generated by a finite dimensional algebra. Choose a number d such that for every m the subvariety W_m is generated by a classical finite dimensional algebra C_m, satisfying the condition: $\dim P(C_m) \leq d$. This number exists by Lemma 4.

Put $q = pd+1$, $m = pq$, and consider the subvariety W_m. Choose a classical finite dimensional algebra C which generates W_m and satisfies condition: $\dim P(C) \leq d$. Let l be the index of nilpotency of $\mathrm{Rad}\,(C)$.

Take a natural number $n \geq q(l-1)$ and consider an arbitrary partial linearization $g = 0$ of the identity $x^n = 0$, whose degrees with respect to every variable are $\leq \frac{n}{q}$. We prove that the algebra A satisfies the identity $g = 0$.

Let k be the maximal integer such that $\frac{n}{q} \geq p^k$. Applying Lemma 2, we can assume that degrees of g with respect to every variable are $\leq p^k$. By the first part of Lemma 4 we can also assume that the polynomial g is the main linearization $h_{n,k}$ of order k.

First of all we prove that the algebra C satisfies the identity $h_{n,k} = 0$. Fix some linear F-basis $c_1, \ldots, c_t$, $t \leq d$ of the semisimple part $P(C)$. If the algebra C does not satisfy the identity $h_{n,k} = 0$, then for some partial linearization $h = h(x_1, \ldots, x_s)$ of the polynomial $h_{n,k}$ and for some elements $a_i \in \{c_1, \ldots c_t\} \cup \mathrm{Rad}\,(C)$ the inequality

$$h(a_1, \ldots, a_s) \neq 0 \tag{14}$$

holds in C. Without loss of generality we may assume that $a_i = c_1$ for $1 \leq i \leq s_1$, $a_i = c_2$ for $s_1 < i \leq s_2, \ldots, a_i = c_t$ for $s_{t-1} < i \leq s_t, a_i \in \mathrm{Rad}\,(C)$ for $s_t < i \leq s$. Make the following substitution into the polynomial h: $x_i = x_1$ for $1 \leq i \leq s_1$, $x_i = x_2$ for $s_1 < i \leq s_2, \ldots, x_i = x_t$ for $s_{t-1} < i \leq s_t, x_i = x_i$ for $s_t < i \leq s$. Denote by u the result of this substitution. By (14) the polynomial u is non-zero. Since the number l is the index of nilpotency of $\mathrm{Rad}\,(C)$, the total degree with respect to the variables $x_{s_t+1}, \ldots, x_s$ is less than l. It follows from this that for some $i \leq t$

$$\deg_{x_i} u \geq \frac{n-l+1}{t} \geq \frac{n-l+1}{d} = \frac{pn}{q} + \frac{n-q(l-1)}{qd} \geq \frac{pn}{q} \geq p^{k+1}. \tag{15}$$

Since the identity $u = 0$ is some partial linearization of $x^n = 0$ and follows from the identity $h_{n,k} = 0$, then by the second part of Lemma 4 $\deg_{x_i} u < p^{k+1}$ for every x_i. This contradicts (15).

It follows that every algebra from the variety W_m satisfies the identity $h_{n,k} = 0$. In particular, the algebra

$$F\langle x_1, \ldots, x_m\rangle / (F\langle x_1, \ldots, x_m\rangle \cap \Gamma)$$

satisfies this identity. It follows from this that

$$h_{n,k} \in F\langle x_1, \ldots, x_m\rangle \cap \Gamma \subseteq \Gamma,$$

because the polynomial $h_{n,k}$ depends on $\leq pq = m$ variables $x_1, x_2, \ldots$. The Theorem is proved.

The other main result actually follows from Theorem 1.

Theorem 2. *Let F_V be a relatively-free algebra of countable rank of a variety V. For any k an arbitrary T-ideal of F_V, generated by some set of polynomials from* $\mathrm{Rad}\,(F_V)$ *depending on $\leq k$ variables, is nil of bounded index.*

Proof. By the definition of the relatively-free algebra $F_V = F\langle X\rangle/\Gamma$, where Γ is the ideal of identities of the variety V. We can assume the ideal Γ is proper. For any non-negative integer s denote by Γ_s the ideal of identities of the full matrix algebra $M_s(F)$. It is well-known that $\mathrm{Rad}\,(F_V) = \Gamma_s/\Gamma$, where the integer s is maximal with property: $\Gamma \subseteq \Gamma_s$. We may assume that $X \subseteq F_V$.

Let I be a T-ideal of F_V generated by some set of polynomials from $\mathrm{Rad}\,(F_V)$, depending on $\leq k$ variables. By Theorem 2 from the paper [4] the ideal I is finitely generated as a T-ideal. Therefore we can assume that the T-ideal I is generated by one polynomial $f = f(x_1, \ldots, x_m)$, because a finite sum of nil ideals of bounded index is a nil ideal of bounded index.

Take an arbitrary $y, z \in X \setminus \{x_1, \ldots, x_m\}$ and consider the set of 4 polynomials $M = \{f, yf, fy, yfz\}$. Denote by M' the set of all the polynomials $uf(u_1, \ldots, u_m)v$, where $u_i \in F\langle X\rangle$, $u, v \in F\langle X\rangle \cup \{1\}$. Every element from the ideal I can be written as a linear combination of the polynomials from the set M'.

Let D be the subalgebra of F_V, generated by the set $M \cup \{x\}$, where $x \in X \setminus \{x_1, \ldots, x_m, y, z\}$. The algebra D is a finitely generated PI-algebra, consequently the radical of D is nilpotent of some index r. Since $M \subseteq \mathrm{Rad}\,(D)$, we have the equalities in F_V

$$gx^{i_1}gx^{i_2}\cdots x^{i_{t-1}}g = 0 \tag{16}$$

for every $g \in M$, $t \geq r$ and for all non-negative integers i_j. The equalities (16) are identities of the algebra F_V. Making the arbitrary substitutions into (16) we obtain the equalities in F_V

$$au^{i_1}au^{i_2}\cdots u^{i_{t-1}}a = 0 \tag{17}$$

for all $a \in M', u \in F\langle X\rangle$.

Take the natural numbers q, n, satisfying the conditions:

1. $\frac{n}{q} \geq r$;
2. The algebra F_V satisfies all the partial linearizations of the identity $x^n = 0$, whose degrees with respect to every variable are $\leq \frac{n}{q}$.

These numbers exist by Theorem 1. We prove that the ideal I satisfies the identity $x^n = 0$. Indeed, take an arbitrary $b \in I$, $b = a_1 + \cdots + a_n$, $a_i \in M'$. The element b^n can be written as a linear combination of the elements of the form

$$h(a_1, \ldots, a_l), \tag{18}$$

where the polynomials h are partial linearizations of the polynomial x^n.

Consider an element e of the form (18). If the degrees of the polynomial h with respect to every variable are $\leq \frac{n}{q}$, then $e = 0$, because $h = 0$ is an identity of the algebra F_V. Assume that the degree with respect to at least one variable is greater than $\frac{n}{q}$. Since $\frac{n}{q} \geq r$, the element e can be written as a linear combination of the elements from the left sides of the equalities (17). By (17) $e = 0$. The Theorem is proved.

Theorem 2 implies immediately

Corollary 1. *If the full matrix algebras over F are finitely based, then the radical of a relatively-free algebra is nil of bounded index.*

Corollary 2. *The radical of a relatively-free algebra of a non-matrix variety is nil of bounded index*

References

[1] A. Kemer, The standard identity in characteristic p. A conjecture of I.B. Volichenko, *Israel J. Math.* **81** (1993), 343–356.

[2] S. Amitsur, A generalization of Hilbert's Nullstellensatz, *Proc. Amer. Math. Soc.* **8** (1957), 649–656.

[3] A. Kemer, Multilinear identities of the algebras over a field of characteristic p, *Intern. J. of Algebra and Computations* **2** (1995), 1–9.

[4] A. Kemer, Identities of finitely generated algebras over an infinite field, *Izv. Akad. Nauk SSSR Ser. Mat.* **54** (1990), 726–753; English transl.; in *Math. USSR Izv.* **32** (1991).

A. Kemer
Department of Mathematics
Ulyanovsk University
Ulyanovsk, 432700
Russia

Canadian Mathematical Society
Conference Proceedings
Volume **22**, 1998

Representations of Finite Dimensional Algebras and Singularity Theory

Helmut Lenzing

1. Introduction

This survey deals with classes of algebras, finite or infinite dimensional over a base field k, arising in the representation theory of finite dimensional algebras:

1. finite dimensional tame hereditary [14, 51] resp. canonical algebras [51, 21, 52, 40] more generally algebras with a separating tubular family [39, 41],
2. preprojective algebras of the path algebra of an extended Dynkin quiver [23], more generally preprojective algebras of a (tame) hereditary algebra [16, 8, 53],
3. surface singularities [12, 22],
4. algebras of automorphic forms [44, 47, 34],
5. two-dimensional factorial algebras [45, 32], see also [46, 56, 42].

We show that a common focus for these classes of rings is provided by the concept of an *exceptional curve*, which generalizes the notion of a weighted projective line from [21], and which for algebraically closed base fields agrees with the latter notion. Roughly speaking, an exceptional curve is a noncommutative curve which is smooth, projective and allows an exceptional sequence of coherent sheaves. Such curves are quite rare since among the nonsingular projective (commutative) curves over an algebraically closed field only the projective line is exceptional. We need to pass to not algebraically closed base fields or to a noncommutative geometric setting in order to get a richer supply of exceptional curves. Note that our choice of terminology is influenced from representation theory, the reader should therefore not confuse exceptional curves as they occur in this

AMS Subject Classification (1991). 16G20, 16G60, 14H20, 14H52, 14H60.

survey with those which traditionally have the same name in algebraic geometry, see [28].

The structure of the paper is as follows. In Section 2 we define the notion of an exceptional curve by specifying — as it is common for noncommutative algebraic geometry [54] — the properties of its abelian category of coherent sheaves, thought to arise from a finite number of module categories (of finitely generated modules) over typically noncommutative noetherian rings by a gluing process. Noncommutativity allows that more than one simple sheaf may be concentrated in a single point; in case this will not happen we call the curve *homogeneous*, again following usage of the term in representation theory. The geometry of a homogeneous exceptional curve is shown in Section 3.3 to be controlled completely by the representation theory of a finite dimensional tame bimodule algebra, an important special case of a finite dimensional tame hereditary algebra (cf. [14, 51]).

In Sections 4 and 5 we show how to pass, in particular, from the homogeneous to the general case by a process called *insertion of weights*. As we are going to show, this process equips a nonsingular curve with a (quasi)-parabolic structure (cf. [58]), but in contrast to the usual treatment of a parabolic structure, which restricts to vector bundles, we extend this concept to all coherent sheaves. Section 7 forms the heart of this report and shows the wide range of applications of the concept of an exceptional curve.

Let k be a field. A k-algebra in this survey will always be associative with a unit element. Modules are usually right modules. We denote by $\mathrm{mod}(A)$ the category of finitely presented and by $\mathrm{Mod}(A)$ the category of all right A-modules. If A is moreover graded by an abelian group H, then $\mathrm{mod}^H(A)$ resp. $\mathrm{Mod}^H(A)$ denote the corresponding categories of H-graded modules with degree zero morphisms.

By a k-category $\mathcal{C}$ we mean an additive category where the morphism spaces are k-spaces and composition is k-bilinear. We call $\mathcal{C}$ *small* if the isomorphism classes of objects from $\mathcal{C}$ form a set. The small categories arising in this article will further be assumed to be *locally finite* over k, meaning that all morphism spaces — and in case of an abelian category also all extension spaces — do have finite k-dimension. Our main examples in that respect are the category $\mathrm{mod}(A)$ for a finite dimensional k-algebra A, the category $\mathrm{coh}(\mathbb{X})$ of coherent sheaves over a non-singular projective variety or occasionally a full subcategory of one of those. An abelian k-category $\mathcal{C}$ is called *connected* if it is not possible to represent $\mathcal{C}$ as a (co)product of non-zero k-categories.

If $\mathcal{C}$ has finite global dimension, the Grothendieck group $\mathrm{K}_0(\mathcal{C})$, formed with respect to short exact sequences, is equipped with a homological bilinear form, called *Euler form*, given on classes of objects by the expression

$$\langle [X],[Y]\rangle = \sum_{i=0}^{\infty}(-1)^i \dim_k \mathrm{Ext}^i(X,Y),$$

where bracket notation $[X]$ refers to the class in the Grothendieck group. An automorphism Φ of $\mathrm{K}_0(\mathcal{C})$, satisfying $\langle y,x\rangle = -\langle x,\Phi y\rangle$ for all x, y is called *Coxeter transformation* for $\mathcal{C}$ or $\mathrm{K}_0(\mathcal{C})$. Such an automorphism always exists if the Euler form is non-degenerate and in this case is further uniquely determined.

For the basic notions of the representation theory of finite dimensional algebras we refer to [5, 20] and [51]. We mention explicitly that the notion of an *almost-*

split sequence $0 \to A \to B \to C \to 0$ makes sense in any small abelian k-category $\mathcal{C}$. It just means that the sequence has indecomposable end terms and does not split, and further that for any indecomposable object X in $\mathcal{C}$ each non-isomorphism $f : X \to C$ lifts to B, equivalently that each non-isomorphism $f : A \to Y$ into an indecomposable object Y extends to B. If almost-split sequences exist, the assignments $C \mapsto A$ and $A \mapsto C$ are called *Auslander-Reiten translations* and denoted τ resp. τ^-. For a module category $\mathrm{mod}(\Lambda)$ over a finite dimensional algebra, almost-split sequences always exist, moreover the *Auslander-Reiten formula* states

$$\mathrm{D}\,\mathrm{Ext}^1(X,Y) = \overline{\mathrm{Hom}}(Y,\tau X) = \underline{\mathrm{Hom}}(\tau^- Y, X)$$

where $\underline{\mathrm{Hom}}$ (resp. $\overline{\mathrm{Hom}}$) refers to morphisms in the stable category modulo projectives (resp. injectives). Note that the formula resembles Serre duality for curves.

A morphism $f : X \to Y$ is called *irreducible* in $\mathcal{C}$ if for any factorization $f = \beta \circ \alpha$ in $\mathcal{C}$ the morphism α is a split monomorphism or β is a split epimorphism. Assuming that almost-split sequences exist for $\mathcal{C}$, the *Auslander-Reiten quiver* $\Gamma_{\mathcal{C}}$ of $\mathcal{C}$ has the isomorphism classes of indecomposable objects as vertices, while the arrows are determined by the existence of irreducible maps. The term *component*, in this context, refers to a connected component of $\Gamma_{\mathcal{C}}$.

For a hereditary abelian category $\mathcal{H}$, *hereditary* means that $\mathrm{Ext}^2_{\mathcal{H}}(-,-) = 0$, its (bounded) *derived category* $\mathrm{D}^b(\mathcal{H})$ can be obtained as follows: For each integer n we take a copy $\mathcal{H}[n]$ of $\mathcal{H}$ with objects denoted $X[n]$, $X \in \mathcal{H}$, then form the union of all $\mathcal{H}[n]$, which is a k-linear category with morphisms given by $\mathrm{Hom}(X[n], Y[m]) = \mathrm{Ext}^{m-n}_{\mathcal{H}}(X,Y)$, and finally close under finite direct sums. Note that we will use the bracket notation $\mathcal{A}[n]$ also for full subcategories. Most derived categories occurring in this paper are equivalent to derived categories of a hereditary category; for general information on derived categories we refer to [24, 25].

2. Exceptional Curves

This section is addressed to people not specialized in algebraic geometry, and intends to cover in a quick hence fragmentary way those aspects which are important to link these concepts with the range of representation theoretic topics mentioned in the introduction. For a proper treatment of the concepts involved we refer the reader for instance to [54, 2].

The main purpose of this section is to introduce the notion of an *exceptional curve* $\mathbb{X}$ and to put this notion into proper context. Roughly speaking we will call a possibly noncommutative curve $\mathbb{X}$ exceptional if it is noetherian, smooth, projective, and admits an exceptional sequence of coherent sheaves. We start to explain these notions, and the concept of an exceptional curve, in more detail.

2.1 Noncommutative Spaces

We define a (possibly) *noncommutative space* $\mathbb{X}$ through an abelian category $\mathcal{C}$ which we are going to interpret as a category of *coherent sheaves* on $\mathbb{X}$. We are only interested in the situation where the category $\mathcal{C}$ is *noetherian*, meaning that each object in $\mathcal{C}$ satisfies the ascending chain condition for subobjects. Slightly deviating from standard terminology, we will in this case say that $\mathbb{X}$ is a *noetherian space*.

Calling $\mathcal{C}$ a category of coherent sheaves is, for the moment, mainly a façon de parler. It is however possible, following the dictionary provided by ([18], chap. VI, see also [54]) for the case of a commutative noetherian (pre)scheme $\mathbb{X}$, to recover the underlying space and its Zariski topology from the category $\mathcal{C} = \mathrm{coh}(\mathbb{X})$ of coherent sheaves on $\mathbb{X}$. For the case of a curve, the case we are mainly interested in, we will specify its underlying set of points in the proper context of Section 2.5. The categories of coherent sheaves $\mathcal{C}$ arising in this article will (with a single exception discussed in Section 7.8) arise by a gluing process from module categories $\mathrm{mod}(A)$ over noetherian rings.

A *basic gluing* ([18], chap. VI) amounts to form the *pull-back* $\mathcal{C}$ of two abelian categories $\mathcal{C}_1$ and $\mathcal{C}_2$

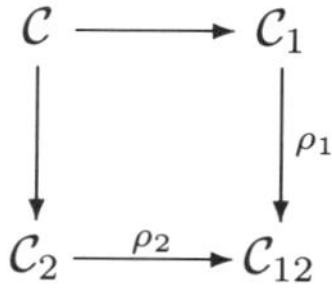

along two exact functors $\rho_1 : \mathcal{C}_1 \to \mathcal{C}_{12}$, $\rho_2 : \mathcal{C}_2 \to \mathcal{C}_{12}$ to a third abelian category $\mathcal{C}_{12}$, where additionally each ρ_i is assumed to induce an equivalence $\mathcal{C}/\ker(\rho_i) \cong \mathcal{C}_{12}$, so that $\mathcal{C}_{12}$ becomes a quotient category of both $\mathcal{C}_1$ and $\mathcal{C}_2$. The pull-back $\mathcal{C}$ is again an abelian category:

The objects of $\mathcal{C}$ are triples (C_1, C_2, α), where C_i is an object of $\mathcal{C}_i$ and α is an isomorphism from $\rho_1 C_1$ to $\rho_2 C_2$. A morphism from (C_1, C_2, α) to (C_1', C_2', α') is a pair (u_1, u_2) of morphisms $u_1 : C_1 \to C_1'$, $u_2 : C_2 \to C_2'$ such that $(\rho_2 u_2) \circ \alpha = \alpha' \circ (\rho_1 u_1)$. Composition of morphisms is componentwise.

In intuitive interpretation, a basic gluing corresponds to an open covering $\mathbb{X} = \mathbb{X}_1 \cup \mathbb{X}_2$ with $\mathcal{C} = \mathrm{coh}(\mathbb{X})$, $\mathcal{C}_1 = \mathrm{coh}(\mathbb{X}_1)$, $\mathcal{C}_2 = \mathrm{coh}(\mathbb{X}_2)$ and $\mathcal{C}_{12} = \mathrm{coh}(\mathbb{X}_1 \cap \mathbb{X}_2)$. Starting with the module categories $\mathrm{mod}(A)$ for noetherian rings A, all categories obtained from those by a finite number of basic gluings are considered to qualify as categories of coherent sheaves (cf. [18], chap. VI for the commutative setting). Even not having specified what $\mathbb{X}$ is, we will write $\mathcal{C} = \mathrm{coh}(\mathbb{X})$ for such a category in order to stay close to geometric intuition.

We illustrate the construction by an example: the k-algebra homomorphisms $r_1 : k[X_1] \to k[X, X^{-1}]$, $r_2 : k[X_2] \to k[X, X^{-1}]$ with $r_1(X_1) = X$, $r_2(X_2) = X^{-1}$ induce by extension of scalars exact functors $\rho_i : \mathrm{mod}(k[X_i]) \to \mathrm{mod}(k[X, X^{-1}])$. The category $\mathcal{C}$ obtained by gluing $\mathrm{mod}(k[X_1])$ and $\mathrm{mod}(k[X_2])$ along ρ_1, ρ_2 is equivalent to the category of coherent sheaves on the projective line $\mathbb{P}_1(k)$, and the gluing data correspond to a an open covering of $\mathbb{P}_1(k)$ by two affine lines.

A space $\mathbb{X}$ as above is called *commutative* if $\mathrm{coh}(\mathbb{X})$ can be obtained by a gluing process starting with categories $\mathrm{mod}(A_i)$ over rings which are commutative, otherwise $\mathbb{X}$ is called *noncommutative*. If $\mathcal{C}$ has two non-isomorphic simple objects S_1, S_2 with $\mathrm{Ext}^1(S_1, S_2) \neq 0$, then $\mathbb{X}$ is forced to be noncommutative and, in fact, that such situations happen is to a large extent typical for a noncommutative setting.

From the technical point of view it is often more convenient to glue full module categories $\mathrm{Mod}(A_i)$, resulting in what then is called a category of *quasi-coherent sheaves*. Also for a proper treatment of homological questions we will need the category $\vec{\mathcal{C}}$ of quasi-coherent sheaves corresponding to $\mathcal{C}$. In the noetherian situation, which we assume, this category can be recovered as the category of all contravariant left exact additive functors from $\mathcal{C}$ to abelian groups [18]. The category $\vec{\mathcal{C}}$ is a Grothendieck category which is locally noetherian, in particular is locally finitely presented. Therefore $\mathcal{C}$ can be identified with the full subcategory of finitely presented objects of $\vec{\mathcal{C}}$, and each object of $\vec{\mathcal{C}}$ is a filtered direct limit of objects from $\mathcal{C}$. As a Grothendieck category $\vec{\mathcal{C}}$ has sufficiently many injective objects, whereas $\mathcal{C}$ may not have any non-zero projectives or injectives.

Notice that the above scheme allows for *different scales of non-commutativity* of a space $\mathbb{X}$: the rings involved in the gluing process can be requested to be commutative, or to be finitely generated as modules over their center, or to satisfy a polynomial identity, or ... In fact, our examples, except the one in Section 7.8, will all belong to the second class, hence stay quite close to commutativity.

2.2 Nonsingular Spaces

Let $\mathbb{X}$ be a noncommutative noetherian space and $\mathcal{C} = \mathrm{coh}(\mathbb{X})$ its noetherian category of coherent sheaves. We define $\mathrm{Ext}^n(C,-)$ as the n-th right derived functor of $\mathrm{Hom}(C,-)$ by means of injective resolutions in $\vec{\mathcal{C}}$. Within $\mathcal{C}$ the resulting values of $\mathrm{Ext}^n(X,Y)$ can alternatively be calculated by means of n-fold Yoneda extensions.

Definition 2.1. A noetherian space $\mathbb{X}$ is called smooth of dimension d if $\mathcal{C} = \mathrm{coh}(\mathbb{X})$ has finite global dimension d, i.e. $\mathrm{Ext}^d_{\mathcal{C}}(-,-) \neq 0$ for $\mathcal{C}$ but $\mathrm{Ext}^{d+1}_{\mathcal{C}}(-,-) = 0$.

For a *smooth curve* ($d = 1$) the category $\mathcal{C} = \mathrm{coh}(\mathbb{X})$ is thus hereditary and noetherian.

For a commutative space $\mathbb{X}$ with a noetherian category of coherent sheaves the above concept of smooth spaces and their dimension yields the usual geometric concept. For the noncommutative spaces, close to commutativity we are going to discuss, the concept also reflects the geometrical meaning quite well. In general situations, further apart from commutativity, a more restrictive concept of smoothness is needed, cf. [2].

2.3 Projective Spaces

Projectivity (relative to a base field k) is an important finiteness condition on a space $\mathbb{X}$ or its category $\mathcal{C}$ of coherent sheaves. It implies, in particular, that $\mathcal{C}$ has morphism and extension spaces which are finite dimensional over k. The definition however is technical, and less obvious than the requirements discussed up to now. The definition to follow is basically modeled after Serre's treatment of projective spaces in [57]. We restrict to the case where $\mathbb{X}$ is smooth.

Let k be a field and let H be a finitely generated abelian group of rank one which may have torsion. We further assume that H is equipped with a (partial) order, compatible with the group structure. We reserve the term *H-graded k-algebra* for the following setting:

R is a k-algebra equipped with a decomposition $R = \bigoplus_{h \in H} R_l$ such that each R_h is a finite dimensional k-space, further $R_h \cdot R_l \subseteq R_{l+h}$ holds for all h, l and finally $R_h \neq 0$ implies $h > 0$ in the given ordering of H. We do not request that $R_0 = k$ or that R is generated — as a k-algebra — by elements of degree one.

Definition 2.2. A smooth space $\mathbb{X}$ is called *projective over* k if there exists a finitely generated noetherian k-algebra $R = \bigoplus_{h \in H} R_h$, graded by an ordered abelian group H of rank one, such that $\operatorname{coh}(\mathbb{X})$ is isomorphic to the quotient category $\operatorname{mod}^H(R)/\operatorname{mod}_0^H(R)$.

Here, $\operatorname{mod}^H(R)$ denotes the category of finitely generated H-graded right R-modules. Therefore each $M \in \operatorname{mod}^H(R)$ is equipped with a decomposition $M = \bigoplus_{h \in H} M_h$ into k-subspaces such that $M_h \cdot R_l \subseteq M_{h+l}$ holds for all h, l in H. Our request on R forces the M_h to be finite dimensional over k. The subindex zero in the expression $\operatorname{mod}_0^H(R)$ refers to the full subcategory of modules of finite length (here the same as finite k-dimension). This subcategory is closed under subobjects, quotients and extensions, hence is a *Serre subcategory* of $\operatorname{mod}^H(R)$. The resulting *quotient category* is formed in the sense of Serre-Grothendieck-Gabriel, see [18] or [49]. It is not difficult to verify the next assertion.

Proposition 2.3. *Assume that R is a finite module over its center C and $C = k[x_1, \dots, x_n]$, where the x_i are assumed to be homogeneous of degree > 0, then the quotient category* $\operatorname{mod}^H(R)/\operatorname{mod}_0^H(R)$ *arises from a finite number of module categories* $\operatorname{mod}(R_i)$ *by gluing, where each R_i is obtained from R by central localization.*

2.4 Exceptional Spaces

An object E in a small abelian k-category is called *exceptional* if the endomorphism algebra of E is a skew field and moreover E has no self-extensions, i.e. $\operatorname{Ext}^n(E, E) = 0$ for all $n > 0$. This notion extends to objects of the bounded derived category, where the Ext-condition needs to be replaced by the request $\operatorname{Hom}(E, E[n]) = 0$ for all $n \neq 0$.

Definition 2.4. A smooth projective space $\mathbb{X}$ is called *exceptional* if the bounded derived category of $\operatorname{coh}(\mathbb{X})$ admits a *complete exceptional sequence* $E_1, \dots, E_n$. We thus request that

1. Each E_i is exceptional;
2. $\operatorname{Ext}^l(E_k, E_i) = 0$ for $i < k$ and for all l;
3. The objects $E_1, \dots, E_n$ generate the bounded derived category of $\operatorname{coh}(\mathbb{X})$ as a triangulated category.

A complete exceptional sequence can be viewed as a system of building blocks for the derived category or for $\operatorname{coh}(\mathbf{X})$ if the E_i lie in $\operatorname{coh}(\mathbf{X})$. In particular, the classes $[E_1], \dots, [E_n]$ form a $\mathbb{Z}$-basis of the Grothendieck group $\mathrm{K}_0(\operatorname{coh}(\mathbb{X}))$. Exceptional spaces accordingly are quite rare especially in the range of commutative spaces. For instance, the only commutative exceptional curve over an algebraically closed field is the projective line $\mathbb{P}_1(k)$. By contrast, and this is in the focus of this survey, there is a rich supply of noncommutative exceptional curves.

Further examples of commutative exceptional spaces are provided by the projective n-space [9], the quadrics [30] and Grassmannians [29]. For additional information in this direction we refer to [55], further to [7] for examples of noncommutative exceptional spaces of dimension ≥ 2.

Closely related to the notion of an exceptional sequence is the notion of a *tilting object* $\bigoplus_{i=1}^n E_i$, where conditions 1. and 2. of Definition 2.4 are replaced by the request $\mathrm{Ext}^l(E_i, E_j) = 0$ for all i, j and all $l \neq 0$. For the situations, studied in this paper, the indecomposable summands E_i of a tilting object can be rearranged to form an exceptional sequence (cf. [27], Lemma 4.1).

2.5 Exceptional Curves

Let k be an arbitrary field. In accordance with the preceding discussion we now fix the notion of an *exceptional curve* $\mathbb{X}$ over k by the following requests on its associated category $\mathcal{H} = \mathrm{coh}(\mathbb{X})$ of coherent sheaves:

1. $\mathcal{H}$ is a connected small abelian k-category with morphism spaces that are finite dimensional over k.
2. $\mathcal{H}$ is hereditary and noetherian, and there exists an equivalence $\tau : \mathcal{H} \to \mathcal{H}$ such that Serre duality $\mathrm{D}\,\mathrm{Ext}^1(X, Y) = \mathrm{Hom}(Y, \tau X)$ holds.
3. $\mathcal{H}$ admits a complete exceptional sequence.

It is possible to skip the request on Serre duality by replacing condition 3. through the related condition "$\mathcal{H}$ has no non-zero projectives and admits a tilting object" (compare [36]).

Let $\mathcal{H}_0$ denote the full subcategory of $\mathcal{H}$ consisting of all objects of finite length, and let $\mathcal{H}_+$ consist of all objects from $\mathcal{H}$ with zero socle. We term the members from $\mathcal{H}_0$ *torsion sheaves* and those of $\mathcal{H}_+$ *vector bundles*.

Combining the techniques of [36], dealing with the case of an algebraically closed base field, and of [41], which treats the related question of a finite dimensional algebra with a separating tubular family over an arbitrary field, we get the following information on $\mathcal{H}$.

1. Each object in $\mathcal{H}$ decomposes into a finite number of indecomposable objects with local endomorphism rings. Moreover each indecomposable object is either a bundle or a torsion sheaf.
2. The category $\mathcal{H}_0$ of torsion sheaves decomposes into a coproduct $\coprod_{x\in\mathbb{X}} \mathcal{U}_x$ of connected uniserial categories $\mathcal{U}_x$, each having a finite number $p(x)$ of simple objects (compare [19]). The latter means that each object of $\mathcal{U}_x$ has a unique composition series in $\mathcal{U}_x$, hence is determined up to isomorphism by its simple $\mathcal{U}_x$ -socle and its length in the category $\mathcal{U}_x$.

   Moreover, finite generation of $\mathrm{K}_0(\mathcal{H})$ implies that $p(x) = 1$ for almost all $x \in \mathbb{X}$.
3. We interpret the index set $\mathbb{X}$ as the *point set of the curve* $\mathbb{X}$ associated with $\mathcal{H}$. The members of $\mathcal{U}_x$ are said to be *concentrated in* x. By definition, the Zariski topology has $\mathbb{X}$ and its finite subsets as the closed sets.
4. Defining w_x as the sum of all classes of simple objects from $\mathcal{U}_x$ in the Grothendieck group $\mathrm{K}_0(\mathcal{H})$, all classes $w_x = [S_x]$ are proportional, hence all are in the same rank one direct factor $\mathbb{Z}.w$ of the Grothendieck group $\mathrm{K}_0(\mathcal{H})$ ([36] Lemma 6).
5. Defining the normalized *rank function* on $\mathrm{K}_0(\mathcal{H})$ as $\frac{1}{\kappa}\langle -, w\rangle$, where $\mathbb{Z}.\kappa = \langle \mathrm{K}_0(\mathcal{H}), w\rangle$, there exists a *line bundle* L, i.e. an indecomposable object from $\mathcal{H}$ of rank one ([36] lemma 6).
6. Each vector bundle E has a filtration $E_0 \subset E_1 \subset \cdots \subset E_r = E$ whose factors E_i/E_{i-1} are line bundles and r equals the rank of E.

7. Because of *Serre duality* $\mathrm{D}\,\mathrm{Ext}^1(X,Y) = \mathrm{Hom}(Y,\tau X)$ the category $\mathcal{H}$ has almost-split sequences where the equivalence τ serves as the Auslander-Reiten translation.
8. $\mathcal{H}$ admits a tilting bundle T whose endomorphism algebra Λ is a canonical algebra in the sense of [52].

Conversely, [41] describes how to associate to each canonical algebra Λ a category $\mathcal{H}(\Lambda)$ of coherent sheaves on an exceptional curve $\mathbb{X}$ which parametrizes the "central" separating tubular family for Λ.

The Grothendieck group $\mathrm{K}_0(\mathcal{H})$ equipped with the Euler form is determined, via a process described in [35], by an invariant of $\mathbb{X}$, called the symbol $\sigma_{\mathbb{X}} = \left(\begin{smallmatrix} p_1,\dots,p_t \\ d_1,\dots,d_t \\ f_1,\dots,f_t \end{smallmatrix} \middle| \epsilon \right)$ of $\mathbb{X}$. The symbol collects the following data:

1. The *weights* $p_i = p(x_i)$ of the finite collection $x_1,\dots,x_t$ of exceptional points x of $\mathbb{X}$ with $p(x) > 1$.
2. $\mathcal{H}$ has a line bundle L, called *special*, such that for each $i = 1,\dots,t$, there is up to isomorphism exactly one simple sheaf S_i concentrated at x_i with the property $\mathrm{Hom}(L,S_i) \neq 0$. Moreover, if S is the direct sum of a τ-orbit of simple sheaves, the middle term of the couniversal extension with L

$$0 \to L^d \to E \to S \to 0, \quad d = \dim_{\mathrm{End}(L)}\mathrm{Ext}^1(S,L)$$

   has the form $E = \bar{L}^r$, where $\bar{L}$ is an indecomposable bundle, whose isomorphism class does not depend on the choice of S.

   We put $\epsilon = \left[\dim_{\mathrm{End}(L)}\mathrm{Hom}(L,\bar{L})\right]^{1/2}$ where $\epsilon > 0$. It follows that ϵ equals 1 or 2, see [35].
3. We put $f_i = \dim_{\mathrm{End}(L)}\mathrm{Hom}(L,S_i)$, $e_i = \dim_{\mathrm{End}(S_i)}\mathrm{Hom}(L,S_i)$ and $d_i = e_i\, f_i$.

From the symbol data one derives a Riemann-Roch formula [35] which determines the *genus* of $\mathbb{X}$ as

$$g_{\mathbb{X}} = 1 + \frac{\epsilon p}{2}\left[\sum_i = 1^t d_i \left(1 - \frac{1}{p_i}\right) - \frac{2}{\epsilon}\right].$$

Here p denotes the least common multiple of the weights $p_1,\dots,p_t$. The genus decides on the representation type:

1. If $g_{\mathbb{X}} < 1$, then $\mathcal{H}$ has a tilting bundle T whose endomorphism algebra is a tame hereditary algebra Σ. See Sections 3.1 and 7.1.
2. If $g_{\mathbb{X}} = 1$, then $\mathcal{H}$ has a tilting bundle T whose endomorphism algebra is a *tubular* canonical algebra Λ. If k is algebraically closed, the indecomposable objects in $\mathcal{H}$ (resp. $\mathrm{mod}(\Lambda)$) are completely classified in [38] (resp. [51]). Moreover, both cases are related since the bounded derived categories of $\mathcal{H}$ and $\mathrm{mod}(\Lambda)$ are equivalent. See also Section 7.6.
3. If $g_{\mathbb{X}} > 1$, then the classification problem for $\mathcal{H}$ is *wild*. We refer to [40] and Section 7.2 for further information.

3. The Homogeneous Case

The sheaf theory of an exceptional curves which is homogeneous is strongly related to, and in fact completely controlled by the representation theory of a tame hereditary algebra of bimodule type whose category of regular modules has the same property of homogenuity. We thus start to discuss the representation theory of tame hereditary (bimodule) algebras.

3.1 Tame Hereditary Algebras

Let $\Lambda = \begin{pmatrix} F & M \\ 0 & G \end{pmatrix}$ be the finite dimensional hereditary k-algebra of a *tame bimodule* ${}_F M_G$, i.e. we request that F, G are finite dimensional skew field extensions of k, and that ${}_F M_G$ is a bimodule such that k acts centrally on F, G, M and $\dim_F M \cdot \dim_G M = 4$ [14]. The pair $(\dim_F M, \dim M_G)$ is called the dimension type of M; tameness therefore is characterized by the dimension types $(2, 2)$, $(1, 4)$ or $(4, 1)$.

If the base field k is algebraically closed, the only choice for M is the bimodule $M = {}_k k \oplus k_k$, and Λ is then called *Kronecker algebra*. The classification of indecomposable modules over this algebra amounts to the classification of pairs of linear maps, a problem solved by Kronecker in 1890. For the base field of real numbers the dimension types $(1, 4)$ and $(4, 1)$ can also be realized, for instance by the bimodules of real quaternions ${}_{\mathbb{H}}\mathbb{H}_{\mathbb{R}}$ and ${}_{\mathbb{R}}\mathbb{H}_{\mathbb{H}}$. The classification of indecomposable modules in these two cases basically amounts to classify real subspaces of a quaternion vector space [15].

The tame bimodule algebras are special cases of connected finite dimensional tame hereditary k-algebras. In this paper we use *tame* in the sense "tame and representation-infinite", where tameness refers to the possibility to classify all indecomposable modules in any given dimension by a finite number of one-parameter families. For a finite dimensional hereditary algebra Λ it is equivalent that the quadratic form $q(x) = \langle x, x \rangle$ associated with the Euler form is positive semi-definite. An indecomposable Λ-module is called *preprojective* (resp. *preinjective*) if for some $n \geq 0$ it has the form $\tau^{-n}P$ (resp. $\tau^n Q$) for an indecomposable projective module P (resp. an indecomposable injective module Q). The remaining indecomposable modules are called *regular*.

Theorem 3.1. (Dlab-Ringel [14]) *Let Λ be a tame bimodule algebra, more generally a connected tame hereditary algebra. The indecomposable Λ-modules fall into the three classes of preprojective, regular, resp. preinjective modules $\mathcal{P}$, $\mathcal{R}$, resp. $\mathcal{I}$, where $\mathcal{R}$ decomposes into a coproduct $\coprod_{x \in \mathbb{X}} \mathcal{U}_x$ of connected uniserial categories. Moreover,* $\mathrm{mod}(\Lambda) = \mathcal{P} \vee \mathcal{R} \vee \mathcal{I}$. □

Here and later we use the notation $\vee$ to denote the closure of the union of $\mathcal{P}$, $\mathcal{R}$, $\mathcal{I}$ with respect to finite direct sums, but also to indicate that in the sequence $\mathcal{P}$, $\mathcal{R}$, $\mathcal{I}$ non-zero morphisms only exist from left to right. Hence there are no non-zero morphisms from objects from $\mathcal{R}$ to $\mathcal{P}$, from $\mathcal{I}$ to $\mathcal{P}$ or from $\mathcal{I}$ to $\mathcal{R}$.

Intuitively speaking, $\mathcal{P}$, $\mathcal{I}$ are discrete families, while $\mathcal{R}$ is a continuous family, thought to be parametrized by the index set of the decomposition $\coprod_{x \in \mathbb{X}} \mathcal{U}_x$. A major aim, in this context, is to understand the geometric structure of the parametrizing space $\mathbb{X}$. This amounts to specify a natural geometric structure on the space of regular Auslander-Reiten components for Λ.

3.2 The Preprojective Algebra

Let Λ be a connected hereditary algebra. We are interested in the preprojective algebras because — at least in the tame case — they yield projectivity of the parameter space $\mathbb{X}$ for the "continuous family" of regular components, conforming to the conventions of Section 2.3. If Λ is tame, the elements x with $\langle x, -\rangle = 0$ form a rank one direct factor $\mathbb{Z}.w$, the *radical* of $\mathrm{K}_0(\Lambda)$. Suitably normalized by a rational factor, the linear form $\langle -, w\rangle$ yields a surjective mapping $\mathrm{rk} : K_0(\Lambda) \to \mathbb{Z}$ called *rank,* which is strictly positive on indecomposable projectives. Moreover, there exists at least one indecomposable projective module of rank one.

The preprojective algebras are a special case of the general construction of the positively $\mathbb{Z}$-graded *orbit algebra* $\mathbb{A}(F; A) = \bigoplus_{n=0}^{\infty} \mathrm{Hom}(A, F^n A)$ associated with a pair consisting of a k-linear endofunctor $F : \mathcal{A} \to \mathcal{A}$ of a small k-category $\mathcal{A}$ and an object A from $\mathcal{A}$ [8, 34]. The product of two homogeneous elements $u : A \to F^n A$ of degree n and $v : A \to F^m A$ of degree m in the orbit algebra is given by the composition $vu = [A \xrightarrow{u} F^n A \xrightarrow{F^m v} F^{n+m} A]$.

Definition 3.2. Let Λ be a connected hereditary algebra and let P denote either the Λ-module Λ or alternatively a projective Λ-module of rank 1 (the latter case requests that Λ is tame). Then the positively $\mathbb{Z}$-graded algebra associated with the Auslander-Reiten translation τ^-

$$\Pi(\Lambda, P) = \bigoplus_{n\in\mathbb{Z}} \mathrm{Hom}(P, \tau^{-n} P)$$

is called a *preprojective algebra* for Λ. We write $\Pi(\Lambda)$ for $\Pi(\Lambda, \Lambda)$.

The preprojective algebra $\Pi(\Lambda)$ was introduced by Gelfand and Panomarev [23] for quivers using a combinatorial definition. Dlab and Ringel [16] extended the notion to the hereditary algebras given as tensor algebras of a species. Our definition as a graded algebra associated with the Auslander-Reiten translation τ^- is taken from [8] and the relationship between the various definitions is studied in [53]. Invoking [11] the following result follows from [8].

Theorem 3.3. *Let* Λ *be a connected* tame hereditary *algebra, and* P *be either* Λ *or an indecomposable projective* Λ*-module of rank one. Then the following assertions hold:*

1. *The preprojective algebra* $\Pi = \Pi(\Lambda, P)$ *is two-sided noetherian and has Krull dimension two, and for* $P = \Lambda$ *also global dimension two. Moreover,* $\Pi(\Lambda)$ *is module-finite over its center which is an affine* k*-algebra.*
2. *The quotient category* $\mathrm{mod}^{\mathbb{Z}}(\Pi)/\mathrm{mod}_0^{\mathbb{Z}}(\Pi)$ *is, in each of the two cases, equivalent to the category* $\mathcal{H}(\Lambda) = \mathcal{I}[-1] \vee \mathcal{P} \vee \mathcal{R}$*, where* $\mathcal{P}$*,* $\mathcal{R}$*,* $\mathcal{I}$ *are the categories of preprojective, regular and preinjective* Λ*-modules, respectively.*
3. $\mathcal{H}(\Lambda)$ *is a connected abelian* k*-category which is hereditary noetherian with Serre duality* $\mathrm{D}\,\mathrm{Ext}^1(X, Y) = \mathrm{Hom}(Y, \tau X)$*.* $\mathcal{H}(\Lambda)$ *has a tilting object* T *with* $\Lambda \cong \mathrm{End}(T)$*.* □

As the notation indicates, $\mathcal{H}(\Lambda)$ is formed in the derived category of $\mathrm{mod}(\Lambda)$. The theorem implies that $\mathcal{H}$ has the interpretation of a category $\mathrm{coh}(\mathbb{X})$ of coherent sheaves on an exceptional curve $\mathbb{X}$. If Λ is the Kronecker algebra over k, then $\mathbb{X}$ equals the projective line $\mathbb{P}_1(k)$, so is a commutative curve.

We illustrate the two kinds of preprojective algebras for the case where Λ is the Kronecker algebra. The preprojective component has the shape

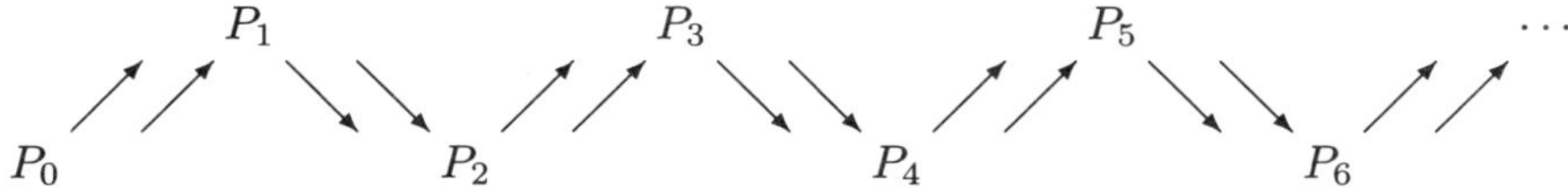

where $\tau^- P_n = P_{n+2}$ and, denoting the arrows from P_n to P_{n+1} consistently by X and Y, the category of indecomposable preprojective Λ-modules is generated by X and Y with relations $XY = YX$, whenever this composition makes sense. Hence $\mathrm{Hom}(P_0, P_n)$ equals R_n the $(n+1)$ -dimensional space of homogeneous polynomials in X, Y of total degree n , and therefore we get isomorphisms

$$\Pi(\Lambda) \cong \bigoplus_{n=0}^{\infty} \begin{pmatrix} R_{2n} & R_{2n-1} \\ R_{2n+1} & R_{2n} \end{pmatrix} \quad \text{and} \quad \Pi(\Lambda, P) \cong \bigoplus_{n=0}^{\infty} R_{2n} = k[X^2, XY, Y^2]$$

as $\mathbb{Z}$-graded algebras.

Remark 3.4. We assume that k is algebraically closed. Then the isomorphism class of $\Pi(\Lambda, P)$ does not depend on the choice of the projective rank one module P. This follows from the fact that the associated curve in this case is a weighted projective line, and therefore the automorphism group of $\mathcal{H}$ acts transitively on the isomorphism classes of line bundles [21].

Moreover, the small preprojective algebra $\Pi' = \Pi(\Lambda, P)$, where P is of rank one, in this case is a *commutative* algebra which is affine over k (see subsection 7.1) and is always of infinite global dimension, in contrast to the fact that the full preprojective algebra $\Pi = \Pi(\Lambda, \Lambda)$ always has global dimension two. The equivalences

$$\mathrm{coh}(\mathbb{X}) \cong \frac{\mathrm{mod}^{\mathbb{Z}}(\Pi)}{\mathrm{mod}_0^{\mathbb{Z}}(\Pi)} \cong \frac{\mathrm{mod}^{\mathbb{Z}}(\Pi')}{\mathrm{mod}_0^{\mathbb{Z}}(\Pi')}$$

can be interpreted in geometric terms: alternatively $\mathbb{X}$ can be obtained from a *singular commutative* surface $\mathbb{F}'$ with $\mathrm{coh}(\mathbb{F}') = \mathrm{mod}(\Pi')$ or from a *nonsingular noncommutative* surface $\mathbb{F}$ with $\mathrm{coh}(\mathbb{F}) = \mathrm{mod}(\Pi)$ as the quotient of a k^*-action.

3.3 Homogeneous Exceptional Curves

Let $\mathbb{X}$ be any noncommutative curve which is exceptional (in particular projective over k and smooth). Accordingly $\mathcal{H} = \mathrm{coh}(\mathbb{X})$ is a connected small abelian category with Serre duality which is noetherian, hereditary and has an exceptional sequence. Let $\mathcal{H}_0$ denote the full subcategory of $\mathcal{H}$ consisting of the objects of finite length. Following the terminology from representation theory we say that $\mathbb{X}$ is *homogeneous* if, in the decomposition of $\mathcal{H}_0 = \coprod_{x\in\mathbb{X}} \mathcal{U}_x$ into connected uniserial subcategories, each $\mathcal{U}_x$ has only one simple object. It is equivalent to say that $\mathrm{Ext}^1(S_1, S_2) = 0$ for each pair of nonisomorphic simple objects. The following result characterizes the exceptional homogeneous curves.

Theorem 3.5. *Let $\mathcal{H} = \operatorname{coh}(\mathbb{X})$ for an exceptional curve $\mathbb{X}$ and assume that $\operatorname{Ext}^1(S_1, S_2) = 0$ for each pair of nonisomorphic simple objects. Then there exists a tame bimodule algebra Λ such that $\mathcal{H} \cong \mathcal{H}(\Lambda)$ in the notations of Theorem 3.3.*

Proof. By assumption $\mathcal{H}_0$ decomposes as a coproduct $\coprod_{x\in\mathbb{X}} \mathcal{U}_x$ of connected uniserial categories, where each $\mathcal{U}_x$ has exactly one simple object S_x. The proof involves several steps, where we use the notations of Section 2.5.

1. Since $\mathbb{X}$ is homogeneous, the class of each simple is a multiple of w, hence $\mathrm{K}_0(\mathcal{H}) = \mathbb{Z}.[L] \oplus \mathbb{Z}.w$.
2. $\mathbb{X}$ has a rational point x, i.e. $[S_x] = w$. Moreover, the couniversal extension $0 \to L^\epsilon \to \bar{L} \to S_x \to 0$, $\epsilon = \dim_{\operatorname{End}(L)} \operatorname{Ext}^1(S_x, L)$, of S_x by L has an indecomposable middle term $\bar{L}$, the companion bundle of L.
3. $T = L \oplus \bar{L}$ is a tilting object in $\mathcal{H}$, $F = \operatorname{End}(L)$, $G = \operatorname{End}(\bar{L})$ are division rings and the (F,G) -bimodule $M = \operatorname{Hom}(L, \bar{L})$ satisfies $\dim_F M \cdot \dim M_G = 4$:

   With respect to the $\mathbb{Z}$-basis $p_1 = [L]$, $p_2 = [\bar{L}]$ of $\mathrm{K}_0(\mathcal{H})$, the Euler form is given by the matrix $C = \begin{pmatrix} \langle p_1,p_1\rangle & \langle p_1,p_2\rangle \\ 0 & \langle p_2,p_2\rangle \end{pmatrix}$. Correspondingly, the Coxeter transformation is given by the matrix $\Phi = -C^{-1}C^{tr}$ which has w as a fixed point, hence admits 1 as an eigenvalue. This in turn implies that $4\langle p_1,p_1\rangle\langle p_2,p_2\rangle = \langle p_1,p_2\rangle^2$, and therefore $\dim_{F_1} M \cdot \dim M_{F_2} = 4$.
4. Since further $\operatorname{Hom}(\bar{L}, L) = 0$, the endomorphism algebra of T is isomorphic to a tame hereditary bimodule algebra Λ. Since Λ equals the endomorphism ring of a tilting object in $\mathcal{H}$ and in $\mathcal{H}(\Lambda)$, we get an equivalence of derived categories $\mathrm{D}^b(\mathcal{H}) \to \mathrm{D}^b(\mathcal{H}(\Lambda))$, preserving the rank, accordingly an equivalence $\mathcal{H} \cong \mathcal{H}(\Lambda)$, cf. [36] for details. □

We refer to [13] for the investigation of related commutative curves.

4. Modification of the Weight Type

Let $\mathbb{X}$ be a smooth noncommutative projective curve, in particular $\mathcal{H} = \operatorname{coh}(\mathbb{X})$ is a hereditary noetherian category with Serre duality. We have already seen, that the genus and thus the complexity of the classification problem for the coherent sheaves on $\mathbb{X}$ largely depends on the weight function of $\mathbb{X}$ attaching to each point of $\mathbb{X}$ the number $p(x)$ of simple sheaves concentrated in x. It is therefore important to dispose of tools changing the weight type of a given $\mathbb{X}$. *Reduction of the weight type* is easily achieved through perpendicular calculus [22]: Let S be a simple sheaf which is concentrated in a point x of weight $p(x) > 1$, then $\operatorname{Ext}^1(S, S) = 0$, hence S is exceptional, and the full subcategory $S^\perp$ of $\mathcal{H}$ consisting of all objects of X satisfying $\operatorname{Hom}(S, X) = 0 = \operatorname{Ext}^1(S, X)$ is again a category of coherent sheaves over an exceptional curve $\mathbb{X}'$ having the same underlying point set as $\mathbb{X}$ but reduced weight. In more detail x has weight $p(x) - 1$ for $\mathbb{X}'$ while the weight of a point $y \neq x$ remains unchanged under the operation. The process can be iterated as long as there still exist exceptional sheaves in the same point x or in a different point, until finally a homogeneous exceptional curve is obtained, where the process of reduction of weights will stop.

For the converse construction which we call *insertion of weights* we need some preparation. First we encode the notion of a point $x \in \mathbb{X}$ in terms of a natural transformation of functors.

Recall that $\mathcal{H}_0 = \coprod_{x\in\mathbb{X}} \mathcal{U}_x$. Each point $x \in \mathbb{X}$ determines, by means of a mutation with respect to the simple objects from $\mathcal{U}_x$, a shift functor $\sigma_x : \mathcal{H} \to \mathcal{H}$, $E \mapsto E(x)$, together with a natural transformation $x : \mathrm{Id} \to \sigma_x$, also denoted by the symbol x (see [41, 38, 43]). For each bundle E , more generally for each sheaf E with $\mathrm{Hom}(\mathcal{U}_x, E) = 0$, these data are given by the $\mathcal{S}_x$-universal extension

$$0 \longrightarrow E \xrightarrow{x_E} E(x) \longrightarrow E_x \longrightarrow 0,$$

where $\mathcal{S}_x$ is the semisimple category consisting of all finite direct sums of simple sheaves concentrated in x and E_x belongs to $\mathcal{S}_x$. If E is an indecomposable torsion object, σ_x acts as follows: If E is concentrated in a point y different from x then we have the $\mathcal{S}_x$-universal extension as above, hence $E(x) = E$ and $x_E = 1_E$. If y equals x, then $E(x) = \tau^- E$, and the kernel of x_E equals the simple socle of E.

4.1 The Category of p-cycles

Let $\mathcal{H}$ be a category of coherent sheaves on a (noncommutative) exceptional curve as before. Fixing a point x of $\mathbb{X}$ and an integer $p \geq 1$ we are going to form the category $\bar{\mathcal{H}}$ of p-cycles in x which may be viewed as a category of coherent sheaves on a curve $\bar{\mathbb{X}}$, having the same underlying point set as $\mathbb{X}$, where the weights for points $y \neq x$ remains unchanged and the weight of x in $\bar{\mathbb{X}}$ equals p times the weight of x in $\mathbb{X}$. Intuitively speaking the effect of the following construction is to form a p-th root of the natural transformation $x_E : E \to E(x)$ corresponding to the point x, and thus relates algebraically to the construction of the Riemann surface of the p-th root function.

Definition 4.1. A *p-cycle E concentrated in x* is a diagram

$$\cdots \longrightarrow E_n \xrightarrow{x_n} E_{n+1} \xrightarrow{x_{n+1}} E_{n+2} \xrightarrow{x_{n+2}} \cdots \longrightarrow E_{n+p} \xrightarrow{x_{n+p}} \cdots$$

which is p-periodic in the sense that $E_{n+p} = E_n(x)$, $x_{n+p} = x_n(x)$ and moreover $x_{n+p-1} \circ \cdots \circ x_n = x_{E_n}$ holds for each integer n.

A morphism $u : E \to F$ of p-cycles concentrated in the same point x is a sequence of morphisms $u_n : E_n \to F_n$ which is *p-periodic*, i.e. $u_{n+p} = u_n$ for each n, and such that each diagram

$$\begin{array}{ccc} E_n & \xrightarrow{x_n} & E_{n+1} \\ \downarrow{\scriptstyle u_n} & & \downarrow{\scriptstyle u_{n+1}} \\ F_n & \xrightarrow{x_n} & F_{n+1} \end{array}$$

commutes. We denote p-cycles in the form $E_0 \xrightarrow{x_0} E_1 \xrightarrow{x_1} \cdots \to E_{p-1} \xrightarrow{x_{p-1}} E_0(x)$ and the category of all p-cycles concentrated in x by $\overline{\mathcal{H}} = \mathcal{H}\left(\begin{smallmatrix} p \\ x \end{smallmatrix}\right)$.

Obviously $\overline{\mathcal{H}}$ is an abelian category, where exactness and formation of kernels and cokernels has a pointwise interpretation. Moreover we have a full exact embedding

$$j : \mathcal{H} \hookrightarrow \overline{\mathcal{H}}, \qquad E \mapsto \bar{E} = E = \cdots = E \xrightarrow{x_E} E(x).$$

We will therefore identify $\mathcal{H}$ with the resulting exact subcategory of $\overline{\mathcal{H}}$. We note that inclusion $j : \mathcal{H} \to \overline{\mathcal{H}}$ has a left adjoint ℓ and a right adjoint r which are both exact functors and are given by

$$\ell\left(E_0 \xrightarrow{x_0} E_1 \xrightarrow{x_1} \cdots \longrightarrow E_{p-1} \xrightarrow{x_{p-1}} E_0(x)\right) = E_{p-1}$$

and

$$r\left(E_0 \xrightarrow{x_0} E_1 \xrightarrow{x_1} \cdots \longrightarrow E_{p-1} \xrightarrow{x_{p-1}} E_0(x)\right) = E_0.$$

Lemma 4.2. *The category $\overline{\mathcal{H}}$ is connected, abelian and noetherian. The simple objects of $\overline{\mathcal{H}}$ occur in two types:*

1. *the simple objects of $\mathcal{H}$ which are concentrated in a point y different from x.*
2. *for each simple object S from $\mathcal{H}$, concentrated in x, the p simples*

$$\begin{array}{rl} S_1: & 0 \to 0 \to \cdots \to 0 \to S \to 0 \\ S_2: & 0 \to 0 \to \cdots \to S \to 0 \to 0 \\ \cdots & \qquad \cdots \\ S_{p-1}: & 0 \to S \to \cdots \to 0 \to 0 \to 0 \\ S_p: & S \to 0 \to \cdots \to 0 \to 0 \to S(x). \end{array}$$

Each S_i is exceptional and $\operatorname{End}(S_i) \cong \operatorname{End}(S) \cong \operatorname{Ext}^1(S_i, S_{i+1})$ *where the indices are taken modulo p.*

If $\mathcal{S} = \{S_1, \ldots, S_{p-1} | S$ simple in $\mathcal{H}$ and concentrated in $x\}$, then the extension closure $\langle \mathcal{S} \rangle$ of $\mathcal{S}$ is localizing in $\overline{\mathcal{H}}$. Moreover

a. *the quotient category $\overline{\mathcal{H}}/\langle \mathcal{S} \rangle \cong \mathcal{H}$ is isomorphic to $\mathcal{H}$, the isomorphism induced by $r : \overline{\mathcal{H}} \to \mathcal{H}$.*
b. *the right perpendicular category $\mathcal{S}^\perp$ formed in $\overline{\mathcal{H}}$ is equivalent to $\mathcal{H}$.*

Proof. Abelianness and noetherianness are obvious by pointwise consideration. It is straightforward from the construction and from the properties of $\mathcal{H}$ that the category $\mathcal{S}^\perp$ right perpendicular to $\mathcal{S}$ consists of exactly those p-cycles $E_0 \xrightarrow{x_0} E_1 \xrightarrow{x_1} \cdots \to E_{p-1} \xrightarrow{x_{p-1}} E_0(x)$ such that $x_0, \ldots, x_{p-2}$ are isomorphisms, hence — up to isomorphism — agree with the objects from $\mathcal{H}$. The remaining properties are straightforward to check. $\square$

We note that $\overline{\mathcal{H}}$ is again equipped with a natural shift automorphism $\overline{\sigma}_x : \overline{\mathcal{H}} \to \overline{\mathcal{H}}$, sending a p-cycle $E_0 \xrightarrow{x_0} E_1 \xrightarrow{x_1} \cdots \to E_{p-1} \xrightarrow{x_{p-1}} E_0(x)$ to $E_1 \xrightarrow{x_1} E_2 \xrightarrow{x_2} \cdots \to E_p \xrightarrow{x_p} E_0(x) \xrightarrow{x_0} E_1(x)$. Note that $\overline{\sigma}_x(S_i) = S_{i+1}$, where the index i is taken modulo p. It is moreover straightforward to define a natural transformation $\mathrm{Id} \to \sigma$ in terms of the "cycle morphisms" $x_0, \ldots, x_{p-1}$.

Theorem 4.3. *The category $\overline{\mathcal{H}}$ shares with $\mathcal{H}$ the properties to be equivalent to a category of coherent sheaves on a smooth projective curve. Moreover, if $\mathcal{H}$ has a tilting object the same holds for $\overline{\mathcal{H}}$.*

Proof. We have already seen that $\overline{\mathcal{H}}$ is an abelian category which is noetherian; it is less obvious that $\overline{\mathcal{H}}$ is also hereditary:

Pass from $\mathcal{H}$ to the Grothendieck category $\vec{\mathcal{H}}$ of left exact functors from $\mathcal{H}^{\text{op}}$ to abelian groups, such that $\mathcal{H}$ becomes the full subcategory of finitely presented objects of $\vec{\mathcal{H}}$. In a straightforward way extend σ_x and the corresponding notion of p-cycles in x from $\mathcal{H}$ to $\vec{\mathcal{H}}$. Invoking Zorn's lemma it is easily checked that a p -cycle E is an injective object if and only if all the E_i are injective objects in $\vec{\mathcal{H}}$ and moreover all the $x_i : E_i \to E$ are epimorphisms. This property is obviously preserved when passing to quotients. Hence the category of p-cycles in $\vec{\mathcal{H}}$, and therefore also $\overline{\mathcal{H}}$ is hereditary.

Note that inclusion $\mathcal{H} \hookrightarrow \overline{\mathcal{H}}$ has a left adjoint $\ell : \overline{\mathcal{H}} \to \mathcal{H}$, given with the previous notations by $\ell(E) = \overline{E}$. If E is projective in $\overline{\mathcal{H}}$ then $\ell(E)$ is projective in $\mathcal{H}$, so is zero, hence E belongs to $\langle \mathcal{S} \rangle$. This implies $E = 0$, and so $\overline{\mathcal{H}}$ has no non-zero projective objects.

Finally, the assumptions imply by a variation of the arguments from [36, 41] the existence of a tilting bundle T for $\mathcal{H}$. For each simple sheaf in $\mathcal{H}$ concentrated in x we form a filtration $0 = F_0(S) \subset F_1(S) \subset \cdots \subset F_p(S) = S$ such that $F_p(S)/F_{p-1}(S) \cong S_i$. Let $T(S) = F_1(S) \oplus \cdots \oplus F_{p-1}(S)$, then the direct sum of T and all $T(S)$, with S simple concentrated at x, is a tilting object in $\overline{\mathcal{H}}$. □

Note that the formation of categories of p-cycles may be iterated and thus the process of insertion of weights may involve any finite set of points of the original curve. We illustrate what is going to happen by an example:

Assuming that k is algebraically closed, the preceding theorem together with the characterization of categories of weighted projective lines from [36] implies the following result. Let $\mathbb{X}$ be the usual projective line over k, and let $\lambda_1, \ldots, \lambda_t$ be a family of pairwise distinct points from $\mathbb{X}$ and $p_1, \ldots, p_t$ be a sequence of integers ≥ 1. Let $\mathbb{X}_0 = \mathbb{X}$ and let inductively denote $\mathbb{X}_i$ be the exceptional curve obtained from $\mathbb{X}_{i-1}$ by inserting weight p_i in λ_i, i.e. forming the category of p_i-cycles in $\operatorname{coh}(\mathbb{X}_{i-1})$ which are concentrated in λ_i. Then $\mathbb{X}_t$ is isomorphic to the weighted projective line corresponding to the above weight data $p_1, \ldots, p_t$ and parameter data $\lambda_1, \ldots, \lambda_t$.

4.2 Parabolic Structure

The base field may again be arbitrary. As in the preceding subsection, we start with a category $\mathcal{H}$ of coherent sheaves on an exceptional curve $\mathbb{X}$. In the notation of the previous section a torsion sheaf in $\overline{\mathcal{H}}$ is exactly a p-cycle $\bar{E} = E_0 \xrightarrow{x_0} E_1 \to \cdots \to E_{p-1} \xrightarrow{x_{p-1}} E_0(x)$ where each E_i is a torsion sheaf. Moreover $\bar{E}$ is a bundle in $\bar{\mathcal{H}}$ if and only if each E_i is a bundle in $\mathcal{H}$, and then each x_i is a monomorphism since each $x_{E_i} : E_i \to E_i(x)$ is a monomorphism. We denote by $\mathcal{S}_x$ the semi-simple category consisting of all finite direct sums of simple sheaves in $\mathcal{H}$ concentrated at x.

In the bundle case we may hence interpret the p-cycle $\bar{E}$ as a filtration

$$E_0 \subseteq E_1 \cdots \subseteq E_{p-1} \subseteq E_0(x),$$

equivalently as a filtration

$$0 = E_0/E_0 \subseteq E_1/E_0 \subseteq \cdots \subseteq E_{p-1}/E_0 \subseteq E_0(x)/E_0$$

of the *fibre* $E_x = E_0(x)/E_0$ of E at x given by the $\mathcal{S}_x$-universal extension

$$0 \to E_0 \xrightarrow{x_{E_0}} E_0(x) \to E_x \to 0.$$

The fibre E_x is a member of the semisimple category $\mathcal{S}_x$ so may be viewed as an r-tuple of finite dimensional vector spaces over a finite skew field extension D of k. Here, D is isomorphic to the endomorphism algebra of any simple sheaf from $\mathcal{H}$ concentrated in x and r denotes the number of such sheaves.

Theorem 4.4. *For any exceptional curve $\mathbb{X}$ and any point x of $\mathbb{X}$ the two concepts of p-cycles of vector bundles, where the p-cycle is concentrated in x, and of a quasi-parabolic structure of filtration length p at x (see* [58] *for the definition) agree.*

Proof. Straightforward from the above discussion. □

In comparison, the concept of p-cycles seems to be advantageous because it applies to torsion sheaves as well, and thus always yields an abelian category.

5. Characterization of Exceptional Curves

In [41] we have associated with each canonical algebra Λ (in the sense of [52]) a connected abelian k-category $\mathcal{H}(\Lambda)$ defining an exceptional curve $\mathbb{X}$ that parametrizes the "central" separating tubular family of $\mathrm{mod}(\Lambda)$.

Theorem 5.1. *Let k be a field. For an abelian k-category $\mathcal{H}$ the following assertions are equivalent:*

a. *$\mathcal{H}$ is equivalent to a category of coherent sheaves on an exceptional curve.*
b. *$\mathcal{H}$ has the form $\mathcal{H}(\Lambda)$ for a canonical algebra Λ.*
c. *$\mathcal{H}$ is equivalent to the category of coherent sheaves on a curve $\mathbb{X}$, arising from a homogeneous exceptional curve $\mathbb{Y}$ by insertion of weights.*

Proof. "$c \Rightarrow a$" see Theorem 4.3.

"$a \Rightarrow b$" see Section 2.5.

"$b \Rightarrow c$" Let $\mathcal{S}$ be a system of exceptional sheaves collecting for each $x \in \mathbb{X}$ all simple sheaves concentrated in x except one. The right perpendicular category $\mathcal{H}' = \mathcal{S}^{\perp}$ then is easily seen to be equivalent to a category of coherent sheaves on a homogeneous curve $\mathbb{Y}$. We fix a tilting bundle for $\mathbb{Y}$, and extend it by the argument from the proof of Theorem 4.3 to a tilting object T on $\mathcal{H}$ and by the same argument realize T as a tilting sheaf on the curve $\overline{\mathbb{Y}}$ arising from $\mathbb{Y}$ by a suitable insertion of weights. There results a rank preserving equivalence of the derived categories $\mathrm{D}^b(\mathrm{coh}(\mathbb{X}))$ and $\mathrm{D}^b(\mathrm{coh}(\overline{\mathbb{Y}}))$ inducing an equivalence between $\mathrm{coh}(\mathbb{X})$ and $\mathrm{coh}(\overline{\mathbb{Y}})$. □

6. Graded Factorial Domains of Dimension Two

Let k be an algebraically closed field. For any sequence $p = (p_1, \ldots, p_t)$ we consider the rank one abelian group $\mathbb{L}(p)$ on generators $\vec{x}_1, \ldots, \vec{x}_t$ with relations

$p_1\vec{x}_1 = \cdots = p_t\vec{x}_t$. For each sequence $\lambda = (\lambda_3, \ldots, \lambda_t)$ of pairwise distinct non-zero elements of k the algebra

$$S(p, \lambda) = k[X_1, \ldots, X_t]/I,$$

where the ideal I is generated by the regular sequence

$$X_i^{p_i} - (X_2^{p_2} - \lambda_i X_1^{p_1}),\ i = 3, \ldots, t,$$

is $\mathbb{L}(p)$-graded by giving the class x_i of X_i degree $\vec{x}_i$. It is easily checked that this algebra is graded factorial [21].

Theorem 6.1. (Kussin [32], Mori [45]) *Let k be an algebraically closed field. An affine k-algebra S of Krull dimension two, positively graded by an ordered rank one abelian group H is graded-factorial if and only if, as a graded algebra, it is isomorphic to the $\mathbb{L}(p)$-graded algebra $S(p, \lambda)$ for some choice of positive integers $p_1, \ldots, p_t$ and pairwise distinct non-zero scalars $\lambda_3, \ldots, \lambda_t$ from k.* □

The result, in its full generality, is due to Kussin [32], who treated the group-graded case. The particular case of a positive $\mathbb{Z}$ -grading is known for some time and due to Mori [45]. Note for this that $\mathbb{L}(p)$ is torsion-free, i.e. $\mathbb{L}(p) \cong \mathbb{Z}$, if and only if $p_1, \ldots, p_t$ are pairwise coprime.

Combined with the characterization of the exceptional curves over an algebraically closed field as the weighted projective lines [36] this yields:

Proposition 6.2. ([36]) *Let k be an algebraically closed field. For a small abelian connected k-category $\mathcal{H}$ the following assertions are equivalent:*

a. *$\mathcal{H}$ is equivalent to the category of coherent sheaves on an exceptional curve.*
b. *$\mathcal{H}$ has the form $\mathrm{mod}^H(S)/\mathrm{mod}_0^H(S)$ for a graded factorial affine k-algebra S of dimension two, graded by a rank one abelian group H.*

Proof. In view of [36] a category of type a is equivalent to a category of coherent sheaves on a weighted projective line. In view of the preceding theorem these are exactly those of type b. □

In geometric terms, graded factoriality of S is (under mild restrictions for S) equivalent to the fact that all line bundles on the associated projective space arise from the structure sheaf by a grading shift.

7. Examples and Applications

It has some tradition to invoke linear algebra methods in the classification of vector bundles for suitable projective varieties. A beautiful result due to Beilinson [9] states that the derived categories of coherent sheaves on the projective n-space and of finite dimensional representations of a certain finite dimensional algebra are equivalent as triangulated categories. This method underlies most applications in this section, even where derived categories are not explicitly mentioned.

7.1 Rational Surface Singularities

Example 7.1. ([22]) Let k be an algebraically closed field, and let $\mathbb{X}$ be a weighted projective line of weight type $(2,3,5)$. Then there exists a tilting sheaf T on $\mathbb{X}$ such that $\Lambda = \mathrm{End}(T)$ is the path algebra over k of the extended Dynkin quiver of type $\widetilde{\mathbb{E}}_8$

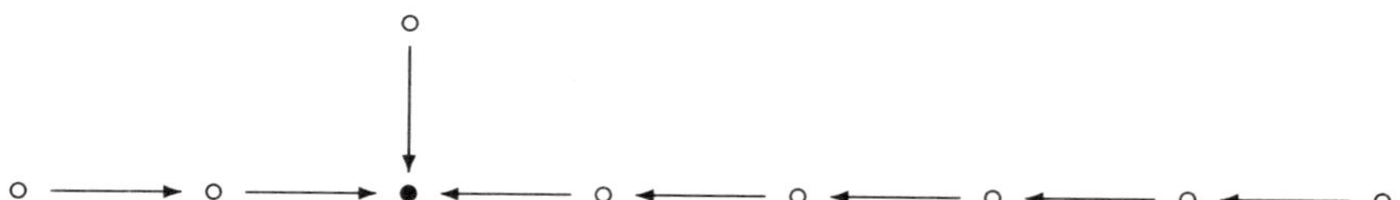

Moreover, if L is a line bundle on $\mathbb{X}$ (resp. a projective Λ-module of rank one) and τ denotes the Auslander-Reiten translation for $\mathrm{coh}(\mathbb{X})$ (resp. for $\mathrm{mod}(\Lambda)$), then there is an isomorphism of graded algebras

$$\bigoplus_{n\geq 0} \mathrm{Hom}(L, \tau^{-n}L) \cong k[X,Y,Z]/(X^2+Y^3+Z^5)$$

where the grading of the algebra on the right is given by $\deg(x,y,z) = (15,10,6)$.

For the base field of complex numbers the equation $X^2+Y^3+Z^5$ describes the rational surface singularity of Dynkin type $\mathbb{E}_8$.

The correspondence between rational surface singularities and the associated small preprojective algebra, illustrated by the preceding example, is valid in general. The attached table is taken from [22].

Theorem 7.2. ([22]) *For any Dynkin diagram $\Delta = (p_1,p_2,p_3)$ let Λ be the path algebra of a quiver of extended Dynkin type $\widetilde{\Delta}$, P a projective Λ-module of rank one, and Π_Δ the preprojective algebra $\Pi(\Lambda, P)$. Then Π_Δ has the form $k[x,y,z] = k[X,Y,Z]/(f_\Delta)$, where the degrees of the generators x,y,z, their degrees, and the relation f_Δ can be seen from the following table:*

Dynkin type Δ	*generators* (x,y,z)	$\mathbb{Z}$*-degrees*	*relations* f_Δ
(p,q)	$(x_0\,x_1, x_1^{p+q}, x_0^{p+q})$	$(1,p,q)$	$X^{p+q} - Y\,Z$
$(2,2,2l)$	$(x_2^2, x_0^2, x_0\,x_1\,x_2)$	$(2,l,l+1)$	$Z^2 + X(Y^2 + Y\,X^l)$
$(2,2,2l+1)$	$(x_2^2, x_0\,x_1, x_0^2\,x_2)$	$(2,2l+1,2l+2)$	$Z^2 + X(Y^2 + Z\,X^l)$
$(2,3,3)$	$(x_0, x_1\,x_2, x_1^3)$	$(3,4,6)$	$Z^2 + Y^3 + X^2\,Z$
$(2,3,4)$	$(x_1, x_2^2, x_0\,x_2)$	$(4,6,9)$	$Z^2 + Y^3 + X^3\,Y$
$(2,3,5)$	(x_2, x_1, x_0)	$(6,10,15)$	$X^2 + Y^3 + X^5$

□

Lists for characteristic zero, usually give the cases $(2,3,3)$ and $(2,2,n)$ in a different form $X^4+Y^3+Z^2$ resp. $X(Y^2-X^n)+Z^2$, then equivalent to the one given here.

7.2 Algebras of Automorphic Forms

Example 7.3. ([34]) Let $\mathbb{X}$ be the weighted projective line over k of weight type $(2,3,7)$. There exists a tilting sheaf T on $\mathbb{X}$ such that $\Lambda = \operatorname{End}(T)$ is the canonical algebra of type $(2,3,7)$ given by the quiver

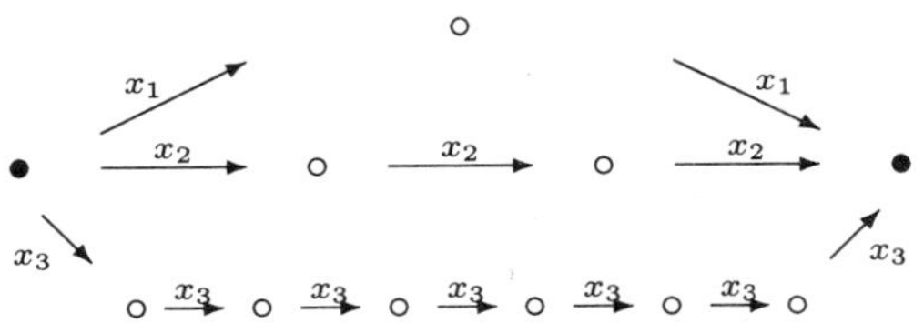

x_1 x_1
x_2 x_2 x_2
x_3 x_3
x_3 x_3 x_3 x_3 x_3

with relations $x_1^2 + x_2^3 + x_3^7 = 0$.

If L is a line bundle on $\mathbb{X}$ (resp. an indecomposable rank one module over Λ not lying in the preprojective component), and τ denotes the Auslander-Reiten translation for $\operatorname{coh}(\mathbb{X})$ (resp. for $\operatorname{mod}(\Lambda)$), then we get an isomorphism

$$\bigoplus_{n\geq 0} \operatorname{Hom}(L, \tau^n L) \cong k[X_1, X_2, X_3]/({X_1}^2 + {X_2}^3 + {X_3}^7)$$

of $\mathbb{Z}$-graded algebras, where the generators $x_1 = [X_1]$, $x_2 = [X_2]$, $x_3 = [X_3]$ for the algebra on the right hand side are homogeneous of degree $21, 14, 6$, respectively.

For the base field of complex numbers, this algebra is isomorphic to the algebra of entire automorphic forms with respect to an action of the triangle group $G = \langle \sigma_1, \sigma_2, \sigma_3 | \sigma_1^2 = \sigma_2^3 = \sigma_3^7 \rangle$ on the upper complex half plane $\mathbb{H}_+$. In fact, this is a special case of a more general setting, treated in [34] in more detail:

It is classical that a Fuchsian group of the first kind, $G = \langle \sigma_1, \ldots, \sigma_t | \sigma_1^{p_1} = \cdots = \sigma_t^{p_t} = 1 = \sigma_1 \cdots \sigma_t \rangle$, satisfying the extra condition $(t-2) - \sum_{i=1}^{t} \frac{1}{p_i} > 0$, can be realized as a discrete group G of automorphisms acting discontinuously on $\mathbb{H}_+$ such that

1. the quotient $\mathbb{H}_+/G$ is isomorphic to the Riemann sphere $\mathbb{P}_1(\mathbb{C})$,
2. only for finitely many orbits $\lambda_1, \ldots, \lambda_t \in \mathbb{P}_1(\mathbb{C})$ the corresponding stabilizer groups are non-trivial and then cyclic of finite order $p_1, \ldots, p_t$, respectively.

The above data $p = (p_1, \ldots, p_t)$ and $\lambda = (\lambda_1, \ldots, \lambda_t)$ define a weighted projective line $\mathbb{X} = \mathbb{X}(p, \lambda)$ and an $\mathbb{L}(p)$-graded factorial algebra $S(p, \lambda)$ as in Section 6. It is easily seen that the character group G^* of G is finite, and moreover that $\mathbb{L}(p)$ can be identified with a subgroup H of $\mathbb{Q} \times G^*$. Then for each pair (a, χ) the $\mathbb{C}$-space $A_{a,\chi}$ of χ-automorphic forms of degree a is finite-dimensional and multiplication of automorphic forms then defines a graded algebra $A = \bigoplus_{(a,\chi)\in\mathbb{Q}\times G^*} A_{a,\chi}$ (see [44, 47] for further details).

Theorem 7.4. ([44, 47, 34]) *The algebra A of automorphic forms on $\mathbb{H}_+$ with respect to G is naturally H-graded, and is isomorphic to the $\mathbb{L}(p)$-graded algebra $S(p, \lambda)$.*

Restricting the grading to the subgroup $\mathbb{Z} \times \{0\}$ yields the algebra R of entire automorphic forms, which is isomorphic to the algebra $\mathbb{A}(\tau; L) = \bigoplus_{n\geq 0} \operatorname{Hom}(L, \tau^n L)$

associated with the Auslander-Reiten translation. Here, L may be chosen as a line bundle on the weighted projective line $\mathbb{X}$ associated with $S(p, \lambda)$ or as a rank one non-preprojective module over the canonical algebra Λ associated with $\mathbb{X}$. □

Note that the algebra $R = \mathbb{A}(\tau; L)$ can be formed over any algebraically closed field k; it is always commutative, affine over k and Cohen-Macaulay of Krull dimension two. Exactly for the minimal wild canonical algebras $(2,3,7)$, $(2,4,5)$, $(3,3,4)$ and their close "neighbours" given in the table below, the algebra R can be generated by three homogeneous elements. In this case R has the form

$$R = k[x, y, z] = k[X, Y, Z]/(F),$$

where the relation F, the degree-triple deg (x, y, z), and deg F are displayed in the table below which is taken from [34].

	deg (x,y,z)	relation F	deg F	
$(2,3,7)$	$(6,14,21)$	$Z^2+Y^3+X^7$	42	
$(2,3,8)$	$(6,8,15)$	$Z^2+X^5+XY^3$	30	
$(2,3,9)$	$(6,8,9)$	$Y^3+XZ^2+X^4$	36	
$(2,4,5)$	$(4,10,15)$	$Z^2+Y^3+X^5Y$	30	
$(2,4,6)$	$(4,6,11)$	$Z^2+X^4Y+XY^3$	22	
$(2,4,7)$	$(4,6,7)$	$Y^3+X^3Y+XZ^2$	18	
$(2,5,5)$	$(4,5,10)$	$Z^2+Y^2Z+X^5$	20	•
$(2,5,6)$	$(4,5,6)$	$XZ^2+Y^2Z+X^4$	16	
$(3,3,4)$	$(3,8,12)$	$Z^2+Y^3+X^4Z$	24	•
$(3,3,5)$	$(3,5,9)$	$Z^2+XY^3+X^3Z$	18	•
$(3,3,6)$	$(3,5,6)$	$Y^3+X^3Z+XZ^2$	15	•
$(3,4,4)$	$(3,4,8)$	$Z^2-Y^2Z+X^4Y$	16	•
$(3,4,5)$	$(3,4,5)$	$X^3Y+XZ^2+Y^2Z$	13	•
$(4,4,4)$	$(3,4,4)$	$X^4-YZ^2+Y^2Z$	12	•

For the base field of complex numbers the 14 equations are equivalent to Arnold's 14 exceptional unimodal singularities in the theory of singularities of differentiable maps [1]. In the theory of automorphic forms they occur as the relations of exactly those algebras of entire automorphic forms having three generators [60]. In the rows marked by • the two references quote a different expression for the singularity F, which — for the base field of complex numbers — is equivalent to the above.

7.3 A Real Exceptional Curve

Proposition 7.5. *The $\mathbb{Z}$-graded $\mathbb{R}$-algebra $\mathbb{R}[X_1, X_2, X_3]/(X_1^2 + X_2^2 + X_3^2)$, where the generators x_i get degree one, is graded-factorial and defines an exceptional curve $\mathbb{X}$, which is commutative and homogeneous and has a tilting bundle whose endomorphism ring is the tame bimodule algebra*

$$\Lambda = \begin{pmatrix} \mathbb{R} & \mathbb{H} \\ 0 & \mathbb{H} \end{pmatrix}.$$

Moreover, each simple sheaf has endomorphism ring $\mathbb{C}$.

Proof. We fix a line bundle L on $\mathbb{X}$. By graded factoriality any line bundle has the form $L(n)$ for some integer n, moreover the Auslander-Reiten translation is given by the grading shift $F \mapsto F(-1)$. This implies that the middle term of the almost-split sequence $0 \to L \to A \to L(1) \to 0$ is indecomposable. It follows that $L \oplus A$ is a tilting bundle with the required properties. □

The following relates the study of $\mathrm{coh}(\mathbb{X})$ to the classification of real subspaces of a quaternion vector space (see [15]).

Corollary 7.6. *The category* $\mathrm{coh}(\mathbb{X})$ *is equivalent to* $\mathcal{H}(\Lambda)$, *and* R *as a graded algebra is isomorphic to the preprojective algebra* $\Pi(\Lambda, P)$, *where* P *is a projective* Λ*-module of rank one.* □

7.4 Factoriality Derived from Representation Theory

It is remarkable that classification results from the representation theory of finite dimensional algebras allow to decide on the factoriality of certain complete local rings. We start with a regular local ring R. An R -algebra S, accordingly an S-module M, is called Cohen-Macaulay (CM for short) if it is finitely generated free as a module over R. We denote by $\mathrm{CM}(S)$ the category of Cohen-Macaulay modules. If R is additionally complete, a theorem of Auslander [3] states that S is an isolated singularity if and only if $\mathrm{CM}(S)$ has almost-split sequences. Here, being an isolated singularity means that each localization of S with respect to any non-maximal prime ideal yields a regular local ring. It is straightforward to define graded analogues of these notions.

Starting with a $\mathbb{Z}$-graded CM-algebra S such that completion $\hat{S}$ is an isolated singularity, Auslander and Reiten have shown [4] that under mild restrictions on S the completion functor $\mathrm{CM}^{\mathbb{Z}}(S) \to \mathrm{CM}(\hat{S})$, $M \mapsto \hat{M}$, preserves indecomposability and almost-split sequences. Moreover, two indecomposable graded CM-modules X, Y do have isomorphic completions if and only if X and Y agree up to a degree-shift, i.e. $Y = X(n)$ for some integer n. Further, the completion $\hat{S}$ is CM-finite, i.e. $\mathrm{CM}(\hat{S})$ has only finitely many isomorphism classes of indecomposable objects, if and only if S is graded CM-finite, i.e. $\mathrm{CM}^{\mathbb{Z}}(S)$ has only finitely many shift-classes of indecomposable objects. Moreover, in this case the completion functor is dense, i.e. surjective on isomorphism classes.

Theorem 7.7. *Each of the isolated surface singularities*

1. $k[[X, Y, Z]]/(X^2 + Y^3 + Z^5)$, *where* k *is a field,*
2. $\mathbb{R}[[X, Y, Z]]/(X^2 + Y^2 + Z^2)$

is a factorial domain.

Proof. Assertion 1 is, for the base field of complex numbers, due to Mumford [46] and to Scheja [56] for the case of an algebraically closed base field of any characteristic. A representation theoretic proof is given below. The line of the argument is the same in both cases.

The path algebra (over k) of the extended Dynkin quiver $\mathbb{E}_8$ has only finitely many preprojective Auslander-Reiten orbits. By means of ([22], Theorem 8.6), see also Example 7.1, this implies that the $\mathbb{Z}$-graded algebra $R = k[X, Y, Z]/(X^2 + Y^3 + Z^5)$, with the grading specified by $\deg(x, y, z) = (15, 10, 6)$ has up to grading shift only a finite number of graded Cohen Macaulay modules, i.e. is graded CM-finite. Since, R is graded factorial, each graded Cohen-Macaulay module over R has the form $R(n)$ for some integer n.

By Auslander and Reiten's completion theorem the completion functor is dense; in particular each rank one CM-module L over $\hat{R}$ is of the form $\hat{M}$, where M is a graded CM-module over R of rank one. By graded factoriality of R, the module M has the form $M = R(n)$ for some integer n, therefore $L \cong \hat{R}$. By a standard argument, this implies that $\hat{R}$ is factorial. The argument is similar with respect to the second algebra. □

7.5 Completion of Cohen-Macaulay Modules

Quite some time ago Auslander has raised the question, whether the completion functor is always dense. The answer is no:

Theorem 7.8. *For the algebra* $S = \mathbb{C}[X, Y, Z]/(X^2 + Y^3 + Z^7)$, *graded by* $\deg(x, y, z) = (21, 14, 6)$ *the following holds:*

1. *The surface singularity* $\hat{S} = \mathbb{C}[[X, Y, Z]]/(X^2 + Y^3 + Z^7)$ *is not factorial.*
2. *The completion functor* $\mathrm{CM}^{\mathbb{Z}}(S) \to \mathrm{CM}(\hat{S})$, $M \mapsto \hat{M}$ *is not dense.*

Proof. S as a graded algebra equals the coordinate algebra of the weighted projective line $\mathbb{X}$ over $\mathbb{C}$ of weight type $(2, 3, 7)$. The curve $\mathbb{X}$ has a wild classification problem for its category $\mathrm{vec}(\mathbb{X})$ of vector bundles [40]. Since the category $\mathrm{CM}^{\mathbb{Z}}(S)$ of graded CM-modules and the category $\mathrm{vec}(\mathbb{X})$ of vector bundles over $\mathbb{X}$ are equivalent, with the shift $X \mapsto X(1)$ for $\mathrm{CM}^{\mathbb{Z}}(S)$ corresponding to the Auslander-Reiten translation for $\mathrm{vec}(\mathbb{X})$ [22], it follows from the properties of the completion functor that $\hat{S}$ has infinite CM-type and hence, in view of Theorem 7.7, is not isomorphic to the algebra $R = \mathbb{C}[[X, Y, Z]]/(X^2 + Y^3 + Z^5)$. A result of Brieskorn [12] (see [42] for an extension to algebraically closed base fields of characteristic different from 2, 3 or 5) states that R is the only complete local two-dimensional factorial algebra with residue class field $\mathbb{C}$, and thus $\hat{S}$ is not factorial. Hence there exists a rank one CM-module M over $\hat{S}$ which is not isomorphic to $\hat{S}$. But for the weight type $(2, 3, 7)$ all line bundles over $\mathbb{X}$ lie in the same Auslander-Reiten orbit [40], hence under completion are all mapped to $\hat{S}$. Therefore, M is not in the image of the completion functor. □

We refer to [59] for a different relationship between CM-modules and representation theory.

7.6 Nonisomorphic Derived-equivalent Curves

We now discuss an interesting weighted variant of the real exceptional curve treated in Section 7.3. Let H be the rank-one abelian group on generators $\vec{x}_1, \vec{x}_2, \vec{x}_3$ with relations $2\vec{x}_1 = 2\vec{x}_2 = \vec{x}_3$.

Example 7.9. (Kussin, 1996) The $\mathbb{R}$-algebra $\mathbb{R}[X_1, X_2, X_3]/(X_1^4 + X_2^4 + X_3^2)$, H-graded by $\deg(x_i) = \vec{x}_i$, defines an exceptional non-commutative curve $\mathbb{X}$ of genus one having a tilting bundle, such that the following holds.

1. There exist tilting bundles T_1, T_2 whose endomorphism algebras Λ_1, Λ_2 are canonical algebras in the sense of [52] with underlying bimodules of type $(1, 4)$ and $(2, 2)$ respectively.

2. There exists a second non-commutative exceptional curve $\mathbb{Y}$ of genus one such that coh($\mathbb{X}$) and coh($\mathbb{Y}$) are not equivalent but such their derived categories are.

Moreover, each simple sheaf on $\mathbb{X}$ has endomorphism ring $\mathbb{C}$ whereas $\mathbb{Y}$ has simple sheaves with endomorphism rings $\mathbb{R}$, $\mathbb{C}$ and $\mathbb{H}$, respectively.

It is therefore not always possible to reconstruct the points of an exceptional curve $\mathbb{X}$ from its derived category $\mathrm{D}^b(\mathrm{coh}(\mathbb{X}))$. By contrast, if $\mathbb{X}$ is an exceptional curve of genus different from one, it is easy to recover coh($\mathbb{X}$) from $\mathrm{D}^b(\mathrm{coh}(\mathbb{X}))$. The example sheds some light on recent work of A. Bondal and D. Orlov [10], who reconstruct the the variety $\mathbb{X}$ from its derived category of coherent sheaves for certain classes of (commutative) spaces.

The species of Λ_1, Λ_2, in particular the underlying bimodules ${}_{\mathbb{R}}\mathbb{H}_{\mathbb{H}}$ and ${}_{\mathbb{C}}\mathbb{C} \oplus \overline{\mathbb{C}}_{\mathbb{C}}$ where the bar indicates a right $\mathbb{C}$-action by conjugation, are depicted below. We do not give the relations.

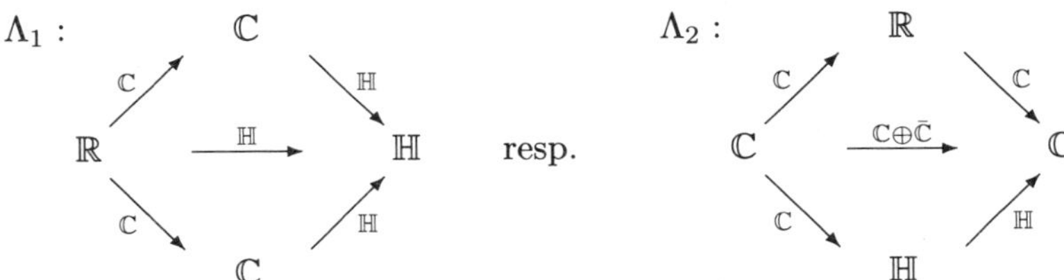

7.7 Infinite Dimensional Modules

We will be very brief here. Let $\mathcal{H}$ be the category of coherent sheaves over an exceptional curve $\mathbb{X}$, and let T be any tilting sheaf and $\Lambda = \mathrm{End}(T)$. By tilting theory, the functor $\mathrm{Hom}(T, -)$ yields in particular a full embedding from the full subcategory of injective sheaves in $\vec{\mathcal{H}}$ into $\mathrm{Mod}(\Lambda)$. The indecomposable injective (quasi-coherent) sheaves fall in two classes

1. the torsion sheaves, coinciding with the injective envelopes of simple sheaves;
2. a single torsion-free sheaf isomorphic to the injective envelope of any line bundle.

For the special situation where Λ is tame hereditary, the quasicoherent sheaves of type 1 (resp. type 2) yield by application of $\mathrm{Hom}(T, -)$ the Prüfer modules (resp. the indecomposable torsionfree divisible or generic Λ-module) from [50]. For the case of a tubular algebra, corresponding to genus one, we refer to [33].

7.8 Curve Attached to a Wild Hereditary Algebra

Assume that Λ is a wild hereditary algebra, for instance the path algebra of the wild quiver ($t \geq 5$)

$\vec{\Delta}$: $\circ \cdots \circ$; x_2, x_{t-1} ; $\circ \xrightarrow{x_1} \bullet \xleftarrow{x_t} \circ$

whose classification problem for indecomposable representations is equivalent to classify the position of t subspaces in a vectorspace. The preprojective component of Λ in terms of generators and relations looks as follows

$$\begin{array}{ccccccccccccc}
 & & \circ & & & & \circ & & & & \circ & & \\
 & {}^{x_1}\nearrow & & \searrow^{y_1} & & {}^{x_1}\nearrow & & \searrow^{y_1} & & {}^{x_1}\nearrow & & \searrow^{y_1} & \cdots \\
\bullet & \xrightarrow{x_2} & \circ & \xrightarrow{y_2} & \bullet & \xrightarrow{x_2} & \circ & \xrightarrow{y_2} & \circ & \xrightarrow{x_2} & \circ & \xrightarrow{y_2} & \bullet \cdots \\
 & {}_{x_t}\searrow & \vdots & \nearrow_{y_t} & & {}_{x_t}\searrow & \vdots & \nearrow_{y_t} & & {}_{x_t}\searrow & \vdots & \nearrow_{y_t} & \cdots \\
 & & \circ & & & & \circ & & & & \circ & &
\end{array}$$

where $\sum_{i=1}^{t} y_i x_i = 0$ and $x_i y_i = 0$ for $i = 1, \ldots, t$.

The characteristic polynomial of the Coxeter transformation Φ is $(T+1)^{t-1} \cdot (T^2 - (t-2)T + 1)$, so for $t \geq 5$ the spectral radius ρ of Φ is > 1. Note that $x_1, \ldots, x_t$ are elements of the path algebra $\Lambda = k[\vec{\Delta}]$ which forms the zero component of the preprojective algebra $\Pi = \Pi(\Lambda)$ given as $\Pi = \Lambda\langle y_1, \ldots, y_t\rangle$, where $\sum_{i=1}^{t} y_i x_i = 0$, $x_i y_i = 0$ for $i = 1, \ldots, t$ and $\deg(y_i) = 1$. In particular Π is a finitely presented k-algebra.

In accordance with [17, 48] we get for the preprojective algebra Π of any connected wild hereditary algebra

$$\lim_{n\to\infty} \frac{\dim_k \Pi_n}{\rho^n} > 0,$$

hence Π has infinite Gelfand-Kirillov dimension. In particular Π does not satisfy a polynomial identity, and also is not noetherian [6]. It is further shown in [61] that the minimal spectral radius (=growth number) of any wild hereditary algebra is given by the largest real root of $T^{10} + T^9 - T^7 - T^6 - T^5 - T^4 - T^3 + T + 1$ which is about 1.176.

It follows from [33] that the quotient category

$$\mathcal{H} = \frac{\mathrm{mod}^{\mathbb{Z}}(\Pi)}{\mathrm{mod}_0^{\mathbb{Z}}(\Pi)}$$

satisfies all requirements for an exceptional curve from Section 2.5, except noetherianness. Moreover $\mathcal{H}$ has a tilting object with endomorphism algebra Λ, which implies that

$$\mathcal{H} \cong \mathcal{H}(\Lambda) = \mathcal{I}(\Lambda)[-1] \vee \mathcal{P}(\Lambda) \vee \mathcal{R}(\Lambda),$$

where $\mathcal{P}(\Lambda)$, $\mathcal{R}(\Lambda)$, $\mathcal{I}(\Lambda)$ denotes the subcategory of preprojective, regular and injective Λ-modules, respectively. Further $\mathcal{H}$ has no simple and therefore no non-zero noetherian or artinian object. We note that abelianness of $\mathcal{H}(\Lambda)$ can alternatively be derived from [26]. Note that it is open whether $\mathcal{H}$ or $\vec{\mathcal{H}}$ can be obtained by gluing from module categories.

It is tempting to view $\mathcal{H}$ as a candidate for a category of "coherent sheaves on a non-noetherian exceptional curve" $\mathbb{X}$. One could think of its geometric meaning to parametrize the regular components of $\mathrm{mod}(\Lambda)$ which would relate to work of Kerner [31] on the regular components of wild hereditary algebras. But this topic needs further investigation.

References

[1] V.I. Arnold, S.M. Gusejn-Zade and A.N. Varchenko, *Singularities of differentiable maps. Volume I*, Monographs in Mathematics, Vol. 82, Birkhauser, Basel, 1985.

[2] M. Artin and J.J. Zhang, Noncommutative projective schemes, *Adv. Math.* **109** (1994), 228–287.

[3] M. Auslander, Isolated singularities and existence of almost split sequences, Notes by Luise Unger. Representation theory II, Groups and orders, Proc. 4th Int. Conf., Ottawa 1984, Lect. Notes Math. **1178** (1986), 194–242.

[4] M. Auslander and I. Reiten, Cohen-Macaulay modules for graded Cohen-Macaulay rings and their completions, Commutative algebra, Proc. Microprogram, Berkeley 1989, Publ., Math. Sci. Res. Inst. **15** (1989), 21–31.

[5] M. Auslander, I. Reiten and S. O. Smalø, *Representation theory of Artin algebras*, Cambridge Studies in Advanced Mathematics. v. 36, Cambridge University Press, Cambridge, 1995.

[6] D. Baer, Homological properties of wild hereditary Artin algebras, Representation theory I, Finite dimensional algebras, Proc. 4th Int. Conf., Ottawa 1984, Lect. Notes Math. **1177** (1986), 1–12.

[7] D. Baer, Tilting sheaves in representation theory of algebras, *Manuscr. Math.* **60** (1988), 323-347.

[8] D. Baer, W. Geigle and H. Lenzing, The preprojective algebra of a tame hereditary Artin algebra, *Comm. Algebra* **15** (1987), 425–457.

[9] A.A. Beilinson, Coherent sheaves on P_n and problems of linear algebra, *Anal. Appl.* **12** (1979), 214–216.

[10] A. Bondal and D. Orlov, Reconstruction of a variety from the derived category and groups of automorphisms, Preprint 1996.

[11] A. Braun and C.R. Hajarnavis, Finitely generated P.I. rings of global dimension two, *J. Algebra* **169** (1994), 587–604.

[12] E. Brieskorn, Rationale Singularitäten komplexer Flächen, *Inventiones Math.* **4** (1968), 336–358.

[13] W.W. Crawley-Boevey, Regular modules for tame hereditary algebras, *Proc. Lond. Math. Soc.* **62** (1991), 490–508.

[14] V. Dlab and C.M. Ringel, Indecomposable representations of graphs and algebras, *Mem. Am. Math. Soc.* **173** (1976).

[15] V. Dlab and C.M. Ringel, Real subspaces of a quaternion vector space, *Can. J. Math.* **30** (1978), 1228–1242.

[16] V. Dlab and C.M. Ringel, The preprojective algebra of a modulated graph, Representation theory II, Proc. 2nd int. Conf., Ottawa 1979, Lect. Notes Math. **832** (1980), 216-231.

[17] V. Dlab and C.M. Ringel, Eigenvalues of Coxeter transformations and the Gelfand-Kirillov dimension of the preprojective algebras, *Proc. Am. Math. Soc.* **83** (1981), 228-232.

[18] P. Gabriel, Des catégories abéliennes, *Bull. Soc. Math. France* **90** (1962), 323-448.

[19] P. Gabriel, Indecomposable representations. II, *Sympos. Math.* **11** (1973), 81–104.

[20] P. Gabriel and A.V. Roiter, *Algebra VIII: Representations of finite-dimensional algebras*, Encyclopaedia of Mathematical Sciences 73, Springer-Verlag, Berlin, 1992.

[21] W. Geigle and H. Lenzing, A class of weighted projective curves arising in representation theory of finite dimensional algebras, Singularities, representations of algebras, and vector bundles, Lect. Notes Math. **1273** (1987), 265–297.

[22] W. Geigle and H. Lenzing, Perpendicular categories with applications to representations and sheaves, *J. Algebra* **144** (1991), 273-343.

[23] I.M. Gelfand and V. A. Ponomarev, Model algebras and representations of graphs, *Funct. Anal. Appl.* **13** (1979), 157–166.

[24] P.P. Grivel, Catégories derivées et foncteurs derivés, in Algebraic D-modules, Perspectives in Math., vol.2, Academic Press, 1987.

[25] D. Happel, *Triangulated categories in the representation theory of finite dimensional algebras*, LMS Lecture Note Series **119**,, Cambridge, 1988.

[26] D. Happel, I. Reiten and S. Smalø, Tilting in abelian categories and quasitilted algebras, *Mem. Am. Math. Soc.* **575** (1996).

[27] D. Happel and C.M. Ringel, Tilted Algebras, *Trans. Am. Math. Soc.* **274** (1982), 399–443.

[28] R. Hartshorne,, *Algebraic geometry*, Springer-Verlag, Berlin-Heidelberg-New York, 1977.

[29] M.M. Kapranov, On the derived category of coherent sheaves on Grassmann manifolds, *Math. USSR, Izv.* **24** (1985), 183–192.

[30] M.M. Kapranov, The derived category of coherent sheaves on a quadric, *Funct. Anal. Appl.* **20** (1986), 141-142.

[31] O. Kerner, Tilting wild algebras, *J. London Math. Soc.* **39** (1989), 29–47.

[32] D. Kussin, Graded factorial algebras of dimension two, *Bull. London Math Soc.* (to appear).

[33] H. Lenzing, Curve singularities arising from the representation theory of tame hereditary algebras. Representation theory I, Finite dimensional algebras, Proc. 4th Int. Conf., Ottawa 1984, Lect. Notes Math. **1177** (1986), 199-231.

[34] H. Lenzing, *Wild Canonical algebras and rings of automorphic forms* (V. Dlab and L.L. Scott, eds.), Kluwer Academic Publishers Finite Dimensional Algebras and Related Topics, 1994, pp. 191-212.

[35] H. Lenzing, A K-theoretic study of canonical algebras, *CMS Conf. Proc.* **18** (1996), 433–454.

[36] H. Lenzing, Hereditary noetherian categories with a tilting complex, *Proc. Amer. Math. Soc.* **125** (1997), 1893–1901.

[37] H. Lenzing, Generic modules over a tubular algebra, Algebra and Model Theory, Proceedings Essen 1994 and Dresden 1995 (to appear).

[38] H. Lenzing and H. Meltzer, Sheaves on a weighted projective line of genus one, and representations of a tubular algebra, Representations of algebras, Sixth International Conference, Ottawa 1992 CMS Conf. Proc. **14** (1993), 313–337.

[39] H. Lenzing and H. Meltzer, Tilting sheaves and concealed-canonical algebras, Representations of algebras. Seventh International Conference, Cocoyoc (Mexico) 1994. CMS Conf. Proc. **18** (1996), 455–473.

[40] H. Lenzing and J. A. de la Peña, Wild canonical algebras, *Math. Z.* **224** (1997), 403-425.

[41] H. Lenzing and J. A. de la Peña, Concealed-canonical algebras and algebras with a separating tubular family, *Proc. London Math. Soc.* (to appear).

[42] J. Lipman, Rational singularities with applications to algebraic surfaces and unique factorization, *Publ. Math. Inst. Hautes Études Scientifiques* **36** (1969), 195–280.

[43] H. Meltzer, Mutations as derived equivalences for sheaves on weighted projective lines of genus one, Preprint, *Paderborn* (1994).

[44] J. Milnor, On the 3-dimensional Brieskorn manifolds M(p, q, r), Knots, Groups, 3-Manif., Pap. dedic. Mem. R. H. Fox (1975), 175–225.

[45] S. Mori, Graded factorial domains, *Japan. J. Math.* **3** (1977), 223–238.

[46] D. Mumford, The topology of normal singularities of an algebraic surface and a criterion for simplicity, Publ. Math. Inst. Hautes Études Scientifiques, vol. 9, 1961, pp. 229–246.

[47] W. Neumann, Brieskorn complete intersections and automorphic forms, *Invent. Math.* **42** (1977), 285–293.

[48] J.A. de la Pena and M. Takane, Spectral properties of Coxeter transformations and applications, *Arch. Math.* **55** (1990), 120–134.

[49] N. Popescu, *Abelian categories with applications to rings and modules*, Academic Press, London, New York, 1973.

[50] C.M. Ringel, Infinite dimensional representations of finite-dimensional hereditary algebras, *Symposia Mat. Inst. Alta Mat.* **23** (1979), 321–412.

[51] C.M. Ringel, *Tame algebras and integral quadratic forms*, Lect. Notes Math. 1099, Springer, Berlin-Heidelberg-New York, 1984.

[52] C.M. Ringel, The canonical algebras with an appendix by William Crawley-Boevey, *Topics in Algebra, Banach Center Publ.* **26** (1990), 407–432.

[53] C.M. Ringel, The preprojective algebra of a quiver, Preprint (1997).

[54] A.L. Rosenberg, *Noncommutative algebraic geometry and representations of quantized algebras*, Kluwer Academic Publishers, Dordrecht-Boston-London, 1995.

[55] A.N. Rudakov, ed., *Helices and vector bundles: seminaire Rudakov*, London Math. Soc. Lect. Note Series 148, Cambridge University Press, 1990.

[56] G. Scheja, Einige Beispiele faktorieller lokaler Ringe, *Math. Annalen* **172** (1967), 124–134.

[57] J.-P. Serre, Faisceaux algébriques cohérents, *Ann. of Math., II. Ser.* **61** (1955), 197-278.

[58] C.S. Seshadri, Fibrés vectoriels sur les courbes algébriques, *Astérisque* **96** (1982).

[59] D. Simson, *Linear representations of partially ordered sets and vector space categories*, Gordon and Breach Science Publishers, Brooklyn, NY, 1992.

[60] P. Wagreich, Algebras of automorphic forms with few generators, *Trans. Am. Math. Soc.* **262** (1980), 367–389.
[61] Ch. Xi, On wild hereditary algebras with small growth numbers, *Commun. Algebra* **18** (1990), 3413–3422.

H. Lenzing
Fachbereich Mathematik
Universität-GH Paderborn
D-33095 Paderborn
Germany
e-mail: helmut@uni-paderborn.de

Canadian Mathematical Society
Conference Proceedings
Volume **22**, 1998

Zassenhaus Conjectures for Infinite Groups*

Zbigniew S. Marciniak and Sudarshan K. Sehgal

Abstract. We report on work concerning an extension of two Zassenhaus conjectures to infinite groups. We show that they have negative solutions when one attempts to prove them literally. However, a modified, stable version is still true for infinite nilpotent groups. Moreover, it holds not only for single units but also for p-groups of matrices over $\mathbb{Z}G$. In particular, it implies that all finite groups of units in $\mathbb{Z}G$ have isomorphic copies inside G. We close with an example of a finitely generated nilpotent group whose unit group is not finitely generated.

0. Introduction

In this paper, we shall report on some of the work on (two) conjectures of Hans Zassenhaus and related problems. First we introduce some notation.

Let $\mathbb{Z}G$ denote the integral group ring of a group G. By $\mathcal{U}\mathbb{Z}G$ we will denote the group of invertible elements (*units*) of $\mathbb{Z}G$.

Obviously, all elements of the form $\pm g$, $g \in G$, belong to $\mathcal{U}\mathbb{Z}G$. We will call them *trivial units*. Usually, the ring $\mathbb{Z}G$ has plenty of other units, see [9] and [15].

The elements of $\mathcal{U}\mathbb{Z}G$ which are of finite order are called *torsion units*. We denote by $T\mathcal{U}(\mathbb{Z}G)$ the set of all torsion units in $\mathbb{Z}G$.

There is a classical result which relates the sets of trivial and torsion units in the case of abelian groups.

AMS Subject Classification (1991). 16S34, 20C05, 20C07.

Keywords and Phrases: group ring, nilpotent group, Zassenhaus Conjecture.

* Supported by Canadian NSERC Grant A-5300 and Polish Scientific Grant KBN 2P30101007.

0.1. Theorem. (Higman (1940), [2]) *If A is an abelian group then $T\mathcal{U}\mathbb{Z}A = \pm A$. In other words, all torsion units are trivial.* □

The *abelian* assumption can be relaxed in the following way.

0.2. Theorem. (See [13, p.46].) *Let G be any group. Torsion units which belong to the centre of $\mathcal{U}\mathbb{Z}G$ are trivial.* □

To proceed further, we introduce the augmentation map: $\varepsilon: \mathbb{Z}G \to \mathbb{Z}$ given by $\varepsilon(\sum \alpha_g g) = \sum \alpha_g$. We say that a ring homomorphism $\sigma: \mathbb{Z}G \longrightarrow \mathbb{Z}G$ is *augmentation preserving* if we have $\varepsilon \circ \sigma = \sigma$.

As ε is a ring homomorphism it maps units to units. In consequence, $\varepsilon(u) = \pm 1$ for any $u \in \mathcal{U}\mathbb{Z}G$.

Let $\mathcal{U}_1\mathbb{Z}G$ denote the group of units of augmentation one. It is easy to check that

$$\mathcal{U}\mathbb{Z}G \approx \mathcal{U}_1\mathbb{Z}G \times \{+1, -1\}.$$

Therefore, it is enough to study the smaller group $\mathcal{U}_1\mathbb{Z}G$. We will denote by $T\mathcal{U}_1\mathbb{Z}G$ the set of torsion units in $\mathcal{U}_1\mathbb{Z}G$.

When the group G is finite then $G \subset T\mathcal{U}_1\mathbb{Z}G$. An easy way to obtain a nontrivial torsion unit is to conjugate a group element $g \in G$ out of G. H. Zassenhaus conjectured that this is essentially the only way in which torsion units can arise.

0.3. Conjecture.
(ZC1) *Each $u \in T\mathcal{U}_1\mathbb{Z}G$ is conjugate in $\mathbb{Q}G$ to a group element $g \in G$.*
(Aut) *If $\sigma \in \operatorname{Aut}(\mathbb{Z}G)$ is an augmentation preserving automorphism then*

$$\sigma = \tau_\gamma \circ \alpha$$

where $\alpha \in \operatorname{Aut}(G)$ and τ_γ is the conjugation by a rational unit $\gamma \in \mathbb{Q}G$. Thus, we have $\sigma(g) = \gamma^{-1} g^\alpha \gamma$ for all $g \in G$.

The Zassenhaus Conjecture **(ZC1)** was recorded in [14] and the conjecture **(Aut)** is a version of the Zassenhaus Conjecture **(ZC2)** described in that paper.

K. Roggenkamp has offered an alternative version of the **(Aut)** part: instead of conjugations by elements of $\mathbb{Q}G$ he considers central automorphisms of $\mathbb{Z}G$. In this way he relates the Zassenhaus Conjecture to the problem of extending the Noether–Skolem Theorem to automorphisms of separable algebras.

From among many results on the above conjectures we would like to quote the following three important theorems.

0.4. Theorem. (Polcino Milies-Ritter-Sehgal (1986), [11]) *Conjecture* **(ZC1)** *holds for metacyclic groups of the type $G = \langle a \rangle \rtimes \langle x \rangle$ with $(|a|, |x|) = 1$.* □

0.5. Theorem. (Roggenkamp-Scott (1987), [12]) *Conjecture* **(Aut)** *holds for finite nilpotent groups.* □

0.6. Theorem. (Weiss (1988–91), [16], [17]) *Conjecture* **(ZC1)** *holds for finite nilpotent groups.* □

In general, there are counterexamples to **(Aut)** due to Roggenkamp–Scott. They construct the required automorphism by finding a suitable invertible $\mathbb{Z}G$-bimodule for a finite group G. The Roggenkamp–Scott group is metabelian of order $2^6 3^2 5$. It is a semidirect product $(C_3 \times C_3 \times C_5) \rtimes H$ where $H < GL_3(\mathbb{F}_4)$ consists of all upper triangular matrices with 1's along the diagonal. Using their ideas, Klingler (1991), [4], has given an explicit construction of the automorphism for a slightly modified group. The proofs require quite involved calculations in representation theory.

1. The (Aut) Conjecture for Infinite Groups

An easy counterexample to **(Aut)** for infinite groups is due to Zalesskii and Sehgal [15, p. 279].

1.1. Lemma. *For all $n > 2$ there exists a unit $u \in \mathcal{U}_1\mathbb{Z}S_n$ such that $\tau_u \neq \tau_g$ for any $g \in S_n$.*

Proof. Suppose $\tau_u = \tau_g$. Then ug^{-1} is central in $\mathcal{U} = \mathcal{U}_1\mathbb{Z}S_n$. Using the fact that $\mathbb{Q}$ is the splitting field for S_n it is easy to prove that the centre $\mathcal{Z}(\mathcal{U})$ of $\mathcal{U}$ is finite. Hence $\alpha = ug^{-1} \in \mathcal{Z}(\mathcal{U})$ has only a finite number of possibilities.

However, $\mathcal{U}$ itself is infinite. Namely, it has elements of infinite order, e.g.,

$$u = 1 + (a-1)b(a+1) \quad \text{where} \quad a = (1\,2),\ b = (1\,2\,3),$$
$$u^k = 1 + k(a-1)b(a+1).$$

Hence there exists a unit u with the desired property. □

1.2. The Example

Fix $n > 6$ and set $G_i = S_n$ for all $i \geq 1$. Consider the restricted direct product $G = \dot{\Pi} G_i$. There exists an augmentation preserving automorphism $\varphi \in \mathrm{Aut}(\mathbb{Z}G)$ such that $\varphi \neq \tau_\gamma \circ \alpha$ for any $\alpha \in \mathrm{Aut}(G)$ and $\gamma \in \mathcal{U}(\mathbb{Q}G)$.

Proof. Pick units $v_i \in \mathcal{U}_1(\mathbb{Z}G_i)$ as in Lemma 1.1 and define the automorphisms

$$\varphi_i = \tau_{v_i} \colon \mathbb{Z}G_i \longrightarrow \mathbb{Z}G_i.$$

Then $\varphi = \Pi\varphi_i$ is the desired automorphism.

In fact, suppose that $\varphi = \tau_\gamma \circ \alpha$ for some $\alpha \in \mathrm{Aut}(G)$ and a rational unit $\gamma \in \mathbb{Q}G$. As γ has finite support, there exists j such that $\gamma \in \mathbb{Q}[\prod_{i \leq j} G_i]$. Then τ_γ centralizes all subgroups G_i for $i > j$.

Fix a number $i_0 > j$ and consider the group automorphism $\alpha = \tau_\gamma^{-1} \circ \varphi$. For $x \in G_{i_0}$ we have $\alpha(x) = \varphi_{i_0}(x) \in G \cap \mathbb{Z}G_{i_0} = G_{i_0}$.

It is well known that $\mathrm{Aut}(S_n) = \mathrm{Inn}(S_n)$ for $n \neq 6$. So $(\alpha_{|G_{i_0}})(x) = g^{-1}xg$ for some $g \in G_{i_0}$. But then $\varphi_{i_0} = \alpha_{|G_{i_0}} = \tau_g$ – a contradiction. □

1.3. Remark. It is clear from the above proof that instead of S_n we could use any group H such that $\mathrm{Aut}(H) = \mathrm{Inn}(H)$ and the group of units $\mathcal{U}\mathbb{Z}H$ is infinite with finite centre.

2. The (ZC1) Conjecture for Infinite Groups

We are going to introduce the notion of a unique trace. Let R be a commutative ring and let G be any group. We denote by $[RG, RG]$ the R-submodule of RG

generated by all Lie commutators $[x, y] = xy - yx$ with $x, y \in RG$. Clearly,

$$[RG, RG] = \langle [g, h] : g, h \in G \rangle_R = \left\{ \sum c_h h : \sum_{h \sim g} c_h = 0 \quad \text{for all fixed} \quad g \in G \right\}.$$

By $h \sim g$ we mean that h is conjugate to g in G.

For $u = \sum u_h h \in RG$ write $\widetilde{u}(g) = \sum_{h \sim g} u_h$. Notice that we have $\widetilde{u}(g) = 0$ for all $g \in G$ whenever $u \in [RG, RG]$. We call the numbers $\widetilde{u}(g)$ *the traces* of $u \in RG$.

Suppose $u = \gamma^{-1} g_0 \gamma$ for some $g_0 \in G$. Then

$$u - g_0 = \gamma^{-1} g_0 \gamma - g_0 = [\gamma^{-1} g_0, \gamma] \in [RG, RG].$$

Consequently, $u = g_0 + \lambda$ and $\widetilde{\lambda}(g) = 0$ for all $g \in G$. Hence

$$\widetilde{u}(g_0) = 1, \quad \widetilde{u}(g) = 0 \text{ for } g \text{ not conjugate to } g_0.$$

In other words: if u is conjugate to a group element then it has only one trace not equal to zero. The next result shows that the above condition is very close to **(ZC1)**.

2.1. Theorem. (See [15, p. 238].) *For an arbitrary group G consider the following two conditions:*

(a) *The* **(ZC1)** *conjecture holds for G.*

(b) *For any torsion unit $u = \sum u_g g \in \mathcal{U}_1 \mathbb{Z}G$ there exists a unique (up to conjugacy) group element g_0 such that $\widetilde{u}(g_0) \neq 0$.*

Condition (a) *always implies condition* (b). *If the group G is finite then the two conditions are equivalent.* □

We extract condition (b) and make a definition.

2.2. Definition. Let G be an arbitrary group. We say that G is a UT-group (*Unique Trace group*) if for every torsion unit $u \in T\mathcal{U}_1 \mathbb{Z}G$ there exists a unique (up to conjugacy) element g_0 of G such that $\widetilde{u}(g_0) \neq 0$.

When we want to extend the positive results about **(ZC1)** to a class of infinite groups we must at least prove that this class consists of UT-groups. Here we have a result of this kind.

2.3. Theorem. (Bovdi-Marciniak-Sehgal (1994), [1]) *Every nilpotent group is a UT-group.*

Sketch of Proof. From Theorems 2.1 and 0.6 it follows that all finite nilpotent groups are UT-groups. Let then G be an infinite nilpotent group. Because we can work with one unit $u \in \mathcal{U}_1 \mathbb{Z}G$ at a time, we can assume that G is finitely generated. Then G is residually finite and conjugacy separable (i.e., every pair of conjugacy classes remains distinct in a finite homomorphic image of G). In particular, G has a finite homomorphic image $\overline{G}$ such that:

(i) the support of u maps injectively to $\overline{G}$;

(ii) the conjugacy classes in G, which intersect the support of u, map to distinct conjugacy classes in $\overline{G}$.

Because the image of u in $\mathbb{Z}\overline{G}$ has one nonzero trace only, the same holds for u, by conditions (i) and (ii). □

2.4. Corollary. *If G is a finitely generated nilpotent group and $u \in T\mathcal{U}_1\mathbb{Z}G$ then u is conjugate to some $g \in G$ in the inverse limit $\varprojlim \mathbb{Q}[G/N]$ where N runs over all finite index normal subgroups of G.* □

So what about **(ZC1)**?

No counterexample is known for finite groups G. For infinite nilpotent groups there is a counterexample due to Marciniak–Sehgal (1996), [8]. We shall describe this group now.

Consider the group automorphism $\sigma: \mathbb{Z} \oplus \mathbb{Z} \longrightarrow \mathbb{Z} \oplus \mathbb{Z}$ given by $\sigma(x_1, x_2) = (x_1 + x_2, x_2)$. Let $H = (\mathbb{Z} \oplus \mathbb{Z}) \rtimes_\sigma \mathbb{Z}$. By this we mean a semi-direct product in which the conjugation by the generator of $\mathbb{Z}$ on the normal subgroup $\mathbb{Z} \oplus \mathbb{Z}$ coincides with σ. Then $H' = \langle (1,0) \rangle = \mathcal{Z}(H)$ so H is nilpotent.

Let $D_8 = \langle x, y : x^4 = 1 = y^2, x^y = x^{-1} \rangle$ be the dihedral group. Then $G = H \times D_8$ is our group.

To describe the unit $u \in \mathcal{U}_1\mathbb{Z}G$ which is not conjugate to any group element we need a few auxiliary elements in $\mathbb{Z}G$. Namely, let $g_i = (i, 1) \in \mathbb{Z} \oplus \mathbb{Z} < H$ for $i = 0, 1, 2, 3$. Thus, $g_i = t^i g_0 t^{-i}$ where t is a fixed generator of the top group $\mathbb{Z} < H$. Further, we define the following elements in $\mathbb{Z}G$:

$$
\begin{aligned}
X_i &= g_i - 1, & i &= 0, 1, 2, 3, \\
\alpha &= 2t^{-2} X_1 X_2, & \gamma &= 2t^{-2} X_0 X_2 + t^{-1} X_0, \\
\beta &= 2t^{-2} X_0 X_3, & \delta &= 2t^{-2} X_1 X_3 - t^{-1} X_2.
\end{aligned}
$$

2.5. Theorem. (See [8].) *Consider the element*

$$u = y + [(\alpha - \beta) + (\alpha + \beta)y + (\gamma + \delta)x + (\gamma - \delta)yx] \cdot (x^2 - 1) \in \mathbb{Z}G.$$

Then $u^2 = 1$ and u is not conjugate to a trivial unit in KG for any field K of characteristic zero.

Sketch of Proof. Let K be any field of characteristic zero. Then we have

$$KG \approx KH \otimes \mathbb{Q}D_8 \xrightarrow{\approx} K\big[H \times D_8/\langle x^2 \rangle\big] \oplus M_2(KH).$$

Our unit $u \in \mathbb{Z}G \subset KG$ is mapped to the pair (y, U) where $y \in D_8$ and

$$U = \begin{bmatrix} 1 - 4\alpha & 4\gamma \\ -4\delta & -1 + 4\beta \end{bmatrix} \in M_2(KH).$$

It is not difficult to calculate that $U^2 = I$ and hence $u^2 = 1$ as well.

For the proof of the second part of the statement notice that all finite order elements of G sit inside D_8 which maps to $M_2(K) \subset M_2(KH)$ in the second component. Therefore it is enough to show that U is not conjugate to any matrix with coefficients in $K < KH$.

In fact, it is enough to prove the same about the idempotent matrix

$$E = (I + U)/2 = \begin{bmatrix} 1 - 2\alpha & 2\gamma \\ -2\delta & 2\beta \end{bmatrix}.$$

To this end we consider the projective KH-module P which is the kernel of the multiplication by E on the right: $KH \oplus KH \xrightarrow{\cdot E} KH \oplus KH$. Our statement about conjugation of E translates to the condition that P is stably free of rank one but not free. This can be verified by a standard identification of P with a left ideal of KH which is not principal. □

2.6. Remark. Unfortunately, it is not possible to find a counterexample to **(ZC1)** in the finite group case by the same line of argument. Namely, from the character theory it easily follows that for finite groups H stably free KH-modules are just free.

Nevertheless, for some infinite groups **(ZC1)** is known to be true.

2.7. Theorem. *In the following cases* **(ZC1)** *is true.*

(i) Levin-Sehgal (1993), [5]:
$G = D_\infty = C_2 * C_2 = \langle t, z : z^2 = 1,\ t^z = t^{-1}\rangle$;

(ii) Bovdi-Marciniak-Sehgal (1994), [1]:
$G = T \times A$ *where* T *is finite nilpotent and* A *is torsion free abelian.* □

Moreover, when $G = C_{m_1} * C_{m_2} * \cdots * C_{m_k}$ the same is true provided we allow the conjugating element to be found in a very large ring containing $\mathbb{Q}G$, see Lichtman–Sehgal (1989), [6].

The D_∞ case is done by representing elements of $\mathbb{Z}G$ as 2×2 matrices over $\mathbb{Z}\langle t\rangle$:

$$\alpha = a(t) + b(t)z \mapsto \begin{pmatrix} a(t) & b(t) \\ b^*(t) & a^*(t) \end{pmatrix} \in (\mathbb{Z}\langle t\rangle)_{2\times 2}.$$

Here, $(\sum c_g g)^* = \sum c_g g^{-1}$. If $u \in T\mathcal{U}_1\mathbb{Z}D_\infty$ then, after a long computation, one gets $u \sim tz$ or z in $\mathbb{Z}\left[\frac{1}{2}\right]D_\infty$.

3. Matrices over $\mathbb{Z}G$

The representation of $\mathbb{Z}G$ as matrices over $\mathbb{Z}A$, where $|G : A| < \infty$, which was used in the proof of Theorem 2.5 is a general "induction" procedure. We consider now a special case.

Let $G = A \rtimes X$, $|X| = n < \infty$. Then $\mathbb{Z}G$ is a left $\mathbb{Z}A$-module and we can write $\mathbb{Z}G = \sum_{i=1}^n \mathbb{Z}Ax_i$ where $X = \{x_1, x_2, \cdots, x_n\}$ and $x_1 = 1$.

As $\mathbb{Z}G$ is also a right $\mathbb{Z}G$-module, we can represent $\mathbb{Z}G$ as matrices over $\mathbb{Z}A$, after picking X as a basis of $\mathbb{Z}G$ over $\mathbb{Z}A$. Thus, e.g., for $a \in A$ we have $x_i a = a^{x_i} x_i$ and so

$$a \mapsto \begin{bmatrix} a & & & \\ & a^{x_2} & & \\ & & \ddots & \\ & & & a^{x_n} \end{bmatrix}.$$

In general

$$x_i\alpha = \sum_{j=1}^n f_{ij}(\alpha)x_j, \quad 1 \le i \le n, \qquad \alpha \mapsto \left[f_{ij}(\alpha)\right]_{n\times n} \in (\mathbb{Z}A)_{n\times n}.$$

If one knows something about torsion in $(\mathbb{Z}A)_{n\times n}$ it may be possible to lift the information back to $\mathbb{Z}G$.

Motivated by the above remark, we would like to study torsion units in $GL_n(\mathbb{Z}G)$. However, this linear group contains $GL_n(\mathbb{Z})$ with its torsion elements, which are not "caused" by G. A way out of this difficulty is to consider a proper subgroup of $GL_n(\mathbb{Z}G)$.

Consider $\varepsilon_*: GL_n(\mathbb{Z}G) \longrightarrow GL_n(\mathbb{Z})$, $\varepsilon_*(\alpha_{ij}) = [\varepsilon(\alpha_{ij})]$. We shall study the kernel of this homomorphism:

$$SGL_n(\mathbb{Z}G) = \{A \in GL_n(\mathbb{Z}G) : \varepsilon_*(A) = I\}.$$

Notice that for $n = 1$ we have $GL_1(\mathbb{Z}G) = \mathcal{U}\mathbb{Z}G$ and $SGL_1(\mathbb{Z}G) = \mathcal{U}_1\mathbb{Z}G$.

3.1. Theorem. (Weiss (1988), [16]) *Let G be a finite p-group and let $U \in SGL_n(\mathbb{Z}G)$ be an element of finite order. Then there exists a matrix $M \in (\mathbb{Q}G)_{n\times n}$ such that $MUM^{-1} = \operatorname{diag}(g_1, \ldots, g_n)$, $g_i \in G$.* □

3.2. Remark. In the case $n = 1$ the above theorem gives **(ZC1)** for p-groups, as stated.

3.3. Remark. All torsion elements of $SGL_n(\mathbb{Z}G)$, G a finite p-group, have p-power order.

3.4. Remark. Theorem 3.1 has been applied to groups $G = A \rtimes X$, A a p-group, X a p'-group, to get information about p-elements in $\mathcal{U}_1\mathbb{Z}G$, for details see [15].

3.5. Question. Can one extend the last result to finite nilpotent groups? Namely, given a finite nilpotent group Π and a torsion matrix $U \in SGL_n(\Pi)$, is it true that $U \sim \operatorname{diag}(g_1, \ldots, g_n)$, $g_i \in \Pi$, in $GL_n(\mathbb{Q}\Pi)$?

This has been recently answered by Weiss.

3.6. Theorem. (Weiss) *Let Π be finite nilpotent. Then $U \sim \operatorname{diag}(g_1, \ldots, g_n)$ for all $U \in SGL_n(\Pi)$ if and only if Π has at most one non cyclic Sylow subgroup.* □

He also gives a partly explicit example which shows that, in general, the answer to Question 3.5 is negative. Namely, consider $\Pi = C_6 \times C_6 = \langle a\rangle \times \langle b\rangle$. Let r be a small integer ≥ 1, (probably 1 or 2). Then there exists $U \in GL_{6r}(\mathbb{Z}\Pi)$, $U^6 = I$ and $\operatorname{Tr}(U) = r(-1+a^3+b^3+a^3b^3+a^4+b^4+a^4b^4+a^2b^4)$. Notice the negative coefficient $r(-1)$. If U were conjugate to a matrix with group elements on the diagonal then its trace would be equal to their sum, hence nonnegative.

The number r is not determined. It is needed because U is "exhibited" semi locally by a $\mathbb{Z}G$-lattice M and $(M_{|N})^{\oplus r}$ is to be free. Here $G = \Pi \times \Gamma$, $N = \Pi \times \langle 1\rangle$, and Γ is cyclic.

Let us return to **(ZC1)**. So far we know:

(i) It is not true (even for p-elements) for infinite nilpotent groups;

(ii) It is not true for $GL_n(\mathbb{Z}G)$, even for G finite abelian.

4. Stable Conjugation in $(KG)_{n\times n}$

Although the closing conclusion of the previous section does not look very encouraging, there are more positive results concerning **(ZC1)** for infinite groups. We need only to adjust it slightly.

Looking for the right adjustment, let us return to Theorem 2.5. We have constructed there a unit $u \in \mathbb{Z}[H \times D_8]$ which is a counterexample to **(ZC1)**. This construction was based on the observation that the group algebra $\mathbb{Q}H$ has a finitely generated projective module P which is not free.

Our group H is a torsion free, finitely generated nilpotent group. It is well known that $\widetilde{K}_0(\mathbb{Q}H) = 0$ for such groups. Therefore our projective module P is *stably free.* In our case it was $P \oplus \mathbb{Q}H \approx \mathbb{Q}H \oplus \mathbb{Q}H$.

4.1. Digression. We do not know any example of a projective $\mathbb{Q}H$-module which is not stably free in the case when H is a torsion free group.

The following result is an easy consequence of the last remark.

4.2. Theorem. (See [8].) *The unit $u \in \mathbb{Z}[H \times D_8]$, constructed in Theorem 2.5, has an additional property:* $\begin{pmatrix} u & 0 \\ 0 & -1 \end{pmatrix} \sim \begin{pmatrix} y & 0 \\ 0 & -1 \end{pmatrix}$ *in $GL_2(\mathbb{Q}G)$.* □

Motivated by this result, we propose the following

4.3. Definition. Let R be a K-algebra over a field K. Two matrices $A, B \in M_n(R)$ are said to be *stably conjugate* if there exist roots of unity $\xi_1, \ldots, \xi_d \in K$ so that

$$A \oplus \operatorname{diag}(\xi_1, \ldots, \xi_d) \sim B \oplus \operatorname{diag}(\xi_1, \ldots, \xi_d) \quad \text{in} \quad M_{n+d}(R).$$

Whenever A, B are stably conjugate we write $A \sim_S B$. For us R will be KG and K will be a field of characteristic zero with enough roots of unity.

For any (associative) K-algebra R we have the *Bass rank map*

$$r : M_n(R) \to R/[R, R], \qquad r(A) = \operatorname{Tr}(A) \,(\operatorname{mod}[R, R]).$$

The map r is K-linear and satisfies the standard trace condition $r(AB) = r(BA)$. When restricted to the set of idempotent matrices it defines a map

$$r : K_0(R) \to R/[R, R].$$

The rank map is related to stable conjugation by the following criterion.

4.4. Theorem. (See [1].) *Let $R = KG$ and K has enough roots of unity. Suppose the rank map $r\colon K_0(R) \to R/[R, R]$ is injective. For a matrix $U \in M_n(R)$ satisfying $U^d = I$ the following conditions are equivalent:*

(i) *$U \sim_S \operatorname{diag}(g_1, \ldots, g_n)$ for some $g_i \in G$,*

(ii) *There exists a diagonal matrix $D = (g_1, \ldots, g_n)$ with $g_i \in G$ so that $r(U^k) = r(D^k)$ for all $k \in N$.* □

We were able to check the injectivity of $r\colon K_0(KG) \to KG/[KG, KG]$ for finitely generated nilpotent groups G. Also, when a matrix U belongs to $SGL_n(\mathbb{Z}G)$ and d is a p-power then it is possible to find a diagonal matrix D satisfying (ii) in the above criterion. Using this, we proved

4.5. Theorem. (Marciniak-Sehgal (1996), [7]) *Let G be nilpotent. Suppose that $U \in SGL_n(\mathbb{Z}G)$ satisfies $U^{p^m} = I$. Then*

$$U \sim_S \operatorname{diag}(g_1, \ldots, g_n), \quad g_i \in G.$$ □

In fact, it is possible to obtain a deeper result. When dealing with a single matrix, we are really working with a cyclic p-subgroup $\langle U\rangle < SGL_n(\mathbb{Z}G)$. It is even better to think of it as of a representation $\phi\colon C_{p^m} \to SGL_n(\mathbb{Z}G)$.

In Definition 4.3 we defined when a pair of matrices is stably conjugate. It can be naturally generalized to pairs of representations $\phi, \psi\colon H \to SGL_n(\mathbb{Z}G)$ for arbitrary finite groups H.

Let $\rho\colon H \longrightarrow GL_d(\mathbb{Z}) \subseteq GL_d(\mathbb{Z}G)$ be the regular representation, where $d = |H|$.

4.6. Definition. Representations ϕ and ψ are *stably conjugate* over KG if there exist a natural number k and a matrix $Y \in GL_{n+kd}(KG)$ such that

$$(\phi \oplus \rho^k)(h) = Y \cdot (\psi \oplus \rho^k)(h) \cdot Y^{-1}$$

holds for all $h \in H$. We write then $\phi \sim_S \psi$.

4.7. Remark. Notice that when the group H is cyclic and the field K contains a primitive d-th root of unity then the regular representation ρ can be diagonalized and we get back the definition of stable conjugation of matrices.

Criterion 4.4 can be extended to

4.8. Proposition. (Marciniak-Sehgal (1996), [7]) *Suppose that the rank map*

$$r\colon K_0(KG) \longrightarrow KG/[KG, KG]$$

is injective. If H is a finite group and K is its splitting field then for any representations $\phi, \psi\colon H \longrightarrow GL_n(KG)$ the following conditions are equivalent:

(i) $r \cdot \phi = r \cdot \psi$,

(ii) $\phi \sim_S \psi$. □

Let $\delta_n(G)$ denote the group of all diagonal matrices $\mathrm{diag}(g_1, \dots, g_n) \in SGL_n(\mathbb{Z}G)$, $g_i \in G$.

4.9. Theorem. (See [7].) *Let G be a nilpotent group and let P be a finite p-group. For each representation $\phi\colon P \longrightarrow SGL_n(\mathbb{Z}G)$ there exists a diagonal representation*

$$\psi\colon P \longrightarrow \delta_n(G) < SGL_n(\mathbb{Z}G)$$

such that ϕ and ψ are stably conjugate over $\mathbb{C}G$. □

The last theorem really extends Theorem 4.5, as can be seen from the following corollary.

4.10. Corollary. *If G is nilpotent then any finite abelian p-group $A < SGL_n(\mathbb{Z}G)$ can be stably diagonalized over $\mathbb{C}G$. It means that there exist:*

— *an embedding $\varphi\colon A \longrightarrow \delta_n(G)$,*

— *a family of linear characters $\xi_i\colon A \longrightarrow \mathbb{C}^*$, $1 \le i \le k$,*

— *a matrix $Y \in GL_{n+k}(\mathbb{C}G)$,*

such that for all $X \in A$ holds

$$Y \cdot \Big(X \oplus \mathrm{diag}\big(\xi_1(X), \dots, \xi_k(X)\big)\Big) \cdot Y^{-1} = \varphi(X) \oplus \mathrm{diag}\big(\xi_1(X), \dots, \xi_k(X)\big).$$

□

Returning to group ring units, we obtain the following "realization" result.

4.11. Corollary. *If G is a nilpotent group then every finite subgroup $H < \mathcal{U}_1\mathbb{Z}G$ is isomorphic to a subgroup of G.*

Sketch of Proof. We first show that every finite subgroup $H < \mathcal{U}_1\mathbb{Z}G$ must be nilpotent. This is verified residually, in sufficiently large finite quotients of G. Thus H is a product of its Sylow p-subgroups: $H = H_{p_1} \times \cdots \times H_{p_r}$. By Theorem 4.9 each group H_{p_i} can be embedded into G. In fact, it embeds into the p_i-Sylow subgroup of the torsion part of G. Hence the direct product of the above embeddings maps H into G. □

5. The Group of Units $\mathcal{U}_1\mathbb{Z}G$ Need Not Be Finitely Generated Even for Finitely Generated Nilpotent Groups

The last corollary tells us that for nilpotent groups G the unit group $\mathcal{U}_1\mathbb{Z}G$ has the same finite subgroups as G. One could wonder how complicated $\mathcal{U}_1\mathbb{Z}G$ might be. For example, let us assume that G is nilpotent and finitely generated. Is $\mathcal{U}_1\mathbb{Z}G$ finitely generated?

It is well known that the group $\mathcal{U}_1\mathbb{Z}G$ is finitely generated (even finitely presentable) for finite groups G, as $\mathcal{U}_1\mathbb{Z}G$ is then an arithmetic group. There is also a class of infinite groups for which the answer is positive.

5.1. Theorem. (See [13, p. 194].) *Let G be a finitely generated nilpotent group and let $t(G)$ denote its torsion subgroup. If the group ring $\mathbb{Z}[t(G)]$ has no nilpotent elements then $\mathcal{U}_1\mathbb{Z}G = \mathcal{U}_1\mathbb{Z}[t(G)] \cdot G$. In particular, it is finitely generated.* □

The above theorem applies for example to all finitely generated abelian groups.

In the literature there is only one example of a finitely generated group G with $\mathcal{U}_1\mathbb{Z}G$ not finitely generated: Mirowicz [10] has proved that it is so for $G = D_\infty$, the infinite dihedral group.

We prove here the following

5.2. Proposition. *Let $G = D_8 \times C_\infty$ where D_8 is the dihedral group of order 8 and C_∞ is the infinite cyclic group. Then the group of normalized units $\mathcal{U}_1\mathbb{Z}G$ is not finitely generated.*

As before, we fix the presentation $D_8 = \langle x, y | x^4 = 1,\ y^2 = 1,\ x^y = x^{-1}\rangle$ and write $z = x^2 \in D_8$. Following [3], we consider the subset $V \subset \mathcal{U}_1\mathbb{Z}G$ consisting of all units u of the form $1 + a(z-1)$ with $a \in \mathbb{Z}G$ such that $\varepsilon(a)$ is even.

5.3. Lemma. $\mathcal{U}_1\mathbb{Z}G = VG$.

Proof. Let $\pi\colon \mathbb{Z}G \longrightarrow \mathbb{Z}[G/\langle z\rangle]$ be the natural map. It is well known [13] that the group ring $\mathbb{Z}[G/\langle z\rangle] = \mathbb{Z}[C_\infty \times C_2 \times C_2]$ has trivial units only. Hence for any unit $u \in \mathcal{U}_1\mathbb{Z}G$ we have $\pi(u) = \pi(g)$ for some $g \in G$. Consider $v = ug^{-1} \in \mathcal{U}_1\mathbb{Z}G$. Then $\pi(v) = 1$, i.e. $v = 1 + a(z-1)$ for some $a \in \mathbb{Z}G$. If $\varepsilon(a)$ is even then $v \in V$ and we are done. Otherwise, write $v = v' \cdot z$, where $v' = z - a(z-1) = 1 + (1-a)(z-1) \in V$. In any case $u \in VG$, as desired. □

5.4. Lemma. $V \cap G = \langle 1 \rangle$.

Proof. Suppose that $g = 1 + a(z-1) \in V$ for some $g \in G$. Then $g(z+1) = z+1$ and hence either $g = 1$ or $g = z$. To eliminate the second possibility, write $z = 1 + b(z-1)$

with $\varepsilon(b)$ even. It follows that $(b-1)(z-1)=0$ and $b-1=c(z+1)$ for some $c \in \mathbb{Z}G$. Then $\varepsilon(b)=1+2\varepsilon(c)$ is odd — a contradiction. □

5.5. Lemma. *V is a normal subgroup of $\mathcal{U}_1\mathbb{Z}G$.*

Proof. It is easy to see that V is closed under multiplication. We shall see that V is also closed under taking inverses. To this end let $v \in V$ and $u=v^{-1} \in \mathcal{U}_1\mathbb{Z}G$. From $\pi(v)=1$ it follows that $\pi(u)=1$ as well, i.e. $u=1+b(z-1)$ for some $b \in \mathbb{Z}G$. We must show that $\varepsilon(b)$ is even.

Suppose the contrary. Then $uz=z-b(z-1)=1+(1-b)(z-1) \in V$ and so $z=v(uz) \in V$ — a contradiction with Lemma 5.4.

If $v=1+a(z-1) \in V$ and w is any unit, then $wuw^{-1}=1+waw^{-1}(z-1) \in V$, as $\varepsilon(waw^{-1})=\varepsilon(a)$ is even. Hence V is normal in $\mathcal{U}_1\mathbb{Z}G$. □

5.6. Lemma. *If $\mathcal{U}_1\mathbb{Z}G$ is finitely generated then V is finitely generated as well.*

Proof. From Lemma 5.5 it follows that VD_8 is a subgroup of $\mathcal{U}_1\mathbb{Z}G$. As C_∞ is central in $\mathcal{U}_1\mathbb{Z}G$, so VD_8 is normal in $VG=\mathcal{U}_1\mathbb{Z}G$. Because also $(VD_8)\cap C_\infty=\langle 1\rangle$, it follows that $\mathcal{U}_1\mathbb{Z}G=(VD_8)\times C_\infty$. Therefore VD_8 is a homomorphic image of $\mathcal{U}_1\mathbb{Z}G$, and hence it is finitely generated. Finally, V is the kernel of the epimorphism $VD_8 \longrightarrow D_8$, hence it is of finite index in VD_8. Therefore V is finitely generated, too. □

Let us write $R=\mathbb{Z}[C_\infty]=\mathbb{Z}[t,t^{-1}]$.

5.7. Lemma. *The group V is isomorphic to a subgroup of $SL_2(R)\cap\begin{bmatrix}1+2R & 4R\\ 2R & 1+2R\end{bmatrix}$.*

Proof. We select a Wedderburn isomorphism

$$\phi\colon \mathbb{Q}D_8 \xrightarrow{\approx} \mathbb{Q}\big[D_8/\langle z\rangle\big] \oplus M_2(\mathbb{Q})$$

as follows. For any $a \in \mathbb{Q}D_8$ we set $\phi(a)=\big(\phi_1(a),\phi_2(a)\big)$ where ϕ_1 is induced by the natural epimorphism $D_8 \longrightarrow D_8/\langle z\rangle$ and ϕ_2 is the linear representation

$$\phi_2(x)=\begin{bmatrix}0 & -1\\ 1 & 0\end{bmatrix},\quad \phi_2(y)=\begin{bmatrix}1 & 0\\ 0 & -1\end{bmatrix}.$$

After tensoring with $\mathbb{Q}C_\infty$, we get

$$\phi\otimes\mathrm{id}\colon \mathbb{Q}G \xrightarrow{\approx} \mathbb{Q}\big[G/\langle z\rangle\big] \oplus M_2(\mathbb{Q}C_\infty).$$

Now we have $V<\mathcal{U}_1\mathbb{Z}G \subset \mathbb{Q}G$. From the definition of V it follows that $\phi_1(V)=\langle 1\rangle$. Therefore we have an embedding $\phi_2\colon V \subset GL_2(\mathbb{Q}C_\infty)$.

We shall describe this embedding explicitly. Any unit $u \in V$ can be written in the form $u=1+(a_0+a_1x+a_2y+a_3xy)(z-1)$ with $a_i \in R$ and $\sum\varepsilon(a_i)$ even. Then

$$\phi_2(u)=\begin{bmatrix}1 & 0\\ 0 & 1\end{bmatrix}+ \\ +\left(a_0\begin{bmatrix}1 & 0\\ 0 & 1\end{bmatrix}+a_1\begin{bmatrix}0 & -1\\ 1 & 0\end{bmatrix}+a_2\begin{bmatrix}1 & 0\\ 0 & -1\end{bmatrix}+a_3\begin{bmatrix}0 & 1\\ 1 & 0\end{bmatrix}\right)\cdot\begin{bmatrix}-2 & 0\\ 0 & -2\end{bmatrix}$$

and hence

$$\phi_2(u) = \begin{bmatrix} 1-2(a_0+a_2) & 2(a_1-a_3) \\ -2(a_1+a_3) & 1-2(a_0-a_2) \end{bmatrix}.$$

In this way we have embedded V into the subgroup $GL_2(R) < GL_2(\mathbb{Q}C_\infty)$. After we conjugate with $\left[\begin{smallmatrix} 1 & 1 \\ 0 & 1 \end{smallmatrix}\right]$ we obtain a new embedding ψ with

$$\psi(u) = \begin{bmatrix} 1-2(a_0+a_1+a_2+a_3) & 4(a_1+a_2) \\ -2(a_1+a_3) & 1+2(-a_0+a_1+a_2+a_3) \end{bmatrix}$$
$$\in \begin{bmatrix} 1+2R & 4R \\ 2R & 1+2R \end{bmatrix}.$$

Let us write $W = \psi(V) < GL_2(R)$. It remains to show that each matrix from W has determinant one.

To this end, take any matrix $A \in W$. After we augment the entries of A, we obtain an invertible integral matrix of the form $\varepsilon_*(A) = \left[\begin{smallmatrix} 1+4\alpha & 4\beta \\ 2\gamma & 1+4\delta \end{smallmatrix}\right]$, as $2 | \sum \varepsilon(a_i)$. Therefore $\det(\varepsilon_*(A)) \equiv 1 \pmod 4$ and $\det(\varepsilon_*(A)) = \pm 1$. It follows that $\det(\varepsilon_*(A)) = 1$.

Let $d \in R$ be the determinant of the matrix A. We have already checked that $\varepsilon(d) = 1$. As d is a unit of R of augmentation 1, it is just a group element: $d \in C_\infty$. It is easy to see from the determinant formula that the coefficient at 1 in d is odd and hence $d = 1$. Thus ψ embeds V into $SL_2(R)$, as desired. $\square$

Again, let us denote by W the image $\psi(V) < SL_2(R)$. We exhibit now a sequence of matrices from W.

5.8. Lemma. *For any integer k the matrix $A_k = \left[\begin{smallmatrix} 1 & 0 \\ -2t^k & 1 \end{smallmatrix}\right]$ belongs to W.*

Proof. Notice that the element $\eta = (x-y)(z-1) \in \mathbb{Z}D_8$ is nilpotent. In fact, $z-1$ is central so $\eta^2 = (x-y)^2(z-1)^2 = (z-xy-yx+1)(2-2z) = -2(z+1-yx^{-1}-yx)(z-1) = -2(1-yx^{-1})(z+1)(z-1) = 0$. Therefore the elements $u_k = 1+t^k\eta$ are units and clearly they belong to V. It is easy to verify that $\psi(u_k) = A_k$. $\square$

Proof of Proposition 5.2. Suppose that the group $\mathcal{U}_1\mathbb{Z}G$ is finitely generated. From Lemma 5.6 it follows that V, and hence W, is finitely generated.

Let $\mu\colon R \longrightarrow \overline{R}$ be the reduction modulo 4, i.e. $\overline{R} = (\mathbb{Z}/4\mathbb{Z})[C_\infty]$. Let $\overline{W}$ be the image $\mu(W) < SL_2(\overline{R})$. Clearly, this group is also finitely generated.

We show that $\overline{W}$ is abelian. In fact, it consists of elements of the form $\left[\begin{smallmatrix} 1+2a & 0 \\ 2c & 1+2d \end{smallmatrix}\right]$ with $a, c, d \in \overline{R}$. Hence, when we multiply two elements of $\overline{W}$ we obtain

$$\begin{bmatrix} 1+2a & 0 \\ 2c & 1+2d \end{bmatrix} \cdot \begin{bmatrix} 1+2a' & 0 \\ 2c' & 1+2d' \end{bmatrix} = \begin{bmatrix} 1+2(a+a') & 0 \\ 2(c+c') & 1+2(d+d') \end{bmatrix}$$

and the result does not depend on the order of matrices.

Thus, $\overline{W}$ is a finitely generated abelian group. However, it contains an infinite elementary abelian 2-subgroup $\langle \mu(A_k) | k \in \mathbb{Z} \rangle < \overline{W}$. This contradiction proves that $\mathcal{U}_1\mathbb{Z}G$ is not finitely generated. $\square$

References

[1] A.A. Bovdi, Z.S. Marciniak and S.K. Sehgal, Torsion units in infinite group rings, *Journal of Number Theory* **47** (1994), 284–299.
[2] G. Higman, The units of group rings, *Proc. Lond. Math. Soc. (2)* **46** (1940), 231–248..
[3] E. Jespers and G. Leal, Describing units of integral group rings of some 2-groups, *Comm. in Algebra* **19** (6) (1991), 1809–1827.
[4] L. Klingler, Construction of a counterexample to a conjecture of Zassenhaus, *Comm. in Algebra* **19** (8) (1991), 2303–2330.
[5] F. Levin and S.K. Sehgal, Zassenhaus conjecture for the infinite dihedral group, Proc. AMS Special Session, Series in Algebra, Vol.1, World Scientific, New York, 1993, pp. 57–68.
[6] A.I. Lichtman and S.K. Sehgal, The elements of finite order in the group of units of group rings of free products of groups, *Comm. in Algebra* **17** (9) (1989), 2223–2253.
[7] Z.S. Marciniak and S.K. Sehgal, Finite matrix groups over nilpotent group rings, *Journal of Algebra* **181** (1996), 565–583.
[8] Z.S. Marciniak and S.K. Sehgal, Zassenhaus conjecture and infinite nilpotent groups, *Journal of Algebra* **184** (1996), 207–212.
[9] Z.S. Marciniak and S.K. Sehgal, Constructing Free Subgroups of Integral Group Ring Units, *Proc. Amer. Math. Soc.* (to appear).
[10] M. Mirowicz, Units in group rings of the infinite dihedral group, *Canad. Math. Bull. Vol.* **34** (1) (1991), 83–89.
[11] C. Polcino Milies, J. Ritter and S.K. Sehgal, On a conjecture of Zassenhaus on units in integral group rings II, *Proc. Amer. Math. Soc.* **97** (2) (1986), 206–210.
[12] K. Roggenkamp and L. Scott, On a conjecture of Zassenhaus for finite group rings, preprint (1987).
[13] S.K. Sehgal, *Topics in group rings*, Marcel Dekker, New York, 1978.
[14] S.K. Sehgal, *Torsion units in integral group rings*, Proc. NATO Inst. on Methods in Ring Theory, Antwerp (1983), D. Riedel, pp. 497–504.
[15] S.K. Sehgal, *Units in integral group rings*, Longman's, Essex, 1993.
[16] A. Weiss, Rigidity of p-adic torsion, *Annals of Math.* **127** (1988), 317–332.
[17] A. Weiss Torsion units in integral group rings, *Journal für Reine und Angewandte Mathematik* **415** (1991), 175–187.

Z. S. Marciniak
Institute of Mathematics
Warsaw University, ul. Banacha 2
02-097 Warszawa
Poland
e-mail: zbimar@mimuw.edu.pl

S. K. Sehgal
Department of Mathematical Sciences
University of Alberta
Edmonton T6G 2G1
Canada
e-mail: S.Sehgal@ualberta.ca

Canadian Mathematical Society
Conference Proceedings
Volume **22**, 1998

Graded Rings — an Approach via Semigroups of Matrices

Jan Okniński

The aim of this paper is to present a general technique that can be applied, in particular, to several problems in ring theory. It is based on the structure theory of linear semigroups, that is subsemigroups of the full linear monoid $M_n(D)$ over a field or, more generally, a division ring D. The significance of the theory of linear groups is well known. Their structure theory and properties are exploited repeatedly in many important applications. On the other hand, while semigroups of endomorphisms show up often in various problems, and are a natural object to study, a systematic structural approach had not been known until recently, except for the Zariski closed semigroups of some special types. Our aim is to present the key ingredients of such an approach and to illustrate it with some applications to graded rings. In Section 1 we start with preliminary motivating remarks on graded rings, leading to semigroups of matrices. The structure of the full linear monoid $M_n(D)$ is described in Section 2. This is the starting point for explaining the structure theorem for an arbitrary subsemigroup $S \subseteq M_n(D)$. The result, its intuition and the resulting philosophy of studying such objects are outlined in Section 3. The link with the powerful theory if linear groups, exemplified with the generalized Tits alternative, is highlighted in Section 4. While the focus of the foregoing sections is on the technique, the remaining material is concerned with examples of applications to problems concerning graded rings. The structure of semilocal and perfect graded rings, the homogeneity of the Jacobson radical of a graded ring, and a description of semigroup algebras which are principal ideal rings are discussed. The emphasis is on the way the presented technique shows up in these problems.

Since a complete bibliography would have to be very long, only the key references are provided, from which the main results come and from which an extensive bibliography can be traced. Our general reference to ring theory is [11], while basics of semigroup theory can be found in [3].

AMS Subject Classification (1991). 16W50, 20M20.
Supported by a KBN research grant.

1. Motivation

We will consider rings R graded by a semigroup S. That is, $R = \bigoplus_{s\in S} R_s$ as additive groups, and $R_sR_t \subseteq R_{st}$ for every $s, t \in S$. $H(R) = \bigcup_{s\in S} R_s$ is the set of homogeneous elements of R. Clearly, $H(R)$ is a multiplicative subsemigroup. If $X \subseteq H(R)$, then by $\mathbf{Z}\{X\}$ we denote the (homogeneous) subring of R generated by X. For a subset $I \subseteq R$ the graded part of I is defined by $I_{\mathrm{gr}} = \bigoplus_{s\in S}(I \cap R_s)$. The set of $s \in S$ such that there exists $a = \sum a_s \in I$ with component $a_s \neq 0$ is denoted by $\mathrm{supp}(I)$.

Group crossed products $R = A * G$ form the simplest class of graded rings, [17]. Here $R_g = Ag$ for $g \in G$, so each component R_g contains a unit of R provided that A is a unital ring. On the other hand, if for a graded ring R the set $H = H(R) \setminus \{0\}$ consists of regular elements, then under many classical finiteness conditions on R, H is an Ore subset of R and one can form the homogeneous localization RH^{-1}. If additionally S is cancellative, then the image $\deg(H)$ of H under the degree map $\deg : H \longrightarrow S$ is an Ore semigroup, so RH^{-1} has a natural structure of a group-graded ring with components $\bigcup_{s,t\in S, s=gt} R_s(H \cap R_t)^{-1}$, for $g \in G = \deg(H)\deg(H)^{-1}$. Since each component contains a unit, it follows easily that RH^{-1} is a group crossed product of the group G. Thus, a natural idea is to study R via RH^{-1}, or more generally to reduce problems on an arbitrary graded ring R to certain crossed products derived from R. Such an approach is quite natural and it turns out to be very fruitful even in the simplest case where R is a domain and $S = \mathbf{Z}$, cf. [2], [26]. What are the most general circumstances this can be applied? For example, if $H(R) \subseteq M_n(D)$ for a division ring D, and F is the group of units of the monoid $eM_n(D)e$ for some idempotent $e \in M_n(D)$, then $H(R) \cap F \subseteq F \cong GL_j(D)$, where $j = \mathrm{rank}(e)$, (if nonempty) is a natural candidate for a 'nice' set of homogeneous elements of the graded subring $\mathbf{Z}\{H(R) \cap F\}$. Our main motivating ideas are built along these lines, aiming at applications to rings with many homomorphic images embeddable into simple Artinian rings. Namely, consider the following situation.

Let P be a prime ideal of R such that R/P is a right Goldie ring. When studying the prime ring R/P, replacing R by R/P_{gr} we may assume that $P \cap H(R) = 0$. We have an embedding $R/P \subseteq M_n(D)$ and the natural map $H(R) \longrightarrow \overline{H(R)} \subseteq R/P$. Choose a minimal nonzero idempotent $e \in M_n(D)$ such that $F = \{a \in eM_n(D)e \mid \mathrm{rank}(a) = \mathrm{rank}(e)\}$ intersects $\overline{H(R)}$. Let T be the inverse image in $H(R)$ of $F \cap \overline{H(R)}$. Note that F is a (maximal) subgroup of the multiplicative semigroup $M_n(D)$. The degree map $\deg : H(R) \setminus \{0\} \longrightarrow S$ restricted to T is a homomorphism $T \longrightarrow \deg(T) \subseteq S$. Consider the subring $R_T = \mathbf{Z}\{T\}$ generated by T. It is an $\deg(T)$-graded ring. Since $P \cap H(R) = 0$, it is easy to see that $H(R_T) = T \cup \{0\}$. In many cases (for example, if R is Noetherian or PI) one can show that T is a right Ore subset of R_T and the homogeneous localization $R' = R_TT^{-1}$ is graded by the group G of quotients of $\deg(T)$. Hence $R' = A * G$, a group crossed product. Moreover, as we will see in Section 3, the image $\overline{R_T}$ of R_T in R/P to a large extent is responsible for the properties of R/P. However, in general, one also needs to consider the intersection of $\overline{H(R)}$ with other maximal subgroups of $M_n(D)$, and not only those coming from the matrices of minimal possible (nonzero) rank . In order to implement the above motivating ideas, we need some structural approach to subsemigroups of $M_n(D)$.

2. Full Linear Monoid

Let K be a field and $M = M_n(K)$ — the monoid of $n \times n$ matrices over K under multiplication. Define $M_j = \{a \in M_n(K) \mid \text{rank}(a) \leq j\}$, $j = 0, 1, \ldots, n$. Then $0 = M_0 \subset \ldots \subset M_n = M$ are the only ideals of the monoid M.

Each of the sets $M_j \setminus M_{j-1}, j \geq 1$, admits an egg-box pattern, which will be crucial for our approach. Namely, let Y be the set of reduced row echelon forms of matrices of rank j, $X = Y^t$, the transpose of Y, and $F = \left\{ \left(\begin{smallmatrix} d & 0 \\ 0 & 0 \end{smallmatrix}\right) \mid d \in GL_j(K) \right\}$. Then each $a \in M_j \setminus M_{j-1}$ is uniquely written as

$$a = xgy, \quad x \in X, \quad y \in Y, \quad g \in F$$

Within this pattern a is interpreted as an element lying in row x, column y and with entry g.

For an abstract semigroup S by θ we will denote the zero element of S. Next, we define a semigroup structure on $T = M_j \setminus M_{j-1} \cup \{\theta\}$ by the rule

$$(xgy) \bullet (x'g'y') = \begin{cases} x(gyx'g')y' & \text{if rank}(yx') = j \\ \theta & \text{otherwise} \end{cases}$$

What we get is an instance of so called semigroup of matrix type over the group $F \cong GL_j(K)$ with the 'sandwich matrix'

$$P = (p_{yx})_{y \in Y, x \in X}, \quad \text{where} \quad p_{yx} = \begin{cases} yx & \text{if } yx \in F \\ \theta & \text{otherwise} \end{cases}$$

We write $T = \mathcal{M}(F, X, Y, P)$. Our T has also the property that each row and each column of P has a nonzero entry. Semigroups of this type are called completely 0-simple. They can be viewed as analogs of simple Artinian rings, [3].

Note that T is isomorphic to the Rees factor M_j/M_{j-1}, so it represents multiplication of matrices of rank j modulo ranks $< j$. There are two types of 'boxes' B. If $a \cdot b \neq \theta$ for $a, b \in B$, then $B \cong GL_j(K)$. This happens exactly when B contains an idempotent. The latter is equivalent to $ker(a) \cap Im(a) = 0$, where a is interpreted as a linear map (acting on the column space K^n by left multiplication). Otherwise $B^2 = \theta$, so B is a zero multiplication box.

Each 'row' (with θ) is a minimal right ideal of T and it consists of linear maps with the same image. Each 'column' consists of maps with the same kernel and corresponds to a minimal left ideal of T. We write $a\mathcal{R}b$ (respectively, $a\mathcal{L}b$) if a, b are in the same row (same column). (These are so called Green's relations.)

One of the important aspects of this semigroup is that it can be given a description totally in terms of the group $GL_n(K)$, and therefore it can be studied via group theoretic methods. Namely, consider the following subgroups of $GL_n(K)$

$$L_j = \begin{pmatrix} * & 0 \\ 0 & * \end{pmatrix}, \quad H_j = \begin{pmatrix} * & 0 \\ 0 & I \end{pmatrix}, \quad P_j = \begin{pmatrix} * & * \\ 0 & * \end{pmatrix}, \quad U_j = \begin{pmatrix} I & * \\ 0 & I \end{pmatrix},$$
$$H_j^* = \begin{pmatrix} I & 0 \\ 0 & * \end{pmatrix}, \quad P_j^- = \begin{pmatrix} * & 0 \\ * & * \end{pmatrix}, \quad U_j^- = \begin{pmatrix} I & 0 \\ * & I \end{pmatrix}.$$

X can be identified with the set $GL_n(K)/P_j$ of left coset representatives of P_j in $GL_n(K)$ and Y with the set $GL_n(K)/P_j^-$ of right coset representatives of P_j^-. If $x \in X$, $y \in Y$, then

$$n_{yx} = \begin{cases} h & \text{if } yx \in U_j^- l U_j \text{ with } l \in L_j,\ h \in H_j,\ h^{-1}l \in H_j^* \\ \theta & \text{if } yx \notin P_j^- P_j \end{cases}$$

defines a sandwich matrix $N = (n_{yx})$ over H_j. It can be easily verified that $T \cong \mathcal{M}(H_j, GL_n(K)/P_j, GL_n(K)/P_j^-, N)$.

In case of the skew linear monoid $M_n(D)$ all the essential features of the egg-box pattern also hold, but the lack of the structure theory in $GL_n(D)$ (compared with $GL_n(K)$) does not always allow us to treat $M_j \setminus M_{j-1}$ efficiently via group theory.

3. Structure Theorem

Keeping the notation of Section 2, we shall identify the nonzero elements of the semigroup M_j/M_{j-1} with the corresponding elements of $M_j \setminus M_{j-1}$.

Definition. Let $T = M_j \setminus M_{j-1} \cup \{\theta\}$. A subsemigroup U of T is called uniform if there exist $X' \subseteq X, Y' \subseteq Y$ such that $U \subseteq T' = \mathcal{M}(F, X', Y', P')$, where P' is the corresponding $Y' \times X'$-submatrix of P, P' has no zero rows or columns (in other words T' is completely 0-simple) and U meets every 'box' of T'.

For a subset X of a group G by $gp(X)$ we denote the subgroup generated by X. The key technical lemma reads as follows.

Lemma 1. *Assume that $U \subseteq T$ is a uniform subsemigroup. Then*

1. *$gp(H_1 \cap U), gp(H_2 \cap U)$ are conjugate in $M_n(K)$ for every maximal subgroups H_1, H_2 of T'.*
2. *there exists the least completely 0-simple semigroup $\hat{U}$ with $U \subseteq \hat{U} \subseteq T'$; moreover $T' \cong \mathcal{M}(F, X', Y', Q) \supseteq \mathcal{M}(H, X', Y', Q) \cong \widehat{U}$, for a sandwich matrix Q over a subgroup H of F isomorphic to $gp(H_1 \cap U)$.*

$\hat{U}$ can be viewed as a 'group approximation' of U. It is called the completely 0-simple closure of U in T. In case $H_1 \cap U$ is an Ore semigroup, $\widehat{U}$ is a semigroup of quotients of U (subject to the denominator set consisting of elements of U contained in maximal subgroups of T).

For convenience, we adopt the convention that $S/I = S$ if $I = \emptyset$.

Theorem 1. ([13]) *Let $S \subseteq M_n(K)$. Put $S_j = S \cap M_j$, $j = 0, 1, \ldots, n$. Then $S_0 \subseteq S_1 \subseteq \ldots \subseteq S_n = S$ are ideals of S, if nonempty. Moreover, for every j*

$$S_j/S_{j-1} = U_1 \cup \ldots \cup U_t \cup N$$

a 0-disjoint union, where U_i are uniform subsemigroups of M_j/M_{j-1} and intersect different 'rows' and 'columns' of M_j/M_{j-1}, $t \leq \binom{n}{j}$, and N is an ideal (if nonempty) with $N^{\binom{n}{j}} = \theta$. In particular $U_i U_k \subseteq N$ for $i \neq k$.

Note that $N = \emptyset$ implies that $S_{j-1} = \emptyset$ and S_j/S_{j-1} consists of one uniform semigroup. All U coming from all possible ranks j (or the corresponding subsets of

S, if convenient) are called the uniform components of S, while all N are called the nilpotent components. Uniform components are easy to describe. Namely, $a \in U$ for some component U if and only if there exist $x, y \in S$ such that the matrices $a, xay, (xay)^2$ have the same ranks (that is, xay lies in a maximal subgroup of $M_n(K)$ in the same 'layer' as a). Moreover, if $\text{rank}(a) = \text{rank}(b)$ and b is contained in a uniform component of S, then $b \in U$ exactly when $a\mathcal{R}v, a\mathcal{L}w, b\mathcal{R}w, b\mathcal{L}v$ for some $v, w \in S$.

We note the following striking analogy with Wedderburn's structure theorem for finite dimensional K-algebras:

$$N \longleftrightarrow \text{nilpotent radical}$$
$$U_i \longleftrightarrow \text{simple blocks modulo the radical}$$

(more precisely, U_i could be viewed as analogs of orders in simple Artinian rings).

If $S \subseteq M_n(D)$ for a division ring D, then the assertion of Theorem 1 holds with two qualitative differences [20]:

— there may be infinitely many uniform components at a given rank j,
— N is a union of nilpotent ideals of index two (but it is not necessarily nilpotent).

The above theorem leads naturally to the following strategy:

> Study $S \subseteq M_n(D)$ via its uniform components and the cancellative semigroups of the form $S \cap H$, for maximal subgroups H of $M_n(D)$.

S leads to at most (up to conjugation) 2^n linear groups G_α (generated by $S \cap H$) called groups associated to S. The left and right multiplication by elements of $S \cap G_\alpha$ determine actions on every component U. We want to study S via the group action of G_α on uniform components U_β that lie 'below' G_α and on the corresponding sandwich matrices. This is necessary in order to understand the interplay between the components.

There are several additional features that are useful when this approach is applied. In particular, using an appropriate modification of the exterior power map it can be seen that each factor S_j/S_{j-1} again is a linear semigroup provided that S is finitely generated. Hence, the interplay between the layers of S often can be approached by induction. Moreover, if U is a uniform component of S such that U (or $U \cup \{0\}$ if $0 \in S$) is an ideal of S, then $S \cup \widehat{U}$ is a subsemigroup of $M_n(D)$.

If $S \subseteq M_n(D)$ and the Z-subalgebra generated by S is an order in $M_n(D)$, for the center Z of D, then it follows that S has a unique uniform component (the set of elements of the least nonzero rank) such that U or $U \cup \{0\}$ is an ideal of S. This U often controls the properties of the entire S. This is essentially what happens in Zelmanov's approach to prime PI semigroup algebras, [25].

The theorem agrees with the theory of irreducible representations of S. Namely, every component U is an ideal of a Rees factor of S and there is a one-to-one correspondence:

irreducible representations of S	$\longleftrightarrow$	irreducible representations of the components of S

The methods of algebraic geometry also fit well into this picture. Namely, $S \subseteq \overline{S} \subseteq M_n(K)$, where $\overline{S}$ is the Zariski closure of S. Let U be a uniform component of S and N a nilpotent component of S. Then:

i) there exists a uniform component U' of $\overline{S}$ such that $U' \cap S = U$,
ii) there exists a nilpotent component N' of $\overline{S}$ such that $N' \cap S = N$.

Note also that, since every nonempty intersection $\overline{S} \cap F$ with a maximal subgroup F of $M_n(K)$ is a group, cf. [18], each uniform component of $\overline{S}$ coincides with its completely 0-simple closure.

4. Status of Cancellative Subsemigroups

When implementing our strategy, in the first place we need to be able to transfer the properties between a subsemigroup T of $GL_j(D)$ and the subgroup $gp(T)$ it generates. This is the first step to handle the extension $U \subseteq \widehat{U}$. It turns out that this is often possible. We illustrate this with an analog of one of the most fundamental results on linear groups.

We start with the following basic group theoretic results:

1. Tits Alternative. If $G \subseteq GL_n(K)$ is a finitely generated group then either G has a free nonabelian subgroup or G is solvable-by-finite (that is, a finite extension of a solvable group), [24].

2. Rosenblatt's Theorem. If G is a finitely generated solvable group, then either G has a free nonabelian subsemigroup or G is nilpotent-by-finite, [19].

It follows that a finitely generated linear group is either nilpotent-by-finite or it has a free nonabelian subsemigroup. It turns out that a similar dichotomy occurs already on the semigroup level. Clearly, the main difficulty in the following result lies in the case $S \subseteq GL_n(K)$.

Theorem 2. ([20])*Let $S \subseteq M_n(K)$ be a subsemigroup. Consider the following conditions:*

1. *S has no free nonabelian subsemigroups,*
2. *the associated linear groups are nilpotent-by-finite,*
3. *S satisfies an identity.*

Then 2) $\Leftrightarrow$ 3) $\Rightarrow$ 1). *Moreover, if the field K is finitely generated, then* 1) $\Rightarrow$ 2).

5. Group-graded Rings Which are Semilocal or Perfect

The remaining sections of the paper will be devoted to examples of applications of the approach presented above. We start with rings satisfying certain classical finiteness conditions. Such conditions have been studied by many authors, but not from a structural point of view. However, most of the known results follow from the structure theorem presented in this section. The first result shows that, roughly speaking, the case where $H(R)$ is a completely 0-simple semigroup is crucial. Recall that a multiplicative semigroup S with zero is left T-nilpotent if for any

$x_i \in S, n = 1, 2, \ldots$, there exists $n \geq 1$ such that $x_1 x_2 \cdots x_n = 0$. It is well known that the Jacobson radical of a left perfect ring has this property.

Proposition 1. *Let R be a left perfect (semilocal, respectively) ring graded by a semigroup S. Then $H(R)$ has a finite ideal chain whose factors are either left T-nilpotent (nil) or completely 0-simple.*

The idea of the proof.

i) first show that $H(R)$ is 'strongly π-regular', that is, a power of every $a \in R$ lies in a subgroup of the semigroup $H(R)$. (This is an extension of the well known fact that a simple Artinian ring has this property). The proof uses the following useful lemma:

$$\begin{array}{l}\text{If a ring } R \text{ graded by a semigroup } S \text{ is semilocal and} \\ a \in R_s \text{ is not nilpotent, then } s \text{ is a periodic element of } S.\end{array} \tag{$*$}$$

ii) next, use the assumption on R together with the fact that a 0-simple semigroup which is strongly π-regular must be completely 0-simple. □

We use the above proposition to reduce the study of R to that of S-graded rings whose homogeneous elements form a semigroup with a nicer structure. The largest homogeneous ideal contained in the Jacobson radical $J(R)$ of $R = \bigoplus_{s \in S} R_s$ will be denoted by $J_{\mathrm{gr}}(R)$. Let N be the maximal nil ideal of $H(R)$ for a left perfect S-graded ring R. Then N is left T-nilpotent. So the subring $\mathbf{Z}\{N\}$ generated by N in R is a left T-nilpotent homogeneous ideal of R. We denote this ring by J_1. It follows that $J_1 = J_{\mathrm{gr}}(R)$. By the proposition, the semigroup $H(R)/N$ has a minimal completely 0-simple ideal I. Let $I_1 = (\mathbf{Z}\{I\} + J_{\mathrm{gr}}(R))/J_{\mathrm{gr}}(R)$ be the image of $\mathbf{Z}\{I\}$ in the S-graded ring $R' = R/J_{\mathrm{gr}}(R)$. Then the image of I in I_1/J_1 is a completely 0-simple ideal in $H(I_1/J_1)$. Continuing this process we obtain a chain

$$I_0 = \{0\} \subseteq J_1 \subseteq I_1 \subseteq J_2 \subseteq I_2 \subseteq \cdots \subseteq J_m \subseteq I_m \subseteq J_{m+1} = R$$

of homogeneous ideals of R, with m less than or equal to the number of simple blocks in $R/J(R)$, and such that each J_i/I_i is left T-nilpotent and each $H(I_i/J_i)$ has a completely 0-simple ideal L that generates I_i/J_i as an additive group, and in particular $\mathrm{supp}(I_i/J_i) = \mathrm{supp}(L)$.

Next, we need the following technical lemma.

Lemma 2. *Let G be a group and let R be a semilocal G-graded ring with $J(R_1)$ nil. If L is a homogeneous left ideal of R with $L \cap R_1 \subseteq J(R_1)$, then $L \subseteq J(R)$.*

In particular, if $R_{g^{-1}} r_g \subseteq J(R_1)$, then $r_g \in J(R)$. Hence $J(R_1) \subseteq J(R)$ and $J(R_1) = J(R) \cap R_1$.

Proof. Using $(*)$, first show that $H(R)$ is nil modulo R_1. Then the assumptions imply that $\bigcup_{g \in G} L_g$ must be nil, and consequently $L \subseteq J(R)$.

If $R_{g^{-1}} r_g \subseteq J(R_1)$, then the above applied to $L = R r_g$ implies that $r_g \in J(R)$. This yields $J(R_1) \subseteq J(R) \cap R_1$ (the converse is well known). □

The main result in the group-graded case shows that the rings in question are built of group crossed products.

Theorem 3. ([5]) *Let G be a group and let R be a G-graded ring with $J(R_1) \subseteq J_{\mathrm{gr}}(R) \neq R$. Then the G-graded ring $R/J_{\mathrm{gr}}(R)$ is semilocal (respectively left perfect, semiprimary, left Artinian) if and only if the following conditions are satisfied*

1. $R_1/J(R_1) \cong M_{n_1}(D_1) \times \cdots \times M_{n_r}(D_r)$, *each D_i a division ring,*
2. *for any complete set of orthogonal idempotents e_u of $R_1/J(R_1)$, $1 \leq u \leq q = n_1 + \cdots + n_r$, the ring $R' = R/J_{\mathrm{gr}}(R)$ is the direct product of matrix rings over crossed products over some periodic subgroups H_i of G*

$$M_{m_1}(D_{(1)} * H_1) \times \cdots \times M_{m_l}(D_{(l)} * H_l),$$

   with $q = m_1 + \cdots + m_l$, and for each $1 \leq i \leq l$,

$$M_{m_i}(D_{(i)} * H_i) = R'e_jR', \quad D_{(i)} = e_j(R_1/J(R_1))e_j.$$

   for some $1 \leq j \leq q$. In particular these matrix rings are homogeneous subrings.
3. *each crossed product $D_{(i)} * H_i$ is semilocal (respectively left perfect, semiprimary, left Artinian).*

Idea of the proof. We can assume that $J_{\mathrm{gr}}(R) = 0$. Then R_1 has a unity and one verifies that this is the unity of R as well. Write $1 = \sum e_i$ for primitive orthogonal idempotents e_i of R_1. Each $e_iR_1e_i$ is a division ring. But $H(e_iRe_i) \setminus \{0\}$ is cancellative and Lemma 2 implies that the grading is non-degenerate ($R_{g^{-1}}r_g \neq 0 \neq r_gR_{g^{-1}}$ for $0 \neq r_g \in R_g, g \in G$). This allows us to show that $e_iRe_i \cong e_iR_1e_i * H_i$ for a group H_i. R semilocal implies that H_i is periodic. Glue the pieces together, writing first R as a generalized matrix ring $\sum_{i,j} e_iRe_j$, to get the structure of R. □

The hypothesis $J(R_1) \subseteq J_{\mathrm{gr}}(R)$ is always satisfied for a left perfect ring R, because of Lemma 2 and Lemma 3 below.

Clearly, the condition $J_{\mathrm{gr}}(R) \neq R$ is satisfied for a group graded ring with unity. For an arbitrary group graded ring R one obtains from the theorem that R is left perfect if and only if $J_{\mathrm{gr}}(R)$ is left T-nilpotent and $R/J_{\mathrm{gr}}(R)$ satisfies conditions (1)–(3) if $R \neq J_{\mathrm{gr}}(R)$.

An extension of Theorem 3 to semigroup gradations is also known, [5]. The following lemma is easy in case of rings with unity, but creates unexpected problems in the general case, which in particular is needed to handle semigroup gradations.

Lemma 3. *Let H be a subgroup of S, R be an S-graded ring. If R is semilocal (left perfect), then so is the ring $R_H = \bigoplus_{h \in H} R_h$.*

6. The Radical of a Graded Ring

The homogeneity problem for the radical of R asks whether $J_{\mathrm{gr}}(R) = J(R)$ under some conditions on R and on S. The theorems of Bergman on $\mathbf{Z}$-graded rings and of Cohen and Montgomery on algebras over a field of characteristic zero graded by a finite group G are the most basic results in this area. For these and various generalizations we refer to [8].

We say that S is a u.p. semigroup if for every nonempty finite subsets $A, B \subseteq S$ there exists an element of AB which has a unique presentation in the form ab, where

$a \in A, b \in B$, cf. [12]. Such an S is cancellative. If it has a group of quotients, then this group must be torsion-free. It is an open problem whether the radical of any ring graded by a u.p. semigroup is homogeneous.

Our main result reads as follows.

Theorem 4. ([9]) *The Jacobson radical of an S-graded ring R is homogeneous if either of the following holds*

1. *S is a cancellative semigroup and R is a PI-algebra over a field of characteristic zero.*
2. *S is a u.p.-semigroup and at least one of the following conditions is satisfied:*
   (a) *all nil subsemigroups of $H(R/J_{\mathrm{gr}}(R))$ are locally nilpotent;*
   (b) *every nil subsemigroup of every right primitive homomorphic image of R is locally nilpotent;*
   (c) *for every minimal prime ideal P of R, the ring R/P is a domain or embeds into a matrix ring over a skew field.*

2) covers in particular the case where R is PI, semilocal, or it is right Goldie modulo the prime radical. The same proof shows also that the prime radical of R is homogeneous in any of the above cases.

We will outline some ideas of the proof. For an element $x \in R$ by $H(x)$ we denote the subsemigroup of $H(R)$ generated by the set of homogeneous components of x. The first lemma, roughly speaking, allows us to consider two cases only: the case of a group gradation and the case where the study of quasi inverses of elements $x \in J(R)$ requires only the ring generated by $H(x)$.

Lemma 4. *Let S be a cancellative semigroup, R an S-graded ring, I the ideal of nonunits of S and $G = S \setminus I$ (if S has no identity, then $S = I$). Assume that $x + y = yx$ for some $x \in R_I, y \in R$. Then $y \in A$, where $A = \mathbf{Z}\{H(x)\}$ is the subring generated by $H(x)$.*

An element $a \in R$ is called rigid if $xay = 0$ implies $xa_ty = 0$ for every $t \in \mathrm{supp}(a)$ and $x, y \in h(R)$. This is a useful notion, derived from the often exploited properties of nonzero elements of shortest length in a given ideal of R. Clearly, the idea is to consider the graded ring $R' = R/J_{\mathrm{gr}}(R)$, and a nonzero element of shortest length in $J(R')$ (supposing $J(R') \neq 0$.)

Lemma 5. *Let S be a cancellative semigroup, R an S-graded ring, r a rigid element of R, and let M be the multiplicative semigroup generated by the set $H(r)$. If M contains 0, then M is nilpotent.*

So, the other case to consider is when $H(r)$ has no zero divisors, and hence deg is a homomorphism on $H(r)$. We continue with the PI case only, which is the easiest to explain.

A semigroup S is said to be *permutational* if there exists $n > 1$ such that, for any n elements $x_1, \dots, x_n$ of S, their product can be rearranged as $x_1 \dots x_n = x_{\sigma(1)} \dots x_{\sigma(n)}$ for a nontrivial permutation σ.

Lemma 6. *Let R be an S-graded PI-ring, and let T be a multiplicative subsemigroup of $H(R)$. If T does not contain zero, then $\mathrm{supp}(T)$ is a permutational subsemigroup of S.*

Proof. Apply a multilinear identity in R to elements of $H(R)$. □

Now, the point is that cancellative permutational semigroups have finite-by-abelian-by-finite groups of quotients, [12], so known results on gradations by such groups (extending the theorems of Bergman and of Cohen and Montgomery, [8]) can be applied.

Gluing the above two cases, in view of the reduction to semigroups of matrices (possible because of the assumptions on R), allows us to apply the ideas of Sections 1, 2, 3. They are used to show that $J(R')$ contains nonzero homogeneous elements, contradicting the definition of R'.

7. Principal Ideal Semigroup Algebras are PI

The next two sections are concerned with semigroup algebras $K[S]$ viewed as S-graded rings with components $Ks, s \in S$. We assume that $K[S]$ has a unity and that every left ideal of $K[S]$ is principal. The main problem is to find a structural description of such algebras and of the underlying semigroups. The group ring case is our point of departure.

Theorem 5. ([16]) *Let G be a group and K a field. The following conditions are equivalent:*

1. *$K[G]$ is a principal left ideal ring;*
2. *if $\operatorname{char} K = 0$ then G is finite or finite-by-(infinite cyclic);*
   if $\operatorname{char} K = p > 0$ then G is (finite p')-by-(cyclic p) or G is (finite p')-by-(infinite cyclic).

Note that such rings are finitely generated K-algebras, they are PI and of Gelfand-Kirillov dimension 1. The following extension to cancellative semigroups has been recently obtained.

Theorem 6. ([7]) *Let T be a cancellative monoid and K a field of characteristic p (not necessarily nonzero). The following conditions are equivalent:*

1. *$K[T]$ is a principal left ideal ring;*
2. *T is a semigroup satisfying one of the following conditions:*
   (a) *T is a group satisfying the conditions of Theorem 5;*
   (b) *T contains a finite p'-subgroup H and a nonperiodic element x such that $xH = Hx$, $T = \bigcup_{i \geq 0} Hx^i$ and the central idempotents of $K[H]$ are central in $K[T]$.*

The first major step of our approach to the general case is the following result.

Theorem 7. ([6]) *Let $K[S]$ be a principal left ideal ring. Then $K[S]$ satisfies a polynomial identity.*

Idea of the proof. For simplicity we discuss only the case where $S \subseteq M_n(D)$ and the K-subalgebra A generated by S is an order in $M_n(D)$. Consider a maximal subgroup $G \cong GL_j(D)$ of $M_n(D)$. General techniques allow to lift (left) ideals of $K[gp(S \cap G)]$ to (left) ideals of $K[S]$ [12], Lemma 7.21. The assumption on $K[S]$ implies then that $K[gp(S \cap G)]$ is left Noetherian. The main difficulty is in showing that this group ring actually is a principal ideal ring. Then it is PI by Theorem 5. The structure of $S \subseteq M_n(D)$ together with the Goldie condition for A allow to show that $K[S]$ is PI as well. □

Corollary 1. *If $K[S]$ is a principal left ideal ring, then S is finitely generated and $K[S]$ embeds into a matrix ring $M_n(F)$ over a field extension F of K. In particular, S is a linear semigroup. The groups associated to S are either finite or finite-by-(infinite cyclic).*

Proof. $K[S]$ is finitely generated because so is every left Noetherian PI semigroup algebra, [12], Theorem 19.14. A theorem of Anan'in implies that a finitely generated left Noetherian PI-algebra embeds into a matrix ring over a (finitely generated) commutative algebra A, [1]. A classical result of Malcev then implies that $K[S]$ embeds into matrices over a field, [10]. Theorem 5 is used to get the remaining assertion. □

The following is an easy consequence.

Corollary 2. *If $K[S]$ is a principal left ideal ring, then the Gelfand-Kirillov dimension of $K[S]$ is equal to its classical Krull dimension and it is 0 or 1. In the former case S is finite. Moreover, every prime Artinian homomorphic image of $K[S]$ is finite dimensional over K.*

8. Principal (Left and Right) Ideal Rings

The following construction plays a crucial role in our final result.

Example 1. Assume that H is a finite group whose order is not divisible by the characteristic of K. Let T be a monoid with group of units H such that $T = \bigcup_{i\geq 0} Hx^i$ for some $x \in T$, and either this union is disjoint or $x^n = \theta$ for some $n \geq 1$. Assume also that $Hx = xH$ and the central idempotents of $K[H]$ commute with x. Then $K[T]$ is a principal ideal ring. In fact, one can check that $K[T] \cong A \oplus B$, where A is a semisimple Artinian ring and B is a finite direct sum of algebras of the type $C[x,\sigma]/(x^k)$ for a semisimple Artinian ring C, an automorphism σ of C and some $k \geq 1$. Note that this extends the construction of 2 b) in Theorem 6.

Let C be a prime homomorphic image of a principal left ideal ring $K[S]$. We use the fact that C is a finitely generated prime PI algebra of classical Krull dimension 0 or 1. First, we claim that, if an ideal J of C is idempotent, then either $J = 0$ or $J = C$. Let $J = Ca$. It is known that $CaCa = Ca^2$, so $J = Ca^2 = Ja$. Therefore $ba = a$ for some $b \in J$. Thus $bax = ax$ for all $x \in C$ and consequently b is a left identity of J. Therefore it is an identity of J because C is prime, so $J = 0$ or $J = C$. Secondly, if J is a nonzero ideal of C, then C/J is of finite dimension over K. Indeed, the classical Krull dimension of C/J is 0. Hence $GK(C/J) = \mathrm{cl}\,K\dim(C/J) = 0$.

The former observation will allow us to decompose certain algebras into a direct product of its simpler blocks. The latter leads to the strategy of considering the finite dimensional case first, and then applying this to the general case.

It is well known that finite dimensional algebras which are principal left ideal rings also are principal right ideal rings. Moreover, they are exactly finite direct sums of matrix rings over local algebras whose radical is a principal ideal, [4], Theorem 9.4.1. This decomposition of $K[S]$ turns out to be a refinement of a decomposition on the semigroup level, and leads to a complete description of $K[S]$ as follows.

Proposition 2. *Let S be a finite semigroup and N a nilpotent ideal of S, or $N = \emptyset$, such that S/N is completely 0-simple. Then the contracted semigroup algebra $K_0[S]$ is a principal ideal ring if and only if one of the following conditions is satisfied*

1. *$S \cong \mathcal{M}(G, n, n, Q)$ for a finite group G satisfying the conditions of Theorem 5 and for a sandwich matrix Q that is invertible over $K[G]$;*
2. *$S \cong \mathcal{M}(T, n, n, Q)$ for a finite monoid T satisfying the conditions of Example 1 and for a sandwich matrix Q that is invertible over $K_0[T]$.*

Proposition 3. *Let S be a finite semigroup. Then $K[S]$ is a principal (left) ideal ring if and only if S has a chain of ideals $I_1 \subset I_2 \subset \cdots \subset I_t = S$ such that I_1 and every factor I_j/I_{j-1} is of the type described in Proposition 2.*

The remaining step is to show that principal ideal rings $K[S]$ can be approached as inverse limits of finite dimensional semigroup algebras. This leads to our main result.

Theorem 8. ([6]) *The following conditions are equivalent:*

1. *$K[S]$ is a principal (left an right) ideal ring;*
2. *there exists an ideal chain*

$$I_1 \subset \cdots \subset I_t = S$$

such that I_1 and every factor I_j/I_{j-1} is of the form $\mathcal{M}(T, n, n, P)$ for an invertible over $K_0[T]$ sandwich matrix P, and one of the following conditions holds:

(a) *T is a group of the type described in Theorem 5;*

(b) *T is a monoid of the type described in Example 1.*

In case the equivalent conditions are satisfied it follows that

$$K_0[S] \cong K_0[I_1] \oplus K_0[I_2/I_1] \oplus \cdots \oplus K_0[I_t/I_{t-1}]$$

and each component is isomorphic to some $M_n(K[T])$. Moreover, $K[S]$ is a finite module over its centre, which is finitely generated.

It may be checked that each of the semigroups I_1, I_j/I_{j-1}, yielding the structural blocks in the above theorem, is of the form $\mathcal{M}(T', n, n, Q)$, where T' is a homomorphic image of a cancellative monoid T satisfying the conditions of Theorem 6. In case T is not a group, one of the conditions for $K[T]$ to be of such a type is that the central idempotents of the semisimple ring $K[H]$ are central in $K[T]$, where H is a finite subgroup of $T = \bigcup_{i \geq 0} Hx^i$. In case $K[H]$ is split (that is, a direct product of matrix rings over K) this means that the centre of $K[H]$ is contained in the centre of $K[G]$, where G is the group of quotients of S. Therefore, this condition can be given an intrinsic formulation in terms of S only: the S-conjugacy class of every element $h \in H$ coincides with the conjugacy class of h in H.

The methods used in the proof of Theorem 8 allow to establish the left-right symmetry of the principal ideal condition for any semiprime algebra $K[S]$. This leads to the following result.

Theorem 9 [6]. *Let S be a semigroup and K a field. Then $K[S]$ is a semiprime principal left ideal ring if and only if there exists an ideal chain*

$$I_1 \subset \cdots \subset I_t = S$$

such that I_1 *and every factor* I_j/I_{j-1} *is of the form* $\mathcal{M}(T,n,n,P)$ *for an invertible over* $K_0[T]$ *sandwich matrix* P *and a monoid* T *such that*

1. *either* T *is an infinite group as in Theorem* 5,
2. *or* $T = \bigcup_i Hx^i$ *is of the type described in Example* 1 *and such that for every primitive central idempotent* $e \in K[H]$, *either* $K[H]ex = 0$ *or* $K[H]ex^i \neq 0$ *for all* $i \geq 1$.

Moreover, if the equivalent conditions are satisfied, then $K[S]$ *is a principal right ideal ring and the direct sum decomposition of Theorem* 8 *occurs.*

If G is as in Theorem 5 and $K[G]$ is prime, then G is trivial or infinite cyclic. If T is as in Example 1 and $K_0[T]$ is prime, then H is trivial. Therefore, the following is an easy consequence of Theorem 9.

Corollary 3. $K_0[S]$ *is a prime principal left ideal ring if and only if*

$$S \cong \mathcal{M}(\{1\},n,n,Q), \quad S \cong \mathcal{M}(\langle x\rangle,n,n,Q)$$

or

$$S \cong \mathcal{M}(\langle x,x^{-1}\rangle,n,n,Q)$$

where the matrix Q *is invertible in* $M_n(K)$, $M_n(K[x])$ *or* $M_n(K[x,x^{-1}])$ *respectively. Hence,* $K_0[S] \cong M_n(K)$, $M_n(K[x])$, *or* $M_n(K[x,x^{-1}])$.

9. Concluding Remarks

A number of other important and easy-to-state problems have been recently treated via representations in simple Artinian rings. In particular, several problems concerning presentability of idempotents as a sum of nilpotents in an algebra over a field of characteristic zero have been solved within the algebras $M_4(D)$ [21]. Clearly, the methods go beyond the multiplicative structure, but they carry a lot of the flavour of Sections 2, 3.

This, together with [20], magnifies the feeling of an abundance of algebraic structures within the algebras $M_n(D)$ (compared to the case where D is a field), providing more motivation for a study of skew linear representations of associative algebras, [22]. Also, the structure theorem for $M_n(D), n \geq 4$, leads to certain new invariants for division rings D, and to open problems concerning them. Namely, to every $j < n$ one can associate the supremum of the cardinalities of the sets of uniform components of rank j of subsemigroups of $M_n(D)$.

The theory of linear algebraic (Zariski closed) semigroups, aiming at a generalization of results known for algebraic groups, has been developed mainly by Putcha and Renner, see [18]. The nicest case of reductive monoids has been recently given a special attention, leading in particular to combinatorics analogous to that on $GL_n(\mathbf{F}_q)$, [23].

The material of Sections 2, 3, 4 is the source of the forthcoming monograph [14].

References

[1] A.Z. Anan'in, An intriguing story about representable algebras, in: Ring Theory 1989, Weizmann, 1989, pp. 31–38.

[2] M. Artin and J.T. Stafford, Noncommutative graded domains with quadratic growth, *Invent. Math.* **122** (1995), 231–276.
[3] A.H. Clifford and G.B. Preston, *Algebraic Theory of Semigroups*, Amer. Math. Soc., 1964.
[4] Ju.A. Drozd and V.V. Kirichenko, *Finite Dimensional Algebras*, Springer-Verlag, 1994.
[5] E. Jespers and J. Okniński, Descending chain conditions and graded rings, *J. Algebra* **178** (1995), 458–479.
[6] E. Jespers and J. Okniński, Semigroup algebras that are principal ideal rings, *J. Algebra* **183** (1996), 837–863.
[7] E. Jespers and P. Wauters, Principal ideal semigroup rings, *Comm. Algebra* **23** (1995), 5057–5076.
[8] G. Karpilovsky, *The Jacobson Radical of Classical Rings*, Longman, 1991.
[9] A.V. Kelarev and J. Okniński, On the radical of graded PI rings and related classes of rings, *J. Algebra* **186** (1996), 818–830.
[10] A.I. Malcev, On representation of infinite algebras, *Mat. Sb.* **13** (1943), 263–285. (Russian)
[11] J.C. McConnell and J.C. Robson, *Noncommutative Noetherian Rings*, Wiley, 1987.
[12] J. Okniński, *Semigroup Algebras*, Marcel Dekker, 1991.
[13] J. Okniński, Linear representations of semigroups, in: Monoids and Semigroups with Applications, World. Sci., 1991, pp. 257–277.
[14] J. Okniński, Semigroups of Matrices, in preparation.
[15] J. Okniński and A. Salwa, Generalized Tits alternative for linear semigroups, *J. Pure Appl. Algebra* **103** (1995), 211–220.
[16] D.S. Passman, Observations on group rings, *Comm. Algebra* **5** (1977), 1119–1162.
[17] D.S. Passman, *Infinite Crossed Products*, Academic Press, 1989.
[18] M.S. Putcha, *Linear Algebraic Monoids*, London Math. Soc. Lect. Note Ser. 133, Cambridge Univ. Press, 1988.
[19] J.M. Rosenblatt, Invariant measures and growth conditions, *Trans. Amer. Math. Soc.* **193** (1974), 33–53.
[20] A. Salwa, Structure of skew linear semigroups, *Int. J. Algebra and Comput.* **3** (1993), 101–113.
[21] A. Salwa, Representing idempotents as a sum of two nilpotents — an approach via matrices over division rings, submitted.
[22] A.H. Schofield, *Representations of Rings over Skew Fields*, London Math. Soc. Lect. Note Ser. 92, Cambridge Univ. Press, 1985.
[23] L. Solomon, An introduction to reductive monoids, in: Semigroups, Formal Languages and Groups, NATO ASI Series, Kluwer, 1995, pp. 295–352.
[24] J. Tits, Free subgroups of linear groups, *J. Algebra* **20** (1972), 250–270.
[25] E.I. Zelmanov, Semigroup algebras with identities, *Sib. Math. J.* **18** (1977), 787–798. (Russian)
[26] J.J. Zhang, On Gelfand-Kirillov transcendence degree, *Trans. Amer. Math. Soc.* (to appear).

J. Okniński
Institute of Mathematics
Warsaw University
e-mail: okninski@mimuw.edu.pl

Canadian Mathematical Society
Conference Proceedings
Volume **22**, 1998

Semiprimitivity of Group Algebras: Past Results and Recent Progress

Donald S. Passman

Abstract. Let K be a field and let G be a multiplicative group. The group ring $K[G]$ is an easily defined, rather attractive algebraic object. As the name implies, its study is a meeting place for two essentially different algebraic disciplines. Indeed, group ring results frequently require a blend of group theoretic and ring theoretic techniques. A natural, but surprisingly elusive, group ring problem concerns the semiprimitivity of $K[G]$. Specifically, we wish to find necessary and sufficient conditions on the group G for its group algebra to have Jacobson radical equal to zero. More generally, we wish to determine the structure of the ideal $\mathcal{J}K[G]$. In the case of infinite groups, this problem has been studied with reasonable success during the past 45 years, and our goal here is to survey what is known. In particular, we describe some of the techniques used, discuss a number of the results which have been obtained, and mention several tantalizing conjectures.

§1. Introduction

Consider the following construction of the polynomial ring in two variables, say x and y, over a field K. To start with, form the set $S = \{x^a y^b \mid a, b = 0, 1, 2, \ldots\}$ of monomials in x and y, and define multiplication in S by $x^a y^b \cdot x^c y^d = x^{a+c} y^{b+d}$. In this way, we see that S becomes an associative semigroup with identity element $1 = x^0 y^0$. Next, let $K[x, y] = K[S]$ be the K-vector space with basis consisting of the elements of S. In other words, every element of $K[x, y]$ is a formal finite sum $\sum k_{a,b} x^a y^b$ with coefficients $k_{a,b} \in K$. Of course, the addition in $K[x, y]$ is the usual vector space addition, and multiplication in $K[x, y]$ is defined distributively using

---

AMS Subject Classification (1991). Primary: 16S24; Secondary: 16N20, 20F24, 20F50.

Keywords and Phrases: Group algebra, semiprimitivity, Jacobson radical.

Research supported in part by NSF Grant DMS-9224662.

the multiplication in S. Since the associative law for multiplication in S clearly carries over to $K[S]$, it follows that $K[x,y]$ is an associative K-algebra. Similarly, we could construct the Laurent polynomial ring $K[x,y,x^{-1},y^{-1}]$ by taking S to be the multiplicative group $S = \{x^a y^b \mid a,b = 0, \pm 1, \pm 2, \dots\}$ and again forming $K[S]$. Indeed, this is our first example of a group ring.

More generally, let K be a field and let G be any multiplicative group. Then the *group algebra* or *group ring* $K[G]$ is a K-vector space with basis consisting of the elements of G. Thus every element of $K[G]$ is a formal finite sum

$$\alpha = \sum_{g \in G} k_g g$$

with coefficients $k_g \in K$. Again, addition in $K[G]$ is the obvious vector space operation, and we define multiplication distributively using the given multiplication of G. In this way, $K[G]$ becomes an associative K-algebra, with structure highly dependent on the nature of G. Basic references for group algebras include the books [MZ], [Pa], [P3], [P10], [Se1] and [Se2].

As is well known, group rings are important tools in both group theory and ring theory. For example, they provide the correct framework to study and understand the ordinary and modular character theory of finite groups. Furthermore, when G is a polycyclic-by-finite group, then $K[G]$ is a right and left Noetherian K-algebra and hence it is a useful testing ground for the rich theory of noncommutative Noetherian rings. In turn, the module theory of the latter group algebra can feed back into group theory to yield information on the structure of abelian-by-polycyclic groups. But, group rings are more than just useful tools. They are easily defined, rather attractive algebraic objects which are worthy of being considered in their own right. Their study is necessarily ring theoretic in nature, but the techniques and proofs exhibit a strong group theoretic flavor. The goal of this paper is to survey the progress made on a rather elusive group ring problem.

If R is an associative ring with 1, then a (right) *R-module* V is just a right R-vector space. Thus V is an additive abelian group which admits right multiplication by R, and such that this scalar multiplication satisfies the usual axioms. Of course, these rules are precisely equivalent to the existence of a natural ring homomorphism $\theta_V \colon R \to \operatorname{End}(V)$, where $\operatorname{End}(V)$ is the ring of endomorphisms of the additive abelian group V. We say that $V \neq 0$ is *irreducible* if V has no proper R-submodule. In other words, the irreducible R-modules are the natural analogs of the 1-dimensional vector spaces over fields. For convenience, we let $\operatorname{Irr}(R)$ denote the set of all such irreducible R-modules.

A ring R is said to be *primitive* if it has a faithful irreducible module. In other words, R is primitive if there exists $V \in \operatorname{Irr}(R)$ with θ_V a one-to-one map. Such rings have a nice, rather natural structure; they are dense sets of linear transformations over division rings. Unfortunately, primitive rings are fairly scarce, so the next best situation is to study the ring R by looking at all its irreducible modules. But there is still a fundamental obstruction here, namely

$$\mathcal{J}R = \bigcap_{V \in \operatorname{Irr}(R)} \ker \theta_V = \{ r \in R \mid Vr = 0 \quad \text{for all} \quad V \in \operatorname{Irr}(R) \}.$$

This characteristic ideal is called the *Jacobson radical* of R, and we say that R is *semiprimitive* precisely when $\mathcal{J}R = 0$. Thus R is semiprimitive if and only if it is a subdirect product of primitive rings. In particular, such rings are reasonably well understood.

It is therefore of some interest and importance to determine those groups G with semiprimitive group algebras $K[G]$. More generally, we would like to describe the structure of the Jacobson radical $\mathcal{J}K[G]$ for any group G. In the case of finite groups, the semiprimitivity problem has the following classical solution, dating from the work of Maschke in 1898.

Theorem 1.1. ([M]) *Let G be a finite group and let K be a field.*

i. *If $\operatorname{char} K = 0$, then $K[G]$ is semiprimitive.*
ii. *If $\operatorname{char} K = p > 0$, then $K[G]$ is semiprimitive if and only if G has no elements of order p.*

The goal now is to extend this result, or some variant of it, to the case of infinite groups, and in this survey, which is a revised and updated version of [P18] and [P19], we will discuss the progress which has been made in this direction.

§2. Fields of Characteristic 0

It is not surprising that the early advances on the semiprimitivity problem for infinite groups concerned fields of characteristic 0, and indeed the field C of complex numbers. The first significant result appeared in 1950, with a proof using analytic methods, including the spectral norm and the auxiliary norm of $C[G]$.

Theorem 2.1. ([R]) *If C is the field of complex numbers, then every group algebra $C[G]$ is semiprimitive.*

This result intrigued a number of ring theorists who rightly felt that it should have an algebraic proof. Thus, for example, the semiprimitivity problem for fields of characteristic 0 appeared in the Ram's Head Inn problem list [K1] (see also [K2]), and an algebraic argument for Theorem 2.1 was quickly discovered. It is instructive to consider some of the ingredients of this new proof. Recall that an ideal I of any ring R is said to be *nil* if all elements of I are nilpotent. Since every nil ideal of R is contained in $\mathcal{J}R$, a first step in proving that $K[G]$ is semiprimitive might be to show that it has no nonzero nil ideal. In this direction we have

Lemma 2.2. *Let K be a subfield of the complex numbers which is closed under complex conjugation. If G is any group, then $K[G]$ has no nonzero nil ideal.*

Proof. Define a map $^*\colon K[G] \to K[G]$ by

$$\Big(\sum_g k_g g\Big)^* = \sum_g \bar{k}_g g^{-1}$$

where $^-$ indicates complex conjugation. It is easy to see that $(\alpha\beta)^* = \beta^*\alpha^*$, $\alpha^{**} = \alpha$, and $(\alpha+\beta)^* = \alpha^* + \beta^*$. Furthermore, if $\alpha = \sum_g k_g g$, then the identity coefficient of $\alpha\alpha^*$ is equal to $\sum_g k_g\bar{k}_g = \sum_g |k_g|^2$. Hence $\alpha\alpha^* = 0$ if and only if $\alpha = 0$.

Let I be a nonzero ideal of $K[G]$ and choose $0 \neq \alpha \in I$. Then, by the above, $\beta = \alpha\alpha^*$ is a nonzero element of I, and β is easily seen to be $*$-symmetric. In other words, any nonzero ideal of $K[G]$ contains a nonzero $*$-symmetric element.

Next, we claim that 0 is the unique $*$-symmetric nilpotent element. Indeed, if γ is $*$-symmetric and nilpotent, then so is any power of γ. Thus it suffices to assume that $\gamma^2 = 0$. But then $0 = \gamma^2 = \gamma\gamma^*$, so $\gamma = 0$ as required, and the result follows immediately from the latter two observations. □

The second ingredient holds over any field. Note that if H is a subgroup of G, then $K[H]$ is naturally embedded in $K[G]$. Indeed, this is just the group ring analog of the obvious polynomial ring inclusion $K[x] \subseteq K[x,y]$. Furthermore, since $K[G]$ is a free right and left $K[H]$-module, using coset representatives for H in G as a free basis, we have

Lemma 2.3. *Let K be any field and let H be a subgroup of G.*

i. *If W is an irreducible $K[H]$-module, then there exists an irreducible $K[G]$-module V with W a submodule of V_H, the restriction of V to $K[H]$.*
ii. $\mathcal{J}K[G] \cap K[H] \subseteq \mathcal{J}K[H]$.

The remainder of the argument is of less interest. To start with, the Hilbert Nullstellensatz asserts that if A is a finitely generated commutative algebra over a field K, then $\mathcal{J}A$ is a nil ideal. Furthermore, recall that there is a trivial proof of this result in case K is nondenumerable. Indeed, the same proof shows, without the commutativity assumption, that if A is a countable dimensional algebra over a nondenumerable field, then $\mathcal{J}A$ is nil. In particular, it follows from this and Lemma 2.2 that if H is a countable group, then the complex group algebra $C[H]$ is semiprimitive. Finally, if G is any group and if $\alpha \in \mathcal{J}C[G]$, then there exists a finitely generated and hence countable subgroup H of G with $\alpha \in C[H]$. But then Lemma 2.3(ii) yields

$$\alpha \in C[H] \cap \mathcal{J}C[G] \subseteq \mathcal{J}C[H] = 0,$$

and Theorem 2.1 is proved.

Much more important is the later work of Amitsur on the behavior of the radical under field extensions. If A is a K-algebra and if F is a field containing K, then we denote the F-algebra $F \otimes_K A$ by A^F. Thus A^F is the largest ring generated by its commuting subrings F and A, with the two copies of K identified.

Theorem 2.4. ([A1]) *Let $F \supseteq K$ be fields and let A be a K-algebra.*

i. *$\mathcal{J}(A^F) \cap A \subseteq \mathcal{J}A$ with equality when F/K is algebraic.*
ii. *If F/K is a finite separable extension, then $\mathcal{J}(A^F) = F \otimes_K \mathcal{J}A$.*
iii. *If F is a nontrivial purely transcendental extension of K, then $\mathcal{J}(A^F) = F \otimes_K I$ for some nil ideal I of A.*

Since $K[G]^F = F \otimes_K K[G] = F[G]$, the preceding result and Lemma 2.2 applied to the field Q of rational numbers yield

Theorem 2.5. ([A2]) *Let K be a field of characteristic 0 so that K contains the rational numbers Q, and let G be an arbitrary group.*

i. *If K/Q is not algebraic, then $K[G]$ is semiprimitive.*
ii. *If K/Q is algebraic, then $\mathcal{J}K[G] = K \otimes_Q \mathcal{J}Q[G]$ and $K[G]$ has no nonzero nil ideal.*

In particular, the semiprimitivity problem for algebraic extensions of Q reduces to Q itself. Presumably $Q[G]$ is always semiprimitive, but unfortunately the above result marks the extent of our knowledge. Indeed, there has been no significant progress on the characteristic 0 problem since Theorem 2.5 appeared in 1959. We remark that the semiprimitivity of $Q[G]$ would follow quite easily if one knew that finitely generated algebras necessarily have nil Jacobson radicals. However, as was shown in [B], this is not always the case.

§3. Fields of Characteristic $p > 0$

Now let us turn to modular fields and assume for the remainder of this paper that $\operatorname{char} K = p > 0$. In view of Theorem 1.1, it is reasonable to suppose $K[G]$ is semiprimitive if and only if G is a *p'-group*, that is a group with no elements of order p. One direction of this is most likely true, but as we will see, the other direction is decidedly false. We begin with an interesting trace argument.

For any group G, let $\operatorname{tr}\colon K[G] \to K$ be the map which reads off the identity coefficient, so that $\operatorname{tr}(\sum k_g g) = k_1$. Then tr is obviously a K-linear functional, and it is easy to see that $\operatorname{tr} \alpha\beta = \operatorname{tr} \beta\alpha$ for all $\alpha, \beta \in K[G]$. Next, we note that if A is any K-algebra and if $\alpha_1, \alpha_2, \ldots, \alpha_s \in A$, then

$$(\alpha_1 + \alpha_2 + \cdots + \alpha_s)^{p^n} = \alpha_1^{p^n} + \alpha_2^{p^n} + \cdots + \alpha_s^{p^n} + \beta$$

for some $\beta \in [A, A]$, where the latter subspace is the span of all Lie products $[\gamma, \delta] = \gamma\delta - \delta\gamma$ with $\gamma, \delta \in A$.

Lemma 3.1. *If G is a p'-group, then $K[G]$ has no nonzero nil ideal.*

Proof. Suppose $\alpha = \sum k_g g \in K[G]$ is nilpotent, and choose n sufficiently large so that $\alpha^{p^n} = 0$. Then by the preceding formula,

$$0 = \alpha^{p^n} = \sum_{g \in G} (k_g)^{p^n} g^{p^n} + \beta$$

for some $\beta \in \big[K[G], K[G]\big]$. In particular, since tr annihilates all Lie products, we have $\operatorname{tr} \beta = 0$ and hence

$$0 = \sum_{g \in G} (k_g)^{p^n} \operatorname{tr} g^{p^n}.$$

But G is a p'-group, so $g^{p^n} = 1$ if and only if $g = 1$, and therefore $\operatorname{tr} g^{p^n} = 0$ for all $g \neq 1$. It follows that $0 = (k_1)^{p^n}$, and we conclude that if α is nilpotent, then $0 = k_1 = \operatorname{tr} \alpha$.

Finally, let I be a nil ideal of $K[G]$ and let $\gamma = \sum c_g g \in I$. Then $\gamma x^{-1} \in I$ is nilpotent for any $x \in G$, so the above yields $0 = \operatorname{tr} \gamma x^{-1} = c_x$. Thus $\gamma = 0$, and hence $I = 0$, as required. □

Since any finitely generated field extension of GF(p) is separably generated, it is a simple matter to translate the argument of Theorem 2.5 to this context. In particular, Theorem 2.4 and Lemma 3.1 yield

Theorem 3.2. *Let K be a field of characteristic $p>0$, write $K_0=\mathrm{GF}(p)$, and let G be a p'-group.*

i. *If K/K_0 is not algebraic, then $K[G]$ is semiprimitive.*
ii. *If K/K_0 is algebraic, then $\mathcal{J}K[G]=K\otimes_{K_0}\mathcal{J}K_0[G]$.*

If G is a p'-group, then $K[G]$ is presumably always semiprimitive. But the converse is certainly not true; there are numerous groups G having elements of order p, but with $\mathcal{J}K[G]=0$. For example, we have

(1) $p=2$ and $G=\langle x,y \mid y^{-1}xy=x^{-1}, y^2=1\rangle$ is infinite dihedral.
(2) $G=Z\wr Z_p$ is the wreath product of the infinite cyclic group Z by the cyclic group Z_p of order p.
(3) $G=Z_p\wr Z$ is again a wreath product and has a normal infinite elementary abelian p-subgroup.
(4) $G=\mathrm{FSym}_\infty$, the countably infinite finitary symmetric group.

Note that (1), which appeared in [Wa], was the first such example, and (4) is a result of [F]. Furthermore, we know that the groups in (3) and (4) have primitive group algebras. The real answer to the semiprimitivity problem is most likely

Conjecture 3.3. *Let K be a field of characteristic $p>0$ and let G be a group. Then $\mathcal{J}K[G]\neq 0$ if and only if G has an element of order p "well placed" in G.*

Of course, before this can be proved, we must first determine what "well placed" means. To do this, it is necessary to compute numerous examples. However, we can get some idea of the possible meaning by considering a slightly different problem. For any ring R, let $\mathcal{N}R$ denote the join of all its nilpotent ideals. Thus $\mathcal{N}R$ is a characteristic nil ideal called the *nilpotent radical* of R. For general rings, it is neither nilpotent nor a radical, but we do have $\mathcal{N}R\subseteq\mathcal{J}R$.

Next, if A and B are subgroups of a group G, then the *finitary centralizer* of B in A is defined by

$$\mathbb{D}_A(B)=\big\{a\in A \,\big|\, |B:\mathbb{C}_B(a)|<\infty\big\}.$$

In other words, $\mathbb{D}_A(B)$ consists of all elements of A which almost centralize B, and consequently it is a subgroup of A normalized by $\mathbb{N}_G(A)\cap\mathbb{N}_G(B)$. Corresponding to this finitary centralizer is a finitary center, the *f.c.* or *finite conjugate center* of G, given by

$$\Delta(G)=\mathbb{D}_G(G)=\big\{x\in G \,\big|\, |G:\mathbb{C}_G(x)|<\infty\big\}.$$

Thus $\Delta(G)$ consists of all elements of G having only finitely many conjugates, and it is easy to see that $\Delta=\Delta(G)$ is a characteristic subgroup of G. Furthermore, we let $\Delta^+(G)$ be the set of torsion elements of Δ, that is the elements of finite order in the group. Surprisingly, $\Delta^+=\Delta^+(G)$ is also a characteristic subgroup of G. Indeed, Δ/Δ^+ is a torsion free abelian group and Δ^+ is the join of all finite normal subgroups of G.

The following result is proved using a powerful coset counting argument known as the Δ*-method.*

Theorem 3.4. ([P1], [P2]) *Let $\mathcal{D}(G)$ denote the set of finite normal subgroups of G, and let $\Delta^+=\Delta^+(G)=\langle D\mid D\in\mathcal{D}(G)\rangle$. If $\operatorname{char}K=p>0$, then*

i. $\mathcal{N}K[G]=\mathcal{J}K[\Delta^+]\cdot K[G]$.
ii. $\mathcal{J}K[\Delta^+]=\bigcup_{D\in\mathcal{D}(G)}\mathcal{J}K[D]$.

iii. $\mathcal{N}K[G] \neq 0$ *if and only if* Δ^+ *contains an element of order* p *and hence if and only if* G *has a finite normal subgroup of order divisible by* p.

Note that (i) asserts that $\mathcal{J}K[\Delta^+]$ is contained in $\mathcal{N}K[G]$ and generates it as a right ideal. Furthermore, (iii) is an immediate consequence of parts (i) and (ii), along with Theorem 1.1. Thus "well placed" for this radical means that the element of order p is contained in $\Delta^+(G)$ or equivalently in some finite normal subgroup of G. We close this section with a simple, but quite useful, observation.

Lemma 3.5. ([V]) *If* H *is a normal subgroup of* G *of finite index* n, *then*

$$\mathcal{J}K[G]^n \subseteq \mathcal{J}K[H]\cdot K[G] \subseteq \mathcal{J}K[G].$$

Furthermore, if p *does not divide* n, *then* $\mathcal{J}K[H]\cdot K[G] = \mathcal{J}K[G]$.

In particular, if $JK[H] = 0$ in the above, then $\mathcal{J}K[G]$ is nilpotent and Theorem 3.4 can come into play. With this observation, it is now a simple exercise to prove that $K[G]$ is semiprimitive when $G = Z \wr Z_p$ or when $p = 2$ and G is infinite dihedral.

§4. Solvable Groups and Linear Groups

This brings us to the early 1970's; it was time to compute some examples. We looked for families of groups which were sufficiently diverse to give us meaningful answers, yet simple enough to be dealt with effectively. Two obvious candidates were the families of solvable groups and linear groups. As it turned out, the solvable case yielded the most information and required the more interesting techniques. Therefore we begin our exposition with these groups. We will ignore some earlier special case considerations and just deal with the general problem.

First, recall that G is said to be an *f.c. group* if $G = \Delta(G)$, or equivalently if all conjugacy classes of G are finite. Next, let G be any group, let H be a subgroup of G, and let I be a nonzero ideal of $K[G]$. Then an *intersection theorem* is a result which guarantees that $I \cap K[H] \neq 0$ under suitable assumptions on H, G/H, or I. There are numerous results of this nature in the literature, and Zalesskiĭ proved a marvelous one for solvable groups. Specifically, he showed

Theorem 4.1. ([Z1]) *If* G *is a solvable group, then* G *has a characteristic f.c. subgroup* $\mathfrak{Z}(G)$ *with the following property. If* K *is any field and if* I *is a nonzero ideal of* $K[G]$, *then* $I \cap K[\mathfrak{Z}(G)] \neq 0$.

This *Zalesskiĭ subgroup* $\mathfrak{Z}(G)$ is the f.c. center of a finitary analog of the Fitting subgroup of a finite solvable group. Of course, if G is solvable and if $\mathcal{J}K[G] \neq 0$, then the preceding theorem implies that $\mathcal{J}K[G] \cap K[\mathfrak{Z}(G)] \neq 0$. Thus, the next step in the solution of the semiprimitivity problem for these groups is to deal with this intersection. For this, we require an interesting general result which is a noncommutative analog of the argument used to prove Theorem 2.4(iii).

Lemma 4.2. ([Wa]) *Let* G *be an arbitrary group, let* $H \triangleleft G$, *and suppose that* $\alpha \in \mathcal{J}K[G] \cap K[H]$. *If* x *is any element of* G *of infinite order modulo* H, *then there exists a positive integer* n *such that*

$$\alpha\alpha^x\alpha^{x^2}\cdots\alpha^{x^n} = 0.$$

Here, of course, $\alpha^y = y^{-1}\alpha y$ for any $y \in G$. Now, if x has infinite order modulo H, then so does x^s for any positive integer s. Thus, each such x gives rise to a family of equations, with varying s and varying $n = n(s)$. These *Wallace equations* are rather unwieldy in general. Nevertheless, we were able to use them effectively when H is a solvable f.c. group.

For any element $\beta = \sum b_g g \in K[G]$, let us write $\operatorname{supp}\beta = \{g \in G \mid b_g \neq 0\}$. In particular, the *support* of β is a finite subset of G which is nonempty when $\beta \neq 0$. Furthermore, let quot β denote the set of *quotients* xy^{-1} with $x, y \in \operatorname{supp}\beta$, and for any prime p let p- quot β denote the set of nonidentity elements of quot β having order a power of p. Finally, if L is any subgroup of G, we write

$$\sqrt{L} = \{x \in G \mid x^n \in L \quad \text{for some} \quad n \neq 0\}.$$

Obviously, $\sqrt{L} \supseteq L$, but this *root set* need not be a subgroup of G.

Proposition 4.3. ([HP]) *Let G be an arbitrary group, let H be a normal solvable f.c. subgroup of G, and let $\alpha \in \mathcal{J}K[G] \cap K[H]$. Then*

$$G = \bigcup_{x \in p\text{- quot}\,\alpha} \sqrt{\mathbb{C}_G(x)}.$$

It remained to translate the latter set theoretic union into a more understandable condition. To start with, notice that $G = \sqrt{\langle 1 \rangle}$ is equivalent to G being a periodic group, and therefore the preceding root set equation is related to the Burnside problem. Fortunately, the Burnside problem is quite simple to deal with when G is solvable, and paper [P4] handled this more general situation. Specifically, it showed that if $G = \bigcup_1^n \sqrt{L_i}$ is a finite union of root sets of subgroups and if G is finitely generated and solvable, then some L_i must have finite index in G. By combining all these ingredients, we obtained

Theorem 4.4. ([HP], [P4], [Z1]) *Let G be a solvable group and let K be a field of characteristic $p > 0$. Then $\mathcal{J}K[G] \neq 0$ if and only if $\mathfrak{Z}(G)$ contains an element x of order p which has only finitely many conjugates under the action of each finitely generated subgroup of G.*

Note that the latter condition on x is equivalent to the assertion that if $x \in H \subseteq G$ with H finitely generated, then $x \in \Delta^+(H)$. In particular, if G is a finitely generated group, then this condition reduces to the assumption that $x \in \Delta^+(G)$, and of course this is precisely equivalent to the nonvanishing of $\mathcal{N}K[G]$. In fact, fairly soon afterwards, Zalesskiĭ built upon the preceding, added an additional intersection theorem of sorts, and proved

Theorem 4.5. ([Z2]) *If G is a finitely generated solvable group and K is a field of characteristic $p > 0$, then $\mathcal{J}K[G] = \mathcal{N}K[G]$.*

In particular, in the above situation, we not only know when $K[G]$ is semiprimitive, we actually know the complete structure of $\mathcal{J}K[G]$ by applying Theorem 3.4. Most of these results have now been extended to groups which have a finite normal series with f.c. factor groups. But these generalizations offer nothing new in the way of ideas or techniques. Now let us move on to consider linear groups over a field F. Here there are actually three different problems according to whether $\operatorname{char} F = 0$, char $F = p =$ char K, or char $F = q > 0$ with $q \neq$ char K. The first two cases were completely settled in the 1970's, but the third was only finished quite recently.

The linear groups in characteristic p actually turned out to be the most interesting of the three possibilities. Here, the proof consisted of a complicated trace argument, along with the solution of another variant of the Burnside problem for linear groups. The answer is quite similar to that for solvable groups and requires that we first define a particular characteristic f.c. subgroup $\mathcal{L}(G)$. This is done in a fairly simple manner, so $\mathcal{L}(G)$ is by no means as important as $\mathfrak{Z}(G)$.

Theorem 4.6. ([P5], [P6]) *Let G be an F-linear group and assume that* char $F =$ $p =$ char K. *Then $\mathcal{J}K[G] \neq 0$ if and only if $\mathcal{L}(G)$ has an element x of order p which has only finitely many conjugates under the action of each finitely generated subgroup of G.*

Now let us assume that G is a finitely generated F-linear group. If char $F \neq p$, then it follows quite easily that G has a normal subgroup H of finite index which is residually a finite q-group for some prime $q \neq p$. Consequently, $\mathcal{J}K[H] = 0$ and Lemma 3.5 implies that $\mathcal{J}K[G]$ is nilpotent. On the other hand, if char $F = p$, then it follows from the preceding theorem and a certain amount of work that $\mathcal{J}K[G]$ is at least a locally nilpotent ideal. In other words, we have

Corollary 4.7. ([P7]) *If G is a finitely generated linear group and* char $K = p > 0$, *then $\mathcal{J}K[G] = \mathcal{N}K[G]$.*

Thus a pattern began to emerge and we were led to

Conjecture 4.8. *If G is any finitely generated group and if* char $K = p > 0$, *then $\mathcal{J}K[G] = \mathcal{N}K[G]$.*

There was even some corroborating evidence which held for arbitrary groups. Recall that the nilpotent radical is not a radical in general. Indeed, there exists a finitely generated K-algebra A with $\mathcal{N}(A/\mathcal{N}A) \neq 0$. But this cannot happen for group rings of finitely generated groups if the preceding conjecture is to hold. Fortunately, we were able to show

Theorem 4.9. ([P7]) *If G is any finitely generated group, then $K[G]$ is a finitely generated K-algebra and $\mathcal{N}(K[G]/\mathcal{N}K[G]) = 0$. Furthermore, if H is a subgroup of finite index in G, then $\mathcal{J}K[H] = \mathcal{N}K[H]$ if and only if $\mathcal{J}K[G] = \mathcal{N}K[G]$.*

We remark that this result, Theorem 4.5, and Corollary 4.7 were all proved using the following quite surprising radical-like property of the Δ^+ operator.

Lemma 4.10. *Let G be a finitely generated group and let H be a normal subgroup of G. If $H \subseteq \Delta^+(G)$, then $\Delta^+(G/H) = \Delta^+(G)/H$.*

It is easy to see that this lemma requires G to be finitely generated, and it does not hold for the Δ operator or indeed for the operator $\mathbb{Z}$, where $\mathbb{Z}(G)$ is the center of G. Unfortunately, this marks the extent of our knowledge of the semiprimitivity question for finitely generated groups. There has been no significant progress made on this problem since the above theorems appeared in 1973 and 1974.

§5. Locally Finite Groups

The obvious next step is to deal with arbitrary groups G under the assumption that we know the answer in the finitely generated case. For convenience, let $\mathcal{F}(G)$ denote the set of finitely generated subgroups of G. Then, motivated by Theorems 4.4 and 4.6, we define a local version of the f.c. center by

$$\Lambda(G) = \{x \in G \mid |H : \mathbb{C}_H(x)| < \infty \quad \text{for all} \quad H \in \mathcal{F}(G)\}.$$

In other words,

$$\Lambda(G) = \bigcap_{H \in \mathcal{F}(G)} \mathbb{D}_G(H)$$

consists of all elements of G which have only finitely many conjugates under the action of each finitely generated subgroup of G. If we also let $\Lambda^+ = \Lambda^+(G)$ be the set of torsion elements of $\Lambda = \Lambda(G)$, then the known structure of Δ and Δ^+ translate to

Lemma 5.1. *Let G be an arbitrary group.*

i. *Λ and Λ^+ are characteristic subgroups of G.*
ii. *Λ/Λ^+ is torsion free abelian, and Λ^+ is a locally finite group.*
iii. *If $H \triangleleft G$ with $H \subseteq \Lambda^+$, then $\Lambda^+(G/H) = \Lambda^+(G)/H$.*

Of course, a group G is *locally finite* if every finitely generated subgroup is finite. For such groups, it follows easily that $\Lambda^+(G) = G$. Thus, the assertion of part (ii) that $\Lambda^+(G)$ is locally finite cannot be further sharpened. Notice also that part (iii) above asserts that the operator Λ^+ exhibits radical-like properties. This is clearly a local version of Lemma 4.10.

Now suppose $\alpha \in \mathcal{J}K[G]$ and let H be any finitely generated subgroup of G with $\operatorname{supp}\alpha \subseteq H$. Then $\alpha \in \mathcal{J}K[G] \cap K[H] \subseteq \mathcal{J}K[H]$ and hence, if we happen to know that $\mathcal{J}K[H] = \mathcal{N}K[H]$, then we can use the structure of $\mathcal{N}K[H]$, as described in Theorem 3.4, to better understand α. Specifically, we obtain

Theorem 5.2. ([P7]) *Let G be an arbitrary group and let K be a field of characteristic $p > 0$. If $\mathcal{J}K[H] = \mathcal{N}K[H]$ for all $H \in \mathcal{F}(G)$, then*

$$\mathcal{J}K[G] = \mathcal{J}K[\Lambda^+(G)]\cdot K[G].$$

In particular, it follows from Theorem 4.5 and Corollary 4.7 that if G is either locally solvable or locally linear, then $\mathcal{J}K[G] = \mathcal{J}K[\Lambda^+(G)]\cdot K[G]$. This is, in fact, how the semiprimitivity problem for characteristic 0 linear groups was settled. Namely, if G is such a group, then $\mathcal{J}K[G]$ is generated by $\mathcal{J}K[\Lambda^+(G)]$, and $\Lambda^+(G)$ is a locally finite characteristic 0 linear group. Thus $\Lambda^+(G)$ is abelian-by-finite and, with this, we can easily obtain a result quite similar to Theorem 4.6.

Notice also that if Conjecture 4.8 holds, then Theorem 5.2 reduces the semiprimitivity problem to the case of locally finite groups. In other words, this result splits the general problem into two parts. Specifically, we must first study the finitely generated case and show that $\mathcal{J}K[G] = \mathcal{N}K[G]$ for such groups. Then we must settle the problem for locally finite groups. In particular, this means that the locally finite case is also of crucial importance, and the remainder of this survey will be devoted to a discussion of this situation.

To start with, let us take another look at Theorems 4.4 and 4.6 in the context of locally finite groups. In each case, we have a normal f.c. subgroup H of G and an element $x \in H$ of order p. Since H is generated by its finite normal subgroups, it follows that x is contained in such a subgroup M. In particular, M is a finite subnormal subgroup of G of order divisible by p, and it began to appear that these finite subnormal subgroups might be the key to the solution. But inclusion in the Jacobson radical is a local property, so a local version of subnormality was really more appropriate.

Let G be a locally finite group and let A be a finite subgroup of G. We say that A is *locally subnormal* in G, and write $A \operatorname{lsn} G$, if A is subnormal in B for all finite subgroups B of G with $A \subseteq B$. For example, if G is locally nilpotent, then every finite subgroup is locally subnormal. Basic properties are as follows.

Lemma 5.3. *Let G be a locally finite group and let K be a field.*

i. *$\mathcal{J}K[G]$ is a nil ideal.*
ii. *If $A \lhd\lhd G$, then $\mathcal{J}K[A] \subseteq \mathcal{J}K[G]$.*
iii. *If $A \operatorname{lsn} G$, then $\mathcal{J}K[A] \subseteq \mathcal{J}K[G]$.*

Proof. We sketch the argument. For part (i), let $\alpha \in \mathcal{J}K[G]$ and choose a finite subgroup H of G which contains the support of α. Then $\alpha \in \mathcal{J}K[G] \cap K[H] \subseteq \mathcal{J}K[H]$ by Lemma 2.3(ii), and $\mathcal{J}K[H]$ is nilpotent since H is finite. Thus α is nilpotent, and $\mathcal{J}K[G]$ is indeed a nil ideal. For part (ii), it suffices to assume that $A \lhd G$, and to show that $\mathcal{J}K[A]\cdot K[G]$ is a nil right ideal of $K[G]$. To this end, let $\gamma \in \mathcal{J}K[A]\cdot K[G]$ and write $\gamma = \sum_1^n \alpha_i\beta_i$ with $\alpha_i \in \mathcal{J}K[A]$ and $\beta_i \in K[G]$. Since G/A is locally finite, there exists a finite subgroup B/A of G/A with $\operatorname{supp}\beta_i \subseteq B$ for all i. Then, by Lemma 3.5, $\gamma = \sum_1^n \alpha_i\beta_i \in \mathcal{J}K[A]\cdot K[B] \subseteq \mathcal{J}K[B]$, and hence we conclude from (i) that γ is nilpotent. Part (iii) follows in a similar manner. □

If K is a field of characteristic $p > 0$, and if P is a locally finite p-group, then it follows from part (iii) above that $\mathcal{J}K[P]$ is the *augmentation ideal* of $K[P]$, namely the kernel of the natural homomorphism $K[P] \to K$ given by $P \mapsto 1$. In particular, if $P = \mathbb{O}_p(G)$ is the largest normal p-subgroup of G, then $\mathcal{J}K[P]\cdot K[G]$ is the kernel of the natural homomorphism $K[G] \to K[G/P]$, and this kernel is contained in $\mathcal{J}K[G]$ by (ii) above. In other words, we have

$$\mathcal{J}K[G]/(\mathcal{J}K[P]\cdot K[G]) \cong \mathcal{J}K[G/P],$$

and obviously $\mathbb{O}_p(G/P) = \langle 1\rangle$. Because of this, it usually suffices to assume that $\mathbb{O}_p(G) = \langle 1\rangle$.

As we will see, if $\mathbb{O}_p(G) = \langle 1\rangle$, then the differences between locally subnormal subgroups, finite subnormal subgroups, and finite subgroups of normal f.c. subgroups essentially disappear. Note that we are interested in the p-elements of such a finite subgroup A, and hence our real concern is with $\mathbb{O}^{p'}(A)$, the characteristic subgroup of A generated by its Sylow p-subgroups. In other words, we can usually assume that $A = \mathbb{O}^{p'}(A)$. In the following definition, $\operatorname{len} A$ denotes the *composition length* of A, namely the common length of all composition series for A. Since A is finite, $\operatorname{len} A$ is certainly finite.

Now for any locally finite group G and fixed prime p, let $\mathbb{S}^p(G)$ be the characteristic subgroup of G generated by all $A \operatorname{lsn} G$ with $A = \mathbb{O}^{p'}(A)$. Furthermore, for each integer $n \geq 1$, let $\mathbb{S}^p_n(G)$ be the subgroup of G generated by all $A \operatorname{lsn} G$ with $A = \mathbb{O}^{p'}(A)$ and $\operatorname{len} A \leq n$. Then we have

Theorem 5.4. ([P8]) *Let G be a locally finite group with $\mathbb{O}_p(G) = \langle 1\rangle$. Then $\mathbb{S}^p(G)$ is the ascending union of its characteristic f.c. subgroups $\mathbb{S}^p_n(G)$.*

Suppose, in the above situation, that A lsn G, $A = \mathbb{O}^{p'}(A)$, and say $\operatorname{len} A = n$. Then $A \subseteq \mathbb{S}^p_n(G)$, and the latter is a normal f.c. subgroup of G. In particular, since $\mathbb{S}^p_n(G)$ is generated by its finite normal subgroups, there exists such a subgroup B with $A \subseteq B \triangleleft \mathbb{S}^p_n(G)$. But $|B| < \infty$, so $A \triangleleft\triangleleft B$ and therefore $A \triangleleft\triangleleft G$. Furthermore, if we take B to be the normal closure of A in $\mathbb{S}^p_n(G)$, then $B = \mathbb{O}^{p'}(B)$ and $B \triangleleft\triangleleft G$ with subnormal depth at most 2. Thus these several concepts all merge into one.

To handle groups having normal p-subgroups, it is natural to define $\mathbb{T}^p(G) \supseteq \mathbb{O}_p(G)$ so that

$$\mathbb{T}^p(G)/\mathbb{O}_p(G) = \mathbb{S}^p(G/\mathbb{O}_p(G)).$$

Then $\mathbb{T}^p(G)$ is a characteristic subgroup of G with a fairly nice structure which can be read off from the preceding theorem. Furthermore, we have

Lemma 5.5. *Let $\operatorname{char} K = p$, and write $T = \mathbb{T}^p(G)$ and $P = \mathbb{O}_p(G)$.*

i. *$\mathcal{J}K[T]{\cdot}K[G] \subseteq \mathcal{J}K[G]$.*
ii. *$\mathcal{J}K[T]/(\mathcal{J}K[P]{\cdot}K[T]) = \mathcal{J}K[\mathbb{S}^p(G/P)] = \bigcup \mathcal{J}K[A]$, where the union is over all A lsn G/P with $A = \mathbb{O}^{p'}(A)$.*
iii. *$\mathcal{J}K[T] \neq 0$ if and only if $T \neq \langle 1\rangle$, or equivalently if and only if G has a locally subnormal subgroup of order divisible by p.*

For a number of reasons, we suspected that the set theoretic inclusion in (i) above might always be an equality. For example, we knew that it held for G a locally finite solvable group or an F-linear group with $\operatorname{char} F = 0$ or p. Furthermore, there was some additional corroborating evidence which will be discussed in the next section. With all of this, we were led to

Conjecture 5.6. *If G is a locally finite group and K is a field of characteristic $p > 0$, then*

$$\mathcal{J}K[G] = \mathcal{J}K[\mathbb{T}^p(G)]{\cdot}K[G].$$

§6. Locally Solvable Groups

Before we proceed further, it is worthwhile to see what the latter two conjectures say about the semiprimitivity problem for group rings of arbitrary groups. To this end, let G be any group and let K be a field of characteristic $p > 0$. If H is a finitely generated subgroup of G, then according to Conjecture 4.8, $\mathcal{J}K[H] = \mathcal{N}K[H]$, and therefore Theorem 5.2 yields $\mathcal{J}K[G] = \mathcal{J}K[\Lambda^+(G)]{\cdot}K[G]$. But $\Lambda^+(G)$ is locally finite, so Conjecture 5.6 implies that $\mathcal{J}K[\Lambda^+(G)] = \mathcal{J}K[\mathbb{T}^p(\Lambda^+(G))]{\cdot}K[\Lambda^+(G)]$, and hence we have

$$\mathcal{J}K[G] = \mathcal{J}K[\mathbb{T}^p(\Lambda^+(G))]{\cdot}K[G].$$

Furthermore, Lemma 5.5 contains an appropriate description of $\mathcal{J}K[\mathbb{T}^p(\Lambda^+(G))]$. In particular, it follows from the above and Lemma 5.5(iii) that $\mathcal{J}K[G] \neq 0$ if and only if $\mathbb{T}^p(\Lambda^+(G)) \neq \langle 1\rangle$, and hence if and only if G has an element x of order p contained in a locally subnormal subgroup of $\Lambda^+(G)$. With this, we now know what "well placed" should mean in Conjecture 3.3.

Of course, neither Conjecture 3.3 nor 4.8 has been proved, and we seem to be quite far from the general solution. Nevertheless, significant progress has been made in the case of locally finite groups, so we return to this situation now. Indeed, until further notice, G will always denote a locally finite group and K will be a field of characteristic $p > 0$. As we remarked, Conjecture 5.6 was shown in [P7] to hold for solvable groups and F-linear groups with char $F = 0$ or p. Furthermore, we have

Theorem 6.1. ([P9]) *Let G be a locally finite group.*

i. *$\mathcal{J}K[\mathbb{T}^p(G)]\cdot K[G]$ is a semiprime ideal of $K[G]$, and it is a prime ideal when $\Delta^+(G/\mathbb{O}_p(G)) = \langle 1 \rangle$.*
ii. *If H is a subgroup of finite index in G, then $\mathcal{J}K[G] = \mathcal{J}K[\mathbb{T}^p(G)]\cdot K[G]$ if and only if $\mathcal{J}K[H] = \mathcal{J}K[\mathbb{T}^p(H)]\cdot K[H]$.*

Of course, an ideal I of a ring R is said to be *semiprime* if $\mathcal{N}(R/I) = 0$, and $\mathcal{J}R$ must necessarily have this property. Thus the above result at least partially corroborates Conjecture 5.6. We remark that Theorem 6.1 was surprisingly difficult to prove. It required intersection theorems from [DZ], and a significant amount of group theory. Specifically, *a generalized Fitting subgroup* $\mathbb{F}^*(G)$ was defined in [P9] and shown to have the following minimax property.

Theorem 6.2. ([P9], [P20]) *Let $G = \mathbb{S}^p(G)$ with $\mathbb{O}_p(G)$ finite, and set $F = \mathbb{F}^*(G)$.*

i. *$G = \mathbb{D}_G(F) = \{g \in G \mid |F : \mathbb{C}_F(g)| < \infty\}$, and hence F is a characteristic f.c. subgroup of G.*
ii. *Suppose $G \triangleleft GB$ where B is a finite group with $|F : \mathbb{C}_F(B)| < \infty$. Then GB is generated by its locally subnormal subgroups.*

In other words, part (i) shows that F is small enough to be almost central in G, while part (ii) implies that it is big enough to control certain types of automorphisms of G. We remark that the definition of $\mathbb{F}^*$ was changed to a more natural one in [P20], and the reference to that paper in the preceding theorem refers to this new formulation. Next, we state and prove the following elementary, but extremely powerful consequence of Theorems 3.4 and 4.9.

Lemma 6.3. ([P11]) *Let $H \triangleleft G$ with $\mathcal{J}K[H] = \mathcal{N}K[H]$. If $D = \mathbb{D}_G(H)$, then*

$$\mathcal{J}K[G] = \mathcal{J}K[D]\cdot K[G].$$

Proof. Since $D \triangleleft G$, Lemma 5.3(ii) implies that $\mathcal{J}K[D]\cdot K[G] \subseteq \mathcal{J}K[G]$. For the reverse inclusion, let $\alpha \in \mathcal{J}K[G]$ and choose any subgroup $B \supseteq H$ with $|B/H| < \infty$ and $\alpha \in K[B]$. Then $\alpha \in \mathcal{J}K[G] \cap K[B] \subseteq \mathcal{J}K[B]$, and $\mathcal{J}K[B] = \mathcal{N}K[B] = \mathcal{J}K[\Delta^+(B)]\cdot K[B]$ by Theorems 4.9 and 3.4(i). But $|B : H| < \infty$ and B is periodic, so $\Delta^+(B) = \mathbb{D}_B(H) = D \cap B$, and hence $\alpha \in \mathcal{J}K[D \cap B]\cdot K[B] \subseteq \mathcal{J}K[D \cap B]\cdot K[G]$. Since this holds for all such B, it follows easily that $\alpha \in \mathcal{J}K[D]\cdot K[G]$. □

The final result of this section deals with locally p-solvable groups. Its proof uses Δ-methods applied to finite subgroups of G, a rather surprising idea, and makes crucial use of Theorem 6.1(ii) and the preceding result applied to $H = \mathbb{O}_{p'}(G)$. In addition, it requires Hall-Higman methods (see [HH]) and a number of observations on p-solvable finitary linear groups.

Theorem 6.4. ([P11]) *If G is a locally finite, locally p-solvable group and if K is a field of characteristic $p > 0$, then*

$$\mathcal{J}K[G] = \mathcal{J}K[\mathbb{T}^p(G)]\cdot K[G].$$

With this result, proved in 1979, we completed an intensive ten year attack on the semiprimitivity problem in characteristic $p > 0$. At this point, it seemed appropriate to move on to other tasks. The general locally finite case would surely require a better understanding of the finite simple groups, and the classification was not to be completed for several more years. But before we leave the 1970's, we should mention two special cases of Conjecture 5.6 which would serve as later test problems. To start with, if G is infinite simple, then it follows easily that $\mathbb{T}^p(G) = \langle 1 \rangle$. Furthermore, if $|G|_p < \infty$, that is if there is a bound on the orders of the finite p-subgroups of G, then $\mathbb{T}^p(G)$ is a finite normal subgroup of G. Thus we were led to

Conjecture 6.5. *Let G be a locally finite group.*

i. *If G is an infinite simple group, then $\mathcal{J}K[G] = 0$.*
ii. *If $|G|_p < \infty$, then $\mathcal{J}K[G]$ is nilpotent.*

These were not considered at all during the decade of the 1980's, but they were solved in the affirmative in the early 1990's using the known structure of infinite simple groups. It turned out that the wait was necessary.

§7. Infinite Simple Groups

Finally, we can begin our discussion of recent progress on semiprimitivity. Again we assume that G is a locally finite group and that K is a field of characteristic $p > 0$. If π is any set of primes, we say that g is a *π-element* if $|g|$, the order of g, has all its prime factors in π. For convenience, we let G_π denote the set of π-elements of G, so that $1 \in G_\pi$ for all π. If X is a finite subset of $G^\# = G \setminus \{1\}$, we say that $z \in G$ is a *π-insulator* of X if $z \in G_\pi$ and $zX \cap G_\pi = \emptyset$. Furthermore, we say that G is *π-insulated* if every finite subset of $G^\#$ has a π-insulator. Note that, if $\pi = \{p\}$ consists of the single prime p, then we use *p-element* and *p-insulated* instead of the more cumbersome $\{p\}$-element and $\{p\}$-insulated. The following is proved by a simple trace argument.

Lemma 7.1. *Let π be a set of primes containing $p = \operatorname{char} K$. If G is π-insulated, then $K[G]$ is semiprimitive.*

Surprisingly, this is all the group ring theory we need to settle Conjecture 6.5(i). The remainder of the long argument is entirely group theoretic in nature and requires a close look at the structure of locally finite simple groups. For our purposes, it suffices to assume that all such groups are countably infinite.

Suppose, for example, that $G = \mathrm{FAlt}_\infty$ is the finitary alternating group on the set of positive integers. If Alt_n denotes the subgroup of G moving points in $\{1, 2, \ldots, n\}$ and fixing the rest, then G is the ascending union of the groups Alt_n with $n \geq 5$, and hence G is an ascending union of finite simple groups. Unfortunately, this property is not always true. More typical is the case where G is the finitary special linear group $\mathrm{FSL}_\infty(F)$ with F a finite field. Here G consists of all countably infinite square F-matrices

$$g = \begin{bmatrix} \bar{g} & 0 \\ 0 & I \end{bmatrix}$$

where $\bar{g} \in \mathrm{SL}_n(F)$ for some n and I is the identity matrix on the remaining rows and columns. Notice that $\mathrm{FSL}_\infty(F)$ contains no nonidentity scalar matrix, so there is no need to form the projective group. Now it is clear that G is the ascending union of the finite subgroups $G_n \cong \mathrm{SL}_n(F)$ with $n \geq 4$, but this time the groups G_n are not simple. Instead, G_n has a normal subgroup M_n, corresponding to the scalar matrices, and $G_n/M_n \cong \mathrm{PSL}_n(F)$ is simple. Furthermore, the combined map

$$G_{n-1} \to G_n \to G_n/M_n \cong \mathrm{PSL}_n(F)$$

is easily seen to be an embedding. This is indicative of the following fundamental result.

Lemma 7.2. ([Ke]) *Let G be a locally finite, countably infinite simple group. Then G has finite subgroups G_i for $i = 0, 1, 2, \ldots$ satisfying*

i. *$G_i \subseteq G_{i+1}$ and $G = \bigcup_{i=0}^{\infty} G_i$,*
ii. *$M_i \lhd G_i$ with $G_i/M_i = S_i$ a nontrivial simple group, and*
iii. *for all $i < j$, the composite map*

$$G_i \to G_j \to G_j/M_j = S_j$$

is an embedding.

In the above situation, we say that G is a *limit of the approximating sequence* $S_0, S_1, \ldots$ and we write $G = \lim_{i\to\infty} S_i$. Of course, G is not uniquely determined by the simple groups S_i, but the approximating sequence does encode a surprising amount of information on the structure of G. To start with, the Classification Theorem (see [G]) asserts that the collection of finite simple groups is divided into finitely many infinite families and finitely many exceptions, the *sporadic groups*. Thus, since any subsequence of the triples (G_i, M_i, S_i) also determines an approximating sequence for G, we can assume that all S_i belong to the same infinite family. Now most of these families have a prime power parameter and all of them have an integer parameter n. Furthermore, it turns out that G is a linear group if and only if the parameter n is uniformly bounded. The nonlinear case was settled first.

Theorem 7.3. ([PZ]) *Let G be a locally finite simple group which is not a linear group. Then G is p-insulated for any prime p, and consequently every group algebra $K[G]$ is semiprimitive.*

One aspect of the proof of this result deals with the maps $G_i \to G_j \to S_j$ which are by no means the obvious inclusions. Fortunately, this difficulty can be overcome with a simple idea implemented in a fairly tedious manner. The more interesting aspect of the argument really concerns the infinite groups FAlt_∞, $\mathrm{FSL}_\infty(F)$, $\mathrm{FSU}_\infty(F)$, $\mathrm{FSp}_\infty(F)$, and $\mathrm{F\Omega}_\infty(F)$ where F is a finite field. Note that the latter four groups correspond to those families of finite simple groups of Lie type for which the integer parameter n can become unbounded. The finitary alternating group had actually been considered by Formanek in 1972, and we sketch his clever argument.

Lemma 7.4. ([F]) *If* $G = \mathrm{FAlt}_\infty$ *or* FSym_∞, *then* G *is* p*-insulated for any prime* p.

Proof. If X is a finite subset of $G^{\#}$, then we can choose an even integer k so that the elements of $X \subseteq \mathrm{FSym}_\infty$ move only points in the set $\{1, 2, \dots, k\}$. Now define

$$z = (1 * \dots *)(2 * \dots *) \dots (k * \dots *)$$

where the $*$'s denote distinct points in $\{k+1, k+2, \dots\}$ and where $(j * \dots *)$ is a cycle of length p^j. Clearly $z \in \mathrm{FSym}_\infty$ is a p-element, and hence $g \in \mathrm{FAlt}_\infty$ if p is odd. On the other hand, if $p = 2$, then z is the product of an even number of odd cycles, so again $z \in \mathrm{FAlt}_\infty \subseteq G$.

Finally, let $x \in X$, and write x as a product of disjoint cycles which, by assumption, involve only the first k points. If $(j_1 j_2 \dots j_r)$ is such a nontrivial cycle in x, then zx (acting on the right) contains the cycle

$$(j_1 * \dots * j_2 * \dots * \dots j_r * \dots *)$$

which is the juxtaposition of the corresponding cycles in z. Since the j_i are distinct, the latter displayed cycle has length $p^{j_1} + p^{j_2} + \cdots + p^{j_r}$ and this is not a power of p. Thus zx is not a p-element, so $zX \cap G_p = \emptyset$ and G is p-insulated. □

The corresponding proof for the infinite size matrix groups is much more complicated. In some sense, these groups divide naturally into the four cases

Case 1:	$G = \mathrm{FSL}_\infty(F)$	$\mathrm{char}\, F \neq p$
Case 2:	$G = \mathrm{FSU}_\infty(F)$, $\mathrm{FSp}_\infty(F)$, $\mathrm{F\Omega}_\infty(F)$	$\mathrm{char}\, F \neq p$
Case 3:	$G = \mathrm{FSL}_\infty(F)$	$\mathrm{char}\, F = p$
Case 4:	$G = \mathrm{FSU}_\infty(F)$, $\mathrm{FSp}_\infty(F)$, $\mathrm{F\Omega}_\infty(F)$	$\mathrm{char}\, F = p$

and these are dealt with in turn. The difficulty increases as we go down the list and reaches a crescendo when we hit the bottom.

Now on to the simple linear groups. Here, we have the following wonderful characterization of such groups based on Lemma 7.2 and the Classification Theorem.

Theorem 7.5. ([Be], [Bo], [HS], [T]) *Let* G *be a locally finite simple group. If* G *is an infinite linear group, then* G *is a group of Lie type over a locally finite field* F.

Of course, the field F is *locally finite* if $\mathrm{char}\, F = q > 0$ and F is contained in the algebraic closure of $\mathrm{GF}(q)$. It follows from the above characterization that G contains a 1-parameter family of q-elements and, using this and the Zariski topology on G, we obtain

Theorem 7.6. ([P14]) *Let* G *be a locally finite simple group. If* G *is an infinite linear group over a locally finite field* F *of characteristic* $q > 0$, *then* G *is* $\{p, q\}$*-insulated for any prime* p. *In particular, every group algebra* $K[G]$ *is semiprimitive.*

Thus Theorems 7.3 and 7.6 settle Conjecture 6.5(i) in the affirmative. Furthermore, with a little more work and a knowledge of the Schur multipliers of the groups of Lie type, we can prove that if G is infinite simple, then any twisted group algebra $K^t[G]$ is semiprimitive. This is not merely of academic interest; the twisted result is actually needed to proceed further.

§8. Extension Problems

The second test problem, namely Conjecture 6.5(ii), turned out to be less important. However it did motivate us to study certain extension problems which must necessarily be part of the general solution. To start with, let G be a locally finite group, and let $|G|_p$ denote the supremum of the orders of its finite p-subgroups. In view of the Sylow theorems, it is clear that $|G|_p < \infty$ if and only if G satisfies the ascending chain condition on finite p-subgroups and hence if and only if G has no infinite p-subgroup. Of course, $|G|_p = 1$ is equivalent to G being a p'-group. Now if $|G|_p < \infty$, then we have a finite parameter to induct on, and by so doing, Lemma 7.2 and Theorem 7.5 yield

Lemma 8.1. *Let G be a locally finite group with $|G|_p < \infty$. Then G has a finite subnormal series*

$$\langle 1\rangle = G_0 \lhd G_1 \lhd \cdots \lhd G_n = G$$

with each quotient $\bar{G}_i = G_i/G_{i-1}$ either

i. *a p'-group,*
ii. *a finite simple group, or*
iii. *an infinite simple group of Lie type.*

In particular, for each subscript i, we know the solution to the semiprimitivity problem for $K[\bar{G}_i]$, and thus we should be able to find the solution for $K[G]$ from the preceding lemma provided we can handle the *extension problem.* To this end, let N be a normal subgroup of the arbitrary group G. Then we know that G is an extension of N by G/N, and therefore $K[G]$ is an extension of $K[N]$ by G/N. As we will see below, this structure is best understood in the context of crossed products.

Let R be any ring and let G be any group. Then a *crossed product* $R{*}G$ of G over R is an associative ring having a copy $\bar{G}$ of G as a left R-basis. In other words, every element α of $R{*}G$ is uniquely a finite sum $\alpha = \sum_{x\in G} r_x\bar{x}$ with coefficients $r_x \in R$ and with *support* defined by $\operatorname{supp}\alpha = \{x \in G \mid r_x \neq 0\}$. Addition in $R{*}G$ is as expected, and multiplication is determined by the rules

$$\bar{x}\bar{y} = \tau(x,y)\,\overline{xy} \qquad \text{for all} \quad x,y \in G, \qquad (\textit{twisting})$$

where τ is a map from $G \times G$ to the group of units $U(R)$ of R, and

$$r\bar{x} = \bar{x}r^{\sigma(g)} \qquad \text{for all} \quad r \in R, \quad x \in G, \qquad (\textit{action})$$

where σ is a map from G to $\operatorname{Aut}(R)$. Note that τ and σ are not group homomorphisms in general. The relations they are assumed to satisfy are precisely equivalent to the associativity of the ring, and we may also suppose that $\bar{1} = 1$ is the identity element of $R{*}G$. Obviously, any group algebra is a crossed product with trivial twisting and action.

Now suppose $K[G]$ is given with $N \lhd G$ and let S be a transversal for N in G. Observe that the elements of S act on $R = K[N]$ via conjugation by $r^s = s^{-1}rs$, and that if $s_1, s_2 \in S$, then there exists $s_3 \in S$ and $u \in N$ with $s_1s_2 = us_3$. Since $K[G] = \oplus\sum_{s\in S} K[N]s$ and since there is a natural one-to-one correspondence between the elements of S and those of G/N, it is now clear that $K[G] = R{*}(G/N)$ is a crossed product of G/N over $R = K[N]$. In fact, there is a more general result here. Namely, if $R{*}G$ is any crossed product and if $N \triangleleft G$, then $R{*}G = (R{*}N){*}(G/N)$ where $R{*}N$ is the uniquely determined sub-crossed product of $R{*}G$ consisting of those elements having support in N (see [P12]).

It would be nice if the extension aspects of the semiprimitivity problem followed directly from crossed product considerations. However, this is not the case, as can be seen from the following example.

Lemma 8.2. *Let G be an arbitrary group containing an element of prime order p. Then there exists a semiprimitive commutative algebra R over a field of characteristic p and a crossed product $R{*}G$, such that $R{*}G$ is not semiprime. In particular, $R{*}G$ is not semiprimitive.*

Proof. Let H be the given subgroup of G of order p and let K be a field of characteristic p. If Ω denotes the set of right cosets of H in G, then G permutes Ω by right multiplication, and we let $\omega_0 \in \Omega$ correspond to the coset H. Consequently, $H = G_{\omega_0} = \{g \in G \mid \omega_0 g = \omega_0\}$.

Now let R be the (complete) direct product $\prod_{\omega\in\Omega} K_\omega$, where each K_ω is a copy of K. Then R is a semiprimitive K-algebra, it is in fact von Neumann regular, and the permutation action of G on Ω extends to an action of G on R. In this way we obtain a homomorphism $\sigma: G \to \mathrm{Aut}(R)$ and we use σ to form the *skew group ring* $R{*}G$. In other words, $R{*}G$ is a crossed product with parameter σ, as above, and with trivial twisting. One knows (see [P12]) that such a construction always leads to an associative ring.

For each $\omega \in \Omega$, let e_ω denote the idempotent of R which has a 1 for its ω-coordinate and zeros elsewhere. Then $\bar{g}^{-1} e_\omega \bar{g} = e_{\omega g}$ and $e_\omega e_{\omega'} = 0$ if $\omega \neq \omega'$. In particular, e_{ω_0} commutes with $\bar{H}$, and if $g \in G \setminus H$ and $r \in R$, then

$$e_{\omega_0}(r\bar{g})e_{\omega_0} = r\bar{g}(\bar{g}^{-1}e_{\omega_0}\bar{g})e_{\omega_0} = r\bar{g}e_{\omega_0 g}e_{\omega_0} = 0.$$

It now follows easily that $e_{\omega_0}(R{*}G)e_{\omega_0} = e_{\omega_0}R{*}H \cong K[H]$ since $e_{\omega_0}R \cong K$ and the twisting is trivial. In particular, since K has characteristic p and $|H| = p$, we conclude that $e_{\omega_0}(R{*}G)e_{\omega_0}$ is not semiprime, and therefore neither is $R{*}G$. $\square$

While crossed product methods are sometimes useful in studying semiprimitivity, it turns out that twisted group algebras are absolutely crucial. Recall that a *twisted group algebra* $K^t[G]$ is a crossed product $K{*}G$ of G over K with trivial action. In particular, $K^t[G]$ is an associative K-algebra with K-basis $\bar{G}$ and with $\bar{x}\bar{y} = \tau(x,y)\,\overline{xy}$ for all $x, y \in G$. Here $\tau: G \times G \to K \setminus \{0\}$ is the twisting function, and associativity is equivalent to τ being a 2-cocycle.

As we will see at the end of this section, twisted group algebras come into play because they are homomorphic images of ordinary group algebras. For example, let Z be a central subgroup of G and let I be an ideal of $K[Z]$ with $K[Z]/I \cong K$. Then $I{\cdot}K[G] \triangleleft K[G]$ and it is easy to see that $K[G]/(I{\cdot}K[G])$ is naturally isomorphic to some twisted group algebra $K^t[G/Z]$ of G/Z.

Now, let us return to Conjecture 6.5(ii). Since the work on this problem actually occurred before Theorem 7.6 was proved, it was necessary to use an earlier special case of the latter result contained in [Z3]. By dealing with the extension problem, we were then able to obtain the affirmative solution

Theorem 8.3. ([P13]) *Let G be a locally finite group with $|G|_p < \infty$. If* $\operatorname{char} K = p > 0$, *then $\mathcal{J}K[G]$ is nilpotent.*

The proof of this result starts with a simple reduction which allows us to assume that G has no finite normal subgroup of order divisible by p, and we are left with the task of showing that $\mathcal{J}K[G] = 0$. Some aspects of the latter semiprimitivity argument will be discussed in the more general context of

Theorem 8.4. ([P14]) *Let $K[G]$ be the group algebra of a locally finite group G over a field K of characteristic $p > 0$. Suppose that G has a finite subnormal series*

$$G_0 \triangleleft G_1 \triangleleft \cdots \triangleleft G_n = G$$

with each quotient G_i/G_{i-1} either

i. *a locally p-solvable group,*
ii. *an infinite nonabelian simple group, or*
iii. *generated by its locally subnormal subgroups.*

If $\mathcal{J}K[G_0] = 0$, then $K[G]$ is semiprimitive if and only if G has no locally subnormal subgroup of order divisible by p.

A brief outline of the proof of semiprimitivity here is as follows. First, we can assume that K is algebraically closed and that $n = 1$. Indeed, by Lemma 6.3, we can suppose that G has a normal subgroup N with $|N : \mathbf{C}_N(g)| < \infty$ for all $g \in G$, and such that $G/N = H$ is a group satisfying condition (i), (ii) or (iii). In particular, N is an f.c. group and therefore, by hypothesis, it must be a p'-group. With this, case (i) now follows from Theorem 6.4, while case (iii) is an immediate consequence of Theorem 3.4 and Lemma 3.5. Finally, let H be an infinite simple group. Then, we may suppose that H is countably infinite and not a p'-group, and that G has no nontrivial f.c. homomorphic images. In other words, the pair (G, N) is what we call a *p'-f.c. cover* of H. Now if N is central in G, then G is a central cover of H and it follows from our previous comments that $K[G]$ is a subdirect product of various twisted group algebras $K^t[H]$. Thus, the twisted analogs of Theorems 7.3 and 7.6 apply here and yield the result.

On the other hand, if N is not central in G, then we show that $H = G/N$ is contained in $\mathrm{FGL}(\mathcal{V})$, the *finitary general linear group* on $\mathcal{V}$, where $\mathcal{V}$ is a vector space over the Galois field $\mathrm{GF}(q)$ for some prime q involved in the subgroup N. Furthermore, with a good deal of effort, this implies that H cannot be a linear group, and therefore the following key result of J. Hall applies.

Theorem 8.5. ([H1], [H2], [H3]) *Let G be a countably infinite, locally finite simple group which is not a linear group, and suppose that $G \subseteq \mathrm{FGL}(\mathcal{V})$, where $\mathcal{V}$ is a vector space over some field F_0.*

i. *If F_0 has characteristic 0, then $G \cong \mathrm{FAlt}_\infty$.*
ii. *If* $\operatorname{char} F_0 = q > 0$, *then G is isomorphic to one of the stable finitary groups FAlt_∞, $\mathrm{FSL}_\infty(F)$, $\mathrm{FSU}_\infty(F)$, $\mathrm{FSp}_\infty(F)$, or $\mathrm{F\Omega}_\infty(F)$, where F is some locally finite field of characteristic q.*

We remark that the uncountable groups have also been classified, but the appropriate analogs of $\mathrm{FSL}_\infty(F)$ are somewhat more complicated to describe. Finally, we define a stronger version of p-insulation and we show that if H is *strongly p-insulated*, then any p'-f.c. cover G of H is p-insulated and hence satisfies $\mathcal{J}K[G] = 0$. Thus all that remains is to prove that the stable groups H, as listed in Theorem 8.5(ii), are strongly p-insulated, and this is achieved in [P14].

§9. The Local Subnormal Closure

In some sense, the results associated with Conjecture 6.5 are all global in nature. Namely, they involve global assumptions on the locally finite group G like being simple or having a particular type of finite subnormal series. Obviously, the next step is to move on to more local assumptions. However, by some strange quirk of fate, this earlier work is not wasted. It turns out that the infinite simple groups and the locally p-solvable groups (of Theorem 6.4) are the critical factors in the general solution. We will consider this phenomenon in more detail in the next section.

For now, let $H \subseteq X$ be finite groups. Since the set of subnormal subgroups of X is closed under intersection, it follows that there is a unique smallest subnormal subgroup S of X which contains H. This is called the *subnormal closure* of H in X, and we denote it by $S = H^{[X]}$. If H^S is the normal closure of H in S, then $H \subseteq H^S \triangleleft S \triangleleft\triangleleft X$, so the minimal nature of S implies that $S = H^S$. In fact, S is characterized by the two properties

i. $H \subseteq S \triangleleft\triangleleft X$, and
ii. $S = H^S$

since (ii) implies that H cannot be contained in a proper normal subgroup of S, and hence it is not in a proper subnormal subgroup of S. In general, subnormal closures do not exist for arbitrary subgroups of infinite groups.

Observe that if $H \subseteq X \subseteq Y$ are all finite, then $H \subseteq H^{[Y]} \cap X \triangleleft\triangleleft X$. Thus the minimal nature of $H^{[X]}$ implies that $H^{[X]} \subseteq H^{[Y]} \cap X \subseteq H^{[Y]}$, and this inclusion allows us to define a *local subnormal closure* for finite subgroups of locally finite groups. Specifically, if H is a finite subgroup of the locally finite group G, then we write

$$H^{[G]} = \bigcup_X H^{[X]}$$

where the union is over all finite subgroups X of G containing H. Note that, if G is finite, then the inclusion $H^{[X]} \subseteq H^{[Y]}$ immediately implies that the two possible meanings for $H^{[G]}$ are, in fact, the same. Some basic properties are as follows.

Lemma 9.1. *Let H be a finite subgroup of G, and set $S = H^{[G]}$.*

i. *S is a subgroup of G containing H.*
ii. *If A lsn S, then A lsn G.*
iii. *$S = H^S$ is the normal closure of H in S.*

Obviously, part (ii) above allows us to reduce semiprimitivity questions from $K[G]$ to $K[S]$, and when we do this, the conclusion $S = H^S$ of (iii) comes into play. Surprisingly, this latter fact turns out to be a rather crucial property. For example, consider the following lovely observation of Wielandt.

Theorem 9.2. ([W2], [W3]) *The only primitive, finitary permutation groups on an infinite set Ω are* FSym_Ω *and* FAlt_Ω.

Then, by adding the hypothesis $G = H^G$, we can quickly extend this result to finitary permutation groups which are not even transitive. Indeed, we have

Lemma 9.3. *Let $G \subseteq \mathrm{Sym}_\Omega$ and suppose that $G = H^G$ for some finite subgroup H. If $H \subseteq \mathrm{FSym}_\Omega$, then G has a finite subnormal series*

$$\langle 1 \rangle = G_0 \lhd G_1 \lhd \cdots \lhd G_n = G$$

with each factor $\bar{G}_i = G_i/G_{i-1}$ either an f.c. group or isomorphic to $\mathrm{FAlt}_{\Lambda_i}$ for some infinite set Λ_i.

Proof. Since $H \subseteq \mathrm{FSym}_\Omega \lhd \mathrm{Sym}_\Omega$, it follows that $G = H^G \subseteq \mathrm{FSym}_\Omega$. Now suppose that H moves k points of Ω. Then H can act nontrivially on at most k orbits of G and thus $G = H^G$ implies that G has at most k nontrivial orbits.

For simplicity, let us just consider the case where G is transitive on the infinite set Ω, and let Γ be a block of imprimitivity for G. If $|\Gamma| > k$, then Γ contains a point fixed by H and hence $\Gamma = \Gamma H$. Furthermore, each conjugate H^g of H also moves k points, so $\Gamma = \Gamma H^g$. Thus Γ is stabilized by $\langle H^g \mid g \in G\rangle = H^G = G$, so Γ is an orbit of G and hence $\Gamma = \Omega$. In other words, all nontrivial blocks have size $\leq k$ and therefore we can choose one, say Γ, of maximal size.

Now if Λ denotes the set $\{\Gamma g \mid g \in G\}$ of distinct translates of Γ, then it follows that $|\Lambda| = \infty$ and that G acts in a primitive manner on Λ. In particular, if N is the kernel of this action, then Theorem 9.2 implies that $G/N \cong \mathrm{FSym}_\Lambda$ or FAlt_Λ. Furthermore, N stabilizes all Γg and acts faithfully on the disjoint union $\Omega = \bigcup \Gamma g$ with $\Gamma g \in \Lambda$. Thus, since $N \subseteq G \subseteq \mathrm{FSym}_\Omega$, it follows that N embeds in the direct sum of the finite symmetric groups $\mathrm{Sym}_{\Gamma g}$ and therefore N is an f.c. group. □

Notice how nicely this fits in with the hypothesis of Theorem 8.4. Similar results hold for finitary automorphism groups. To start with, we say that G acts in a *finitary manner* on the group V if $|V : \mathbf{C}_V(x)| < \infty$ for all $x \in G$. Furthermore, G acts in a *strongly finitary manner* if the action is finitary and if all G-stable subgroups of V are normal in V. In particular, both of these concepts include the usual notion of a finitary action of a group G on a vector space V over a finite field. Notice that we do not assume, at this point, that G acts faithfully on V. Note further that if G is strongly finitary on V and if W is a G-stable subgroup of V, then G acts in a strongly finitary manner on both W and V/W. Of course, G acts *irreducibly* on V if and only if V has no proper G-stable normal subgroup.

Lemma 9.4. *Let G act in a strongly finitary manner on the group V, and assume that $G = H^G$ is the normal closure of some finite subgroup H. Then V has a finite chain*

$$\langle 1 \rangle = V_0 \subseteq V_1 \subseteq \cdots \subseteq V_n = V$$

of G-stable normal subgroups such that, for each i, either G acts irreducibly on $\bar{V}_i = V_i/V_{i-1}$ or it acts trivially on this quotient.

Now suppose, in addition, that V is a locally finite f.c. group, and assume that G acts irreducibly on the infinite quotient $\bar{V}_i = V_i/V_{i-1}$. If $\bar{V}_i$ is nonabelian, then it follows easily from the f.c. property that it is a *semisimple group*, namely isomorphic to a (weak) direct product of finite nonabelian simple groups. Furthermore, G permutes these direct factors transitively, and therefore Theorem 9.2 enables us to describe $\bar{G}_i = G/\mathbf{C}_G(\bar{V}_i)$. On the other hand, if $\bar{V}_i$ is abelian, then it is an elementary abelian q-group for some prime q, and again we can describe $\bar{G}_i$ if the representation is imprimitive. Fortunately, when the representation is primitive, we can apply the following key result of Phillips.

Theorem 9.5. ([Ph1], [Ph2]) *Let G be a primitive, locally finite subgroup of* $\mathrm{FGL}(\mathcal{V})$*, where $\mathcal{V}$ is an infinite dimensional vector space. Then G contains a normal infinite simple subgroup D, such that G/D is solvable of derived length ≤ 6.*

As a consequence, we obtain

Lemma 9.6. *Let G act faithfully and in a strongly finitary manner on the locally finite f.c. group V. If G is the normal closure of a finite subgroup H, then G has a finite subnormal series*

$$\langle 1 \rangle = G_0 \triangleleft G_1 \triangleleft \cdots \triangleleft G_m = G$$

with each quotient $\bar{G}_i = G_i/G_{i-1}$ either infinite simple or an f.c. group. Furthermore, each such infinite simple quotient is a finitary linear group over a finite field $\mathrm{GF}(q)$ *for some prime q involved in V.*

In particular, by using Theorem 8.5(ii), we can obtain a precise automorphism group analog of Lemma 9.3.

§10. Local Results

The recent series of local results began in [P16] where it was shown that if G has no nonidentity locally subnormal subgroup, then $\mathcal{J}K[G] = 0$. Obviously, this was close to the precise necessary and sufficient conditions for semiprimitivity. Indeed, all that was missing was the relationship between the orders of the locally subnormal subgroups and the characteristic of the field. To proceed further, it was again necessary to work in the more general context of twisted group algebras. In particular, Theorem 6.4 had to be extended to this context, and then the methods used to prove the preceding local result were generalized to yield

Theorem 10.1. ([P16], [P17]) *Let G be a locally finite group and let K be a field of characteristic $p > 0$. Then $K[G]$ is semiprimitive if and only if G has no locally subnormal subgroup of order divisible by p.*

This is, of course, the semiprimitivity consequence of Conjecture 5.6. As we mentioned in the last section, the earlier results on locally p-solvable groups and on infinite simple groups, as encapsulated in Theorem 8.4, are crucial to the proof of Theorem 10.1, and we give some indication below of this phenomenon. We begin by discussing an ultraproduct argument suggested by the work of [H2] and [H3].

Let $G_1 \subseteq G_2 \subseteq \cdots$ be finite subgroups of G with $G = \bigcup_{i=1}^{\infty} G_i$ and let $\mathcal{N} = \{1, 2, \ldots\}$ be the set of natural numbers. Then we can choose an ultrafilter $\mathcal{F}$ on $\mathcal{N}$ containing the cofinite subsets, and we note that all members of $\mathcal{F}$ are infinite. Now suppose that each G_i acts as permutations (on the right) on a set Ω_i with kernel N_i. Then the ultraproduct $\prod_{\mathcal{F}} G_i$ acts on $\Omega = \prod_{\mathcal{F}} \Omega_i$ via $\otimes_i w_i \cdot \otimes_i g_i = \otimes_i w_i g_i$. Furthermore, we can define a homomorphism $\theta : G \to \prod_{\mathcal{F}} G_i$ by $\theta(g) = \otimes_i \theta_i(g)$ where $\theta_i(g) = g$ if $g \in G_i$ and $\theta_i(g) = 1$ otherwise. In this way, we obtain a permutation action of G on Ω which satisfies

Lemma 10.2. *Let G and Ω be as above.*

i. *If N is contained in the kernel of the action of G on Ω, then there exists a subsequence $\mathcal{M} \subseteq \mathcal{N}$ such that N is the ascending union of the subgroups $N \cap N_i$ with $i \in \mathcal{M}$.*
ii. *If $g \in G$ and if $\theta_i(g)$ moves at most k points of Ω_i for each i, then g moves at most k points of Ω and hence is finitary on Ω.*

Proof. (i) Suppose $x \in N$ and let $\mathcal{S}(x) = \{i \mid \theta_i(x) \text{ acts nontrivially on } \Omega_i\}$. For each $i \in \mathcal{S}(x)$ choose $w_i \in \Omega_i$ moved by $\theta_i(x)$, and if $i \notin \mathcal{S}(x)$ let $w_i \in \Omega_i$ be arbitrary. Then $w = \otimes_i w_i \in \Omega$ and, since $x \in N$, we have

$$\otimes_i w_i \equiv_{\mathcal{F}} (\otimes_i w_i)x = \otimes_i w_i\theta_i(x),$$

where $\equiv_{\mathcal{F}}$ indicates that the two elements of the (complete) direct product agree on a member of $\mathcal{F}$. In other words, $w_i = w_i\theta_i(x)$ almost everywhere and consequently $\mathcal{S}(x)$ has measure 0, that is $\mathcal{S}(x) \notin \mathcal{F}$. Furthermore, if X is any finite subset of N, then $\mathcal{S}(X) = \bigcup_{x\in X} \mathcal{S}(x)$ also has measure 0 and therefore the complement of $\mathcal{S}(X)$ is contained in $\mathcal{F}$ and is infinite. Thus we can choose $i \in \mathcal{N} \setminus \mathcal{S}(X)$ sufficiently large so that $X \subseteq G_i$. But then $\theta_i(X) = X$ acts trivially on Ω_i and hence $X \subseteq N_i$. It follows that every finite subset of N is contained in some $N \cap N_i$ and, since each such N_i is finite, the subsequence $\mathcal{M}$ is easily seen to exist.

(ii) Suppose for example that $k = 3$ so that $\theta_i(g)$ moves at most 3 points of Ω_i. Then, for each i, we can choose $a_i, b_i, c_i \in \Omega_i$, not necessarily distinct, with $\theta_i(g)$ fixing the remaining points. Now let $a = \otimes_i a_i$, $b = \otimes_i b_i$ and $c = \otimes_i c_i$ be the elements of Ω determined by these choices. We claim that these are the only possible points moved by g. To this end, let $w = \otimes_i w_i \in \Omega$ and define $A = \{i \mid w_i = a_i\}$, $B = \{i \mid w_i = b_i\}$, $C = \{i \mid w_i = c_i\}$, and $D = \{i \mid w_i \neq a_i, b_i, c_i\}$. Then $A \cup B \cup C \cup D = \mathcal{N}$ and hence at least one of these four sets must have measure 1. Now, if $A \in \mathcal{F}$, then $w = \otimes_i w_i \equiv_{\mathcal{F}} \otimes_i a_i = a$ and similarly $B \in \mathcal{F}$ yields $w = b$ and $C \in \mathcal{F}$ yields $w = c$. Finally, if $D \in \mathcal{F}$, then since $\theta_i(g)$ acts trivially on $\Omega_i \setminus \{a_i, b_i, c_i\}$, we have $wg = \otimes_i w_i\theta_i(g) \equiv_{\mathcal{F}} \otimes_i w_i = w$ and g fixes w. □

Now, what might the groups G_i act on? To understand our choice, let us first assume that $G_i = W$ is a finite group with no nonidentity solvable normal subgroup. Let $S = \operatorname{soc} W$ be the *socle* of W, so that S is generated by the minimal normal subgroups of W. Since any two distinct minimal normal subgroups commute, it follows that $\operatorname{soc} W$ is the direct product of certain of these subgroups. Furthermore, any minimal normal subgroup is either an elementary abelian q-group for some prime q, or it is *semisimple*, namely a direct product of nonabelian simple groups. This proves (i) below and, of course, parts (ii) and (iii) are routine consequences.

Lemma 10.3. *Let W be a finite group with no nonidentity solvable normal subgroup and set $S = \operatorname{soc} W$.*

i. *$S = M_1 \times M_2 \times \cdots \times M_k$ is a finite direct product of the nonabelian simple groups M_i. Thus S is semisimple.*
ii. *$\mathbb{C}_W(S) = \langle 1 \rangle$, so W acts faithfully as automorphisms on S.*
iii. *The groups M_i are precisely the minimal normal subgroups of S. Thus W permutes the set $\Omega = \{M_1, M_2, \ldots, M_k\}$ by conjugation.*
iv. *If N is the kernel of the action of W on Ω, then $S = N^{(4)}$ where the latter is the fourth derived subgroup of N.*

Proof. (iv) Note that $N = \bigcap_i \mathbb{N}_W(M_i)$, so $N \supseteq S$ and $N^{(4)} \supseteq S^{(4)} = S$. Furthermore, since $\mathbb{C}_W(S) = \langle 1 \rangle$, it follows that N embeds in $\prod_i \mathrm{Aut}(M_i)$. But under this embedding, S corresponds to $\prod_i \mathrm{Inn}(M_i)$, so N/S embeds in $\prod_i \mathrm{Out}(M_i)$. Finally, the precise version of the Schreier conjecture (see [G]), using the Classification of Finite Simple Groups, implies that each outer automorphism group $\mathrm{Out}(M_i)$ is solvable of derived length ≤ 4, and hence $N^{(4)} \subseteq S$, as required. □

If W is an arbitrary finite group, we let $\mathrm{sol}\, W$ denote the unique largest normal solvable subgroup of W. Then $\bar{W} = W/\mathrm{sol}\, W$ has no nonidentity solvable normal subgroup, so the above lemma applies to this group. In particular, if we define $\mathrm{rad}\, W \supseteq \mathrm{sol}\, W$ by $\mathrm{rad}\, W/\mathrm{sol}\, W = \mathrm{soc}\, \bar{W}$, then $\mathrm{rad}\, W$ is *solvable-by-semisimple* and W permutes the set $\Omega(W)$ of simple factors of $\mathrm{rad}\, W/\mathrm{sol}\, W$ by conjugation. For convenience, we call $|\Omega(W)|$ the *width* of W.

Now let us turn to the proof of Theorem 10.1. In view of Lemma 9.1, it suffices to assume that $G = H^G$ for some finite subgroup H of G. Furthermore, we may suppose that G is countably infinite. In particular, we can write $G = \bigcup_{i=1}^{\infty} G_i$ where the G_i are finite subgroups of G satisfying $H \subseteq G_1 \subseteq G_2 \subseteq \cdots$. Now, as we indicated above, each G_i acts as permutations on the set $\Omega_i = \Omega(G_i)$ of simple factors of $\mathrm{rad}\, G_i/\mathrm{sol}\, G_i$. Indeed, if N_i is the kernel of this action, then Lemma 10.3(iv) implies that $N_i^{(4)}$ is a normal subgroup of $\mathrm{rad}\, G_i$ and hence it is solvable-by-semisimple. Furthermore, if we choose the ultrafilter $\mathcal{F}$ on $\mathcal{N}$ to contain the cofinite subsets, then G acts as permutations on the ultraproduct $\Omega = \prod_{\mathcal{F}} \Omega_i$ and Lemma 10.2 comes into play. If N denotes the kernel of G on Ω, then we study the structure of G by considering N and $\bar{G} = G/N \subseteq \mathrm{Sym}_{\Omega}$ in turn.

To start with, Lemma 10.2(i) implies that there exists a subsequence $\mathcal{M}$ of the natural numbers $\mathcal{N} = \{1, 2, \ldots\}$ such that $L = N^{(4)}$ is the ascending union of its finite subgroups $L \cap N_i^{(4)}$ with $i \in \mathcal{M}$. Furthermore, note that $(L \cap N_i^{(4)}) \triangleleft N_i^{(4)}$ and that $N_i^{(4)}$ is solvable-by-semisimple. Thus $L \cap N_i^{(4)}$ is also solvable-by-semisimple, and $N^{(4)} = L = \bigcup_{i \in \mathcal{M}} (L \cap N_i^{(4)})$ is *locally solvable-by-semisimple*. There are now two cases to consider according to whether the widths which occur here are bounded or not. For the bounded case, we have

Lemma 10.4. *Let L be the ascending union of the finite subgroups $L_1 \subseteq L_2 \subseteq \cdots$ and suppose that each L_i is solvable-by-semisimple. If the widths of the various subgroups L_i are uniformly bounded, then L has a finite subnormal series*

$$\langle 1 \rangle = M_0 \triangleleft M_1 \triangleleft \cdots \triangleleft M_n = L$$

with each factor M_{i+1}/M_i either simple or locally solvable.

This follows easily by induction on the given upper bound for the widths. For example, if all L_i are solvable, which occurs when all widths are equal 0, then L is certainly locally solvable. On the other hand, if each L_i is a simple group, then clearly the same is true of L.

Using this lemma and Theorem 8.4, we can easily settle the semiprimitivity problem for $N^{(4)} = L$ in the case of bounded widths. The unbounded case builds upon this, but also requires some techniques from the proof of Theorem 7.3 to construct a particular p-insulator.

Finally, consider $\bar{G} = G/N \subseteq \mathrm{Sym}_{\Omega}$, and notice that $\bar{G} = \bar{H}^{\bar{G}}$. Again, there are two cases to deal with according to the nature of the action of H on the various Ω_i. Suppose first that H moves a bounded number of points in each Ω_i. Then

Lemma 10.2(ii) implies that $\bar{H} \subseteq \mathrm{FSym}_\Omega$ and we conclude from Lemma 9.3 that $\bar{G}$ has a finite subnormal series with factors which are either f.c. groups or isomorphic to FAlt_∞. In particular, the result again follows from Theorem 8.4.

The last case, where H moves arbitrarily large numbers of points in its various actions, requires an entirely new approach based on the representation theory of finite wreath products. Nevertheless, it should be clear from the above remarks that Theorem 8.4 does indeed play a crucial role in this proof.

§11. The Conjecture

Approximately 20 years after it was posed, Conjecture 5.6 was finally solved in the affirmative. Specifically, we have

Theorem 11.1. ([P20]) *If G is a locally finite group and K is a field of characteristic $p > 0$, then*

$$\mathcal{J}K[G] = \mathcal{J}K[\mathbb{T}^p(G)]\cdot K[G]$$

where $\mathbb{T}^p(G)/\mathbb{O}_p(G) = \mathbb{S}^p(G/\mathbb{O}_p(G))$ is the subgroup of $\bar{G} = G/\mathbb{O}_p(G)$ generated by those locally subnormal subgroups $\bar{A}$ with $\bar{A} = \mathbb{O}^{p'}(\bar{A})$.

In particular, in view of Lemma 5.5(ii), this yields a precise description of $\mathcal{J}K[G]$. As usual, the proof of the above result requires that we work in the more general context of twisted group algebras. Obviously, Theorem 10.1 is needed here, and several new ideas also come into play. To start with, we mention another application of the subnormal closure in finite groups.

Let $K[G]$ be given, and recall that if $\alpha = \sum a_x x \in K[G]$, then the support of α is the finite subset of G given by $\operatorname{supp}\alpha = \{x \in G \mid a_x \neq 0\}$. In addition, we call $H = \langle \operatorname{supp}\alpha \rangle$ the *supporting subgroup* of α. Clearly H is the smallest subgroup of G with $\alpha \in K[H]$ and, since G is locally finite, H is finite. We say that $\beta \in K[G]$ is a *truncation* of α if $\beta = \sum' a_x x$, where $\sum'$ indicates a partial sum of the terms of α. Thus $\operatorname{supp}\beta \subseteq \operatorname{supp}\alpha$, and the coefficients of α and of β agree on the smaller set. Of course, β is a *proper truncation* if $\beta \neq 0$ or α.

Note that, if D is any subgroup of G, then there is a natural $K[D]$-bimodule projection map $\pi_D\colon K[G] \to K[D]$ given by

$$\pi_D\colon \sum_{x\in G} a_x x \mapsto \sum_{x \in D} a_x x.$$

Thus π_D is the linear extension of the map $G \to D \cup \{0\}$ which is the identity on D and zero on $G \setminus D$. Clearly, $\pi_D(\alpha)$ is a truncation of α.

Now let $I \triangleleft K[G]$. We say that $0 \neq \alpha$ is a *minimal element* of I if $\alpha \in I$ but no proper truncation of α is contained in I. It is easy to see that I is the linear span of its minimal elements, and that I is the right (or left) ideal generated by those minimal elements having 1 in their support.

Lemma 11.2. *Suppose $\mathcal{J}K[G] \neq 0$ and let α be a minimal element of this ideal having 1 in its support. Then there exists a finite subgroup H of G containing the supporting subgroup $\langle \operatorname{supp}\alpha \rangle$ such that*

i. $H = \langle \operatorname{supp} \alpha \rangle^H$ *and* $H = \mathbb{O}^{p'}(H)$.
ii. α *is a minimal element of* $\mathcal{J}K[H]$.
iii. *If* N *is any subgroup of* G *normalized by* H *and if* $\mathcal{J}K[N] = \mathcal{N}K[N]$, *then* $H \subseteq \mathbb{D}_G(N)$.
iv. *If* N *is any subgroup of* G *normalized by* H *which satisfies both* $\mathbb{O}_p(N) = \langle 1 \rangle$ *and* $\mathcal{J}K[N] = \mathcal{J}K[\mathbb{S}^p(N)]{\cdot}K[N]$, *then* $H \subseteq \mathbb{S}^p(NH)$ *and* $H \subseteq \mathbb{D}_G(F^*)$, *where* $F^* = \mathbb{F}^*(N)$.

Proof. Let $\beta_1, \beta_2, \ldots, \beta_k$ be the finitely many proper truncations of α. By definition, no β_i is contained in $\mathcal{J}K[G]$, and hence the right ideals $\beta_i K[G]$ are not nil. In other words, we can choose elements $\gamma_i \in K[G]$ with $\beta_i\gamma_i$ not nilpotent. Now G is locally finite, so there exists a finite subgroup $L \subseteq G$ which contains $\langle \operatorname{supp} \alpha \rangle$ and the supports of all γ_i. In particular, $\beta_i \notin \mathcal{J}K[L]$ since $\mathcal{J}K[L]$ is nilpotent.

Now let $H = \langle \operatorname{supp} \alpha \rangle^{[L]}$ be the subnormal closure of $\langle \operatorname{supp} \alpha \rangle$ in L. Then $H = \langle \operatorname{supp} \alpha \rangle^H$ and $\alpha \in \mathcal{J}K[G] \cap K[H] \subseteq \mathcal{J}K[H]$. Furthermore, since $H \lhd\lhd L$, Lemma 5.3(ii) implies that $\mathcal{J}K[H] \subseteq \mathcal{J}K[L]$. Thus, since $\beta_i \notin \mathcal{J}K[L]$, we have $\beta_i \notin \mathcal{J}K[H]$ and therefore (ii) is proved. Note that $\mathcal{J}K[H] = \mathcal{J}K[\mathbb{O}^{p'}(H)]{\cdot}K[H]$ by Lemma 3.5, so $\pi_{\mathbb{O}^{p'}(H)}(\alpha) \in \mathcal{J}K[\mathbb{O}^{p'}(H)] \subseteq \mathcal{J}K[H]$. Moreover, $1 \in \operatorname{supp} \alpha$, so $\pi_{\mathbb{O}^{p'}(H)}(\alpha)$ is a nonzero truncation of α contained in $\mathcal{J}K[H]$, and consequently $\pi_{\mathbb{O}^{p'}(H)}(\alpha)$ must equal α. In other words, $\langle \operatorname{supp} \alpha \rangle \subseteq \mathbb{O}^{p'}(H) \lhd H$ and, since $H = \langle \operatorname{supp} \alpha \rangle^H$, it follows that $H = \mathbb{O}^{p'}(H)$.

For part (iii), suppose that N is any subgroup of G normalized by H with $\mathcal{J}K[N] = \mathcal{N}K[N]$. If $X = NH$, then N is a normal subgroup of X of finite index, so Theorem 4.9 implies that $\mathcal{J}K[X] = \mathcal{N}K[X]$. In particular, by Theorem 3.4(i), we have $\mathcal{J}K[X] = \mathcal{J}K[D]{\cdot}K[X]$ where $D = \Delta(X)$. Now $\alpha \in \mathcal{J}K[G] \cap K[X] \subseteq \mathcal{J}K[X]$ and therefore $\pi_D(\alpha)$ is a nonzero truncation of α contained in $\mathcal{J}K[D] \subseteq \mathcal{J}K[X]$. Thus $\pi_D(\alpha) \in \mathcal{J}K[X] \cap K[H] \subseteq \mathcal{J}K[H]$, so the minimal nature of α implies that $\pi_D(\alpha) = \alpha$. In other words, $\langle \operatorname{supp} \alpha \rangle \subseteq D \lhd X$ and therefore $H = \langle \operatorname{supp} \alpha \rangle^H \subseteq D$. But $N \subseteq X$, so the definition of D implies that $|N : \mathbb{C}_N(h)| < \infty$ for all $h \in H$, and consequently $H \subseteq \mathbb{D}_G(N)$, as required.

Finally, suppose H normalizes a group N which satisfies both $\mathbb{O}_p(N) = \langle 1 \rangle$ and $\mathcal{J}K[N] = \mathcal{J}K[\mathbb{S}^p(N)]{\cdot}K[N]$. If $X = NH$ then Theorem 6.1(ii) easily implies that $\mathcal{J}K[X] = \mathcal{J}K[S]{\cdot}K[X]$ where $S = \mathbb{S}^p(X)$. Again, $\alpha \in \mathcal{J}K[G] \cap K[X] \subseteq \mathcal{J}K[X]$ and therefore $\pi_S(\alpha)$ is a nonzero truncation of α contained in $\mathcal{J}K[S] \subseteq \mathcal{J}K[X]$. Thus $\pi_S(\alpha) \in \mathcal{J}K[X] \cap K[H] \subseteq \mathcal{J}K[H]$, and the minimal nature of α implies that $\pi_S(\alpha) = \alpha$. In other words, $\langle \operatorname{supp} \alpha \rangle \subseteq S \lhd X$, and consequently $H = \langle \operatorname{supp} \alpha \rangle^H \subseteq S = \mathbb{S}^p(NH)$. Furthermore, since $N \lhd NH$, we have $F^* = \mathbb{F}^*(N) \subseteq \mathbb{F}^*(NH)$. Thus, since $\mathbb{O}_p(NH)$ is finite, Theorem 6.2(i) applied to the group $\mathbb{S}^p(NH)$ shows that H acts in a finitary manner on $\mathbb{F}^*(NH)$ and hence also on F^*. □

Now let us consider some aspects of the proof of Theorem 11.1. Of course, we can assume that $\mathbb{O}_p(G) = \langle 1 \rangle$, and the goal is to show that $\mathcal{J}K[G] \subseteq \mathcal{J}K[\mathbb{S}^p(G)]{\cdot}K[G]$. In view of our previous comments, it therefore suffices to show that all minimal elements of $\mathcal{J}K[G]$ having 1 in their support are contained in $\mathcal{J}K[\mathbb{S}^p(G)]{\cdot}K[G]$. To this end, let α be such an element and let $H \supseteq \langle \operatorname{supp} \alpha \rangle$ be given by Lemma 11.2. Furthermore, let $L = H^{[G]}$ be the local subnormal closure of H in G. Note that $\alpha \in \mathcal{J}K[L]$ and, by Lemma 9.1(ii), it suffices to prove that $\mathcal{J}K[L] = \mathcal{J}K[\mathbb{S}^p(L)]{\cdot}K[L]$.

Let $V = \mathbb{F}^*(\mathbb{S}^p(L))$ and note that V is an f.c. group by Theorem 6.2(i) since $\mathbb{O}_p(L) \subseteq \mathbb{O}_p(G) = \langle 1 \rangle$. Indeed, since H normalizes V and $\mathcal{J}K[V] = \mathcal{N}K[V]$,

Lemma 11.2(iii) implies that $H \subseteq \mathbb{D}_L(V) \triangleleft L$. But $L = H^L$, so $L = \mathbb{D}_L(V)$ and therefore $L \supseteq V$ acts in a strongly finitary manner on V. In particular, if we let $C = \mathbb{C}_L(V)$, then it follows from Lemma 9.6 that L has a finite subnormal series

$$C = L_0 \triangleleft L_1 \triangleleft \cdots \triangleleft L_m = L$$

with each quotient $\bar{L}_i = L_i/L_{i-1}$ either an infinite simple finitary linear group or an f.c. group.

Furthermore, since $\mathbb{S}^p(C) \subseteq \mathbb{S}^p(L)$ centralizes V, it follows easily that $\mathbb{S}^p(C)$ is contained in $\mathbb{F}(\mathbb{S}^p(L))$, the *Fitting subgroup* of $\mathbb{S}^p(L)$. But the latter group is a p'-group since $\mathbb{O}_p(L) = \langle 1 \rangle$, and certainly $\mathbb{S}^p(C)$ is generated by p-elements. Thus $\mathbb{S}^p(C) = \langle 1 \rangle$ and Theorem 10.1 implies that $0 = \mathcal{J}K[C] = \mathcal{J}K[\mathbb{S}^p(C)]\cdot K[C]$. In other words, we need only climb the chain $C = L_0 \triangleleft L_1 \triangleleft \cdots \triangleleft L_m = L$ and show that the condition $\mathcal{J}K[L_i] = \mathcal{J}K[\mathbb{S}^p(L_i)]\cdot K[L_i]$ lifts from L_{i-1} to L_i. Of course, $\mathbb{O}_p(L_i) = \langle 1 \rangle$ and therefore Theorem 6.1(ii) easily handles the case where $\bar{L}_i$ is an f.c. group. Thus all that remains is to settle this particular extension problem when $\bar{L}_i$ is infinite simple.

Let us completely change notation and just consider the latter extension problem. By applying Lemma 11.2(iv) and the local subnormal closure, we are quickly faced with the following group theoretic structure. For a fixed prime p, we say that (G, C, H) is a *critical triple* if

1. $C \triangleleft G$ and G/C is an infinite simple group.
2. $H = \mathbb{O}^{p'}(H)$ is a finite subgroup of G with $G = H^G$.
3. $H \subseteq \mathbb{S}^p(CH)$, $\mathbb{O}_p(G) = \langle 1 \rangle$ and $G = \mathbb{D}_G(\mathbb{F}(G))$.

Then, with a good deal of work, and by using Theorems 5.4 and 9.5, we obtain

Lemma 11.3. *If (G, C, H) is a critical triple for the prime p, then there exists a subgroup $\tilde{G} \triangleleft G$ having the following numerous properties. To start with, $\tilde{G}$ has finite index in $G^{(6)}$, the 6th derived subgroup of G. Furthermore, if $\tilde{C} = C \cap \tilde{G}$, then $\tilde{G}/\tilde{C} \cong G/C$ is infinite simple and either (1) $\tilde{C}$ is a nilpotent p'-group, or (2) $G/C = \mathrm{FAlt}_{\mathcal{I}}$ for some infinite set $\mathcal{I}$, and G has normal subgroups $\tilde{D} \subseteq \tilde{X} \subseteq \tilde{L} \subseteq \tilde{C} \subseteq \tilde{G}$ satisfying*

- i. *$\tilde{L} = \mathbb{O}^{p'}(\tilde{C})$, so that $\tilde{C}/\tilde{L}$ is a p'-group.*
- ii. *$\tilde{D}$ is a finite abelian p'-group which is central in $\mathbb{O}^{p'}(\tilde{G})$.*
- iii. *$\tilde{L}$ is an f.c. group, and $\tilde{L}/\tilde{X}$ is an abelian p-group.*
- iv. *There exist finite normal subgroups $\tilde{X}_i$ of $\tilde{C}$, for all $i \in \mathcal{I}$, with $(\tilde{X}_i)^g = \tilde{X}_{ig}$ where ig is the image of $i \in \mathcal{I}$ under the permutation $Cg \in \mathrm{FAlt}_{\mathcal{I}}$.*
- v. *$\tilde{D} \subseteq \tilde{X}_i \subseteq \tilde{X}$ and $\tilde{X}/\tilde{D}$ is the (weak) direct product $\prod_{i\in\mathcal{I}}(\tilde{X}_i/\tilde{D})$.*
- vi. *$\tilde{X}_j/\tilde{D} \subseteq \mathbb{O}^{p'}(\mathbb{C}_{\tilde{L}/\tilde{D}}(\tilde{X}_i/\tilde{D}))$ for all distinct $i, j \in \mathcal{I}$.*

This is unfortunately as far as the group theory goes. We must now deal directly with the groups G as described above, compute $\mathcal{J}K[G]$ and verify that the conclusion of Theorem 11.1 is satisfied here. To do this, it suffices to determine $\mathcal{J}K[\tilde{G}]$, since the extension from $\tilde{G}$ to G is easy to handle. Now if case (1) holds and $\tilde{C}$ is a nilpotent p'-group, then $\mathcal{J}K[\tilde{G}] = 0$ by Theorem 8.4 and the result follows quite simply. On the other hand, if case (2) holds, then the only option is to compute $\mathcal{J}K[\tilde{G}]$ by brute force. The ad hoc argument here is fairly long and painful. It does, however, use some interesting crossed product techniques along with the following lemma which allows Theorem 10.1 to again come into play.

Lemma 11.4. *Let Ω be an infinite set and let G be a subgroup of the finitary symmetric group* FSym_Ω. *If the stabilizer $G_\Delta = \{g \in G \mid \Delta g = \Delta\}$ of every finite subset $\Delta \subseteq \Omega$ has only infinite orbits on the complementary set $\Omega \setminus \Delta$, then G has no nonidentity locally subnormal subgroups. In particular, this applies when $G \supseteq \mathrm{FAlt}_{\Omega_1} \times \mathrm{FAlt}_{\Omega_2} \times \cdots \times \mathrm{FAlt}_{\Omega_k}$, where $\Omega = \Omega_1 \cup \Omega_2 \cup \cdots \cup \Omega_k$ is a disjoint union of infinite sets.*

§12. Burnside Groups

To proceed further, we must obviously return to the case of finitely generated groups. Indeed, the next candidates for study should most likely be the finitely generated p-groups, that is the groups associated with the Burnside problem. A natural question here is whether $\mathcal{J}K[G]$ can equal the augmentation ideal $\mathcal{A}K[G]$ of $K[G]$, namely the kernel of the natural epimorphism $K[G] \to K[G/G] = K$. If Conjecture 4.8 is to hold, then we must have

Conjecture 12.1. *Let G be a finitely generated p-group and let* $\operatorname{char} K = p > 0$. *Then $\mathcal{J}K[G] = \mathcal{A}K[G]$ if and only if G is finite.*

This is easily seen to be equivalent to the assertion that $\mathcal{J}K[G] = \mathcal{A}K[G]$ if and only if G is a locally finite p-group with $p = \operatorname{char} K$. So far, the only real evidence here of any generality is the following lovely argument of Lichtman.

Lemma 12.2. ([L]) *Let G be an infinite finitely generated p-group and let K be a field of characteristic $p > 0$. If $\mathcal{J}K[G] = \mathcal{A}K[G]$, then G has an infinite residually finite homomorphic image and, in particular, $G \neq G'$.*

Proof. Let H be the intersection of all normal subgroups of G of finite index. Then G/H is a residually finite homomorphic image of G, and the goal is to show that this factor group is infinite. Suppose, by way of contradiction, that $|G : H| < \infty$. Since G is finitely generated, it follows that $H = \langle h_1, h_2, \ldots, h_n\rangle$ is also finitely generated, and consequently $I = \mathcal{A}K[H] = \sum_{i=1}^{n}(1-h_i)K[H]$ is a finitely generated right ideal of $K[H]$ and hence a finitely generated right $K[H]$-module. Furthermore, $I \neq 0$ since otherwise we would have $H = \langle 1\rangle$ and $|G| < \infty$. Nakayama's lemma now implies that $I{\cdot}\mathcal{J}K[H]$ is properly contained in I.

By assumption, $\mathcal{J}K[G] = \mathcal{A}K[G]$, and consequently

$$\mathcal{J}K[H] \supseteq \mathcal{J}K[G] \cap K[H] = \mathcal{A}K[G] \cap K[H] = \mathcal{A}K[H].$$

It follows that $\mathcal{J}K[H] = \mathcal{A}K[H] = I$ and, by our previous remarks, I properly contains $I{\cdot}\mathcal{J}K[H] = I^2$. Now consider the homomorphism $^{-}\colon K[H] \to K[H]/I^2$. Since $\overline{K[H]} = K + \bar{I}$ and $\bar{I}^2 = 0$, it is clear that this image is a commutative K-algebra properly larger than K. Thus since $\overline{K[H]}$ is spanned over K by $\bar{H}$, we see that $\bar{H}$ is a nontrivial abelian homomorphic image of H and consequently $H \neq H'$. In other words, H/H' is a nonidentity finitely generated abelian p-group, so $1 < |H/H'| < \infty$ and H' is a normal subgroup of G of finite index properly contained in H. This, of course, contradicts the definition of H. □

A slight generalization of the above argument shows that every maximal subgroup of G is normal of index p. Note that $\mathcal{J}K[G] = \mathcal{A}K[G]$ if and only if $K[G]$

has precisely one irreducible module, namely the *principal module.* Thus Conjecture 12.1 can be paraphrased as asserting that if G is an infinite finitely generated p-group, then $K[G]$ has a nonprincipal module. For example, if G is a Tarski monster of period p, as constructed in [O], then certainly $G = G'$ and the preceding lemma implies that a nonprincipal irreducible module exists in this case.

On the other hand, many of the remaining Burnside counterexamples are residually finite. One such is the Gupta-Sidki group which is described in [GuS] and [S1] as a certain subgroup of the automorphism group of a 1-rooted regular tree of degree p. For this group, we nevertheless have

Theorem 12.3. ([S2]) *Let G be a Gupta-Sidki p-group and let* $\operatorname{char} K = p$. *Then $K[G]$ has a nonprincipal irreducible module.*

Actually, this result is stated in [S2] only for $p = 3$ and for $K = \mathrm{GF}(3)$, but it does hold in the above generality with the same proof. Using this as a starting point, it was then shown in [PT] that $K[G]$ has infinitely many nonisomorphic irreducible modules when the field K is sufficiently large.

Finally, the Golod groups G are described in [Go] and [GoS] as finitely generated subgroups of the group of units of a Golod-Shafarevitch algebra $A = K \oplus N$, where N is an infinite dimensional nil ideal. Furthermore, $\bigcap_{k=0}^{\infty} N^k = 0$, so these groups are residually finite. Now there is a natural epimorphism $\tilde{}: K[G] \to A$ which maps $\mathcal{A}K[G]$ onto N. In particular, if $\tilde{}$ is an isomorphism, then $\mathcal{A}K[G]$ is nil and we have a counterexample to Conjecture 12.1. Fortunately, it was shown in [Si] that $\tilde{}$ is not an isomorphism, at least when the construction parameters satisfy certain fairly natural conditions.

This is essentially all that is known about the semiprimitivity problem for Burnside groups. Obviously, much remains to be done.

References

[A1] S. A. Amitsur, The radical of field extensions, *Bull. Res. Council Israel* **7F** (1957), 1–10.

[A2] S. A. Amitsur, On the semi-simplicity of group algebras, *Mich. Math. J.* **6** (1959), 251–253.

[B] K. I. Beidar, On radicals of finitely generated algebras, *Uspekhi Mat. Nauk* **36** (1981), 203–204. (Russian)

[Be] V. V. Belyaev, Locally finite Chevalley groups, Studies in Group Theory, Urals Scientific Centre of the Academy of Sciences of USSR, Sverdlovsk, 1984, pp. 39–50. (Russian)

[Bo] A. V. Borovik, Periodic linear groups of odd characteristic, *Soviet Math. Dokl.* **26** (1982), 484–486.

[DZ] Z. Z. Dyment and A. E. Zalesskiĭ, On the lower radical of a group ring, *Dokl. Akad. Nauk BSSR* **19** (1975), 876–879. (Russian)

[F] E. Formanek, A problem of Passman on semisimplicity, *Bull. London Math. Soc.* **4** (1972), 375–376.

[Go] E. S. Golod, On nil algebras and finitely approximable groups, *Izv. Akad. Nauk. SSSR Ser. Mat.* **28** (1964), 273–276. (Russian)

[GoS] E. S. Golod and I. R. Shafarevitch, On towers of class fields, *Izv. Akad. Nauk. SSSR Ser. Mat.* **28** (1964), 261–272. (Russian)

[G] D. Gorenstein, *Finite Simple Groups*, Plenum, New York, 1982.

[GuS] N. Gupta and S. Sidki, On the Burnside problem for periodic groups, *Math. Z.* **182** (1983), 385–388.

[H1] J. I. Hall, Infinite alternating groups as finitary linear transformation groups, *J. Algebra* **119** (1988), 337–359.

[H2] J. I. Hall, Locally finite simple groups of finitary linear transformations, Finite and Locally Finite Groups, Kluwer, Dordrecht, 1995, pp. 147–188.

[H3] J. I. Hall, Periodic simple groups of finitary linear transformations (to appear).

[HH] P. Hall and G. Higman, The p-length of a p-soluble group and reduction theorems for Burnside's problem, *Proc. London Math. Soc. (3)* **6** (1956), 1–42.

[HP] C. R. Hampton and D. S. Passman, On the semisimplicity of group rings of solvable groups, *Trans. Amer. Math. Soc.* **173** (1972), 289–301.

[HS] B. Hartley and G. Shute, Monomorphisms and direct limits of finite groups of Lie type, *Quart. J. Math. Oxford (2)* **35** (1984), 49–71.

[K1] I. Kaplansky, Problems in the theory of rings, NAS-NRC Publ. 502, Washington, pp. 1–3.

[K2] I. Kaplansky, "Problems in the theory of rings" revisited, *Amer. Math. Monthly* **77** (1970), 445–454.

[Ke] O. H. Kegel, Über einfache, lokal endliche Gruppen, *Math. Z.* **95** (1967), 169–195.

[L] A. I. Lichtman, On group rings of p-groups, *Izv. Akad. Nauk. SSSR Ser. Mat.* **27** (1963), 795–800. (Russian)

[M] H. Maschke, Über den arithmetischen Charakter der Coefficienten der Substitutionen endlicher Substitutionsgruppen, *Math. Ann.* **50** (1898), 492–498.

[MZ] A. V. Mihalev and A. E. Zalesskiĭ, Group Rings, Modern Problems of Mathematics 2, VINITI, Moscow, 1973. (Russian)

[O] A. Yu. Ol'shanskiĭ, *Geometry of Defining Relations in Groups*, Kluwer, Dordrecht, 1991.

[Pa] I. B. S. Passi, *Group Rings and their Augmentation Ideals*, Lecture Notes in Math. vol. 715, Springer-Verlag, Berlin, 1979.

[P1] D. S. Passman, Nil ideals in group rings, *Mich, Math. J.* **9** (1962), 375–384.

[P2] D. S. Passman, Radicals of twisted group rings, *Proc. London Math. Soc. (3)* **20** (1970), 409–437.

[P3] D. S. Passman, *Infinite Group Rings*, Marcel Dekker, New York, 1972.

[P4] D. S. Passman, Some isolated subsets of infinite solvable groups, *Pacific J. Math.* **45** (1973), 313–320.

[P5] D. S. Passman, On the semisimplicity of group rings of linear groups, *Pacific J. Math.* **47** (1973), 221–228.

[P6] D. S. Passman, On the semisimplicity of group rings of linear groups II, *Pacific J. Math.* **48** (1973), 215–234.

[P7] D. S. Passman, A new radical for group rings?, *J. Algebra* **28** (1974), 556–572.

[P8] D. S. Passman, Subnormality in locally finite groups, *Proc. London Math. Soc. (3)* **28** (1974), 631–653.

[P9] D. S. Passman, Radical ideals in group rings of locally finite groups, *J. Algebra* **33** (1975), 472–497.

[P10] D. S. Passman, *The Algebraic Structure of Group Rings*, Wiley-Interscience, New York, 1977.

[P11] D. S. Passman, The Jacobson radical of a group ring of a locally solvable group, *Proc. London Math. Soc.* **38** (1979), 169–192, (ibid **39** (1979), 208–210).

[P12] D. S. Passman, *Infinite Crossed Products*, Academic Press, Boston, 1989.

[P13] D. S. Passman, Semiprimitivity of group algebras of locally finite groups, Infinite Groups and Group Rings, World Scientific, Singapore, 1993, pp. 77–101.

[P14] D. S. Passman, Semiprimitivity of group algebras of infinite simple groups of Lie Type, *Proc. Amer. Math. Soc.* **121** (1994), 399–403.

[P15] D. S. Passman, Semiprimitivity of group algebras of locally finite groups II, *J. Pure Appl. Algebra* **107** (1996), 271–302.

[P16] D. S. Passman, The semiprimitivity problem for group algebras of locally finite groups, *Israel J. Math.* **96** (1996), 481–509.

[P17] D. S. Passman, The semiprimitivity problem for twisted group algebras of locally finite groups, *Proc. London Math. Soc. (3)* **73** (1996), 323–357.

[P18] D. S. Passman, Semiprimitivity of group algebras: a survey, Atas da XII Escola de Álgebra, IMECC-UNICAMP, 1995, pp. 168–187.

[P19] D. S. Passman, Ultraproducts and the group ring semiprimitivity problem, *Resenhas IME-USP* **2** (1996), 253–261.

[P20] D. S. Passman, The Jacobson radical of group rings of locally finite groups, *Trans. Amer. Math. Soc.* (to appear).

[PT] D. S. Passman and W. V. Temple, Representations of the Gupta-Sidki group, *Proc. Amer. Math. Soc.* **124** (1996), 1403–1410.

[PZ] D. S. Passman and A. E. Zalesskiĭ, Semiprimitivity of group algebras of locally finite simple groups, *Proc. London Math. Soc.* **67** (1993), 243–276.

[Ph1] R. E. Phillips, Finitary linear groups: a survey, Finite and Locally Finite Groups, Kluwer, Dordrecht, 1995, pp. 111–146.

[Ph2] R. E. Phillips, Primitive, locally finite, finitary linear groups (to appear).

[R] C. Rickart, The uniqueness of norm problem in Banach algebras, *Ann. Math.* **51** (1950), 615–628.

[Se1] S. K. Sehgal, *Topics in Group Rings*, Marcel Dekker, New York, 1978.

[Se2] S. K. Sehgal, *Units in Integral Group Rings*, Longman Scientific, Essex, 1993.

[Si] P. N. Siderov, Group algebras and algebras of Golod-Shafarevich, *Proc. Amer. Math. Soc.* **100** (1987), 424–428.

[S1] S. Sidki, On a 2-generated infinite 3-group: subgroups and automorphisms, *J. Algebra* **110** (1987), 24–55.

[S2] S. Sidki, A primitive ring associated to a Burnside 3-group, *J. London Math. Soc.* **55** (1997) (to appear).

[T] S. Thomas, The classification of simple linear groups, *Archiv der Mathematik* **41** (1983), 103–116.

[V] O. E. Villamayor, On the semisimplicity of group algebras, *Proc. Amer. Math. Soc.* **9** (1958), 621–627.

[Wa] D. A. R. Wallace, The Jacobson radicals of the group algebras of a group and of certain normal subgroups, *Math. Z.* **100** (1967), 282–294.

[W1] H. Wielandt, Eine Verallgemeinerung der invarianten Untergruppen, *Math. Z.* **45** (1939), 209–244.

[W2] H. Wielandt, *Unendliche Permutationsgruppen*, Lecture Notes, Mathematisches Institut der Universität, Tübingen, 1960.

[W3] H. Wielandt, *Mathematical Works*, Volume 1, de Gruyter, Berlin, 1994.

[Z1] A. E. Zalesskiĭ, On the semisimplicity of a modular group algebra of a solvable group, *Soviet Math.* **14** (1973), 101–105.

[Z2] A. E. Zalesskiĭ, The Jacobson radical of the group algebra of a solvable group is locally nilpotent, *Izv. Akad. Nauk SSSR, Ser. Mat.* **38** (1974), 983–994. (Russian)

[Z3] A. E. Zalesskiĭ, Semiprimitivity of group algebras of linear groups, private communication, 1992.

[Za] H. Zassenhaus, *The Theory of Groups*, Second Edition, Chelsea, New York, 1958.

D. S. Passman
Department of Mathematics
University of Wisconsin
Madison, Wisconsin 53706
USA
e-mail: passman@math.wisc.edu

Canadian Mathematical Society
Conference Proceedings
Volume **22**, 1998

Asymptotics of Codimensions of some P.I. Algebras

Amitai Regev*

Abstract. Recent results on the asymptotics of the multilinear codimensions of P.I. algebras are reviewed here.

Introduction

Assume throughout that F is a field of characteristic zero and all algebras are F algebras. We review here a body of — mostly recent — results, about the asymptotics of the codimensions of P.I. algebras. Among the algebras reviewed are the verbally prime P.I. algebras, as well as algebras whose T-ideal of identities is either a product or an intersection of the ideals of identities of verbally prime algebras. Kemer's structure theory [Kem2] emphasizes the importance of these algebras.

After recalling some general results in §1, and Kemer's structure theory in §2, we review in §3, the computation of the asymptotics of the codimensions $c_n(M_k(F))$ of $M_k(F)$, the $k \times k$ matrices over F [Rgv3]. The Procesi–Razmyslov theory of trace identities [Pro], [Raz1], together with Formanek's work on the various Poincaré series of $M_k(F)$ [Frm1], [Frm2], are essential here.

In §4 we review some recent results about the codimensions of the other verbally prime P.I. algebras. In §5 and §6 we review the corresponding — recent — results for products and for intersections of such T ideals.

The asymptotics of the involutions codimensions of $M_k(F)$ was recently determined [Brl.Gmb.Rgv] — and is discussed in §7. Some additional P.I. algebras are reviewed in §8.

AMS Subject Classification (1991). 16R10.
* Partially supported by an NSF grant number DMS 94-01197.

§1. General Results on Cocharacters and Codimensions

Let A be a P.I. algebra, i.e. an F algebra satisfying polynomial identities. $F < x >= F < x_1, x_2, \cdots >$ are the associative and non commutative polynomials, and $I(A) \subseteq F < x >$ is the subset of all the polynomial identities of A. This $I(A)$ is a T-ideal, i.e. a two sided ideal which is closed under all substitutions of variables in $F < x >$.

The process of linearization (or polarization, i.e. constructing $f(x_1, \cdots) \to g(x, y, \cdots) = f(x+y, \cdots) - f(x, \cdots) - f(y, \cdots)$ when $\deg_{x_1} f \geq 2$) can be reversed (i.e. $g(x, y, \cdots) \to g(x, x, \cdots)$), since $\operatorname{char}(F) = 0$. This implies that it suffices to study only the multilinear identities, since these determine all other identities.

Let $V_n = V_n(x_1, \cdots, x_n)$ denote the F-vector space of the multilinear polynomials of degree n in $x_1, \cdots, x_n$. V_n is naturally a left FS_n module (isomorphic to the regular representation of S_n), with $I(A) \cap V_n \subseteq V_n$ a submodule. The S_n character of the quotient module is the n-th cocharacter of A:

$$\chi_n(A) = \chi_{S_n}\big(V_n/(I(A) \cap V_n)\big);$$
$$c_n(A) = \deg\big(\chi_n(A)\big)$$

is the n-th codimension of A.

The irreducible S_n characters are given by the Young-Frobenius theory: they are in a one-to-one correspondence with $\operatorname{Par}(n)$, the partitions of n; given $\lambda \in \operatorname{Par}(n)$, it corresponds (in an explicit way) to the irreducible S_n-character χ_λ. The degree $\deg \chi_\lambda = d_\lambda$ is given by either the Young–Frobenius or by the "hook" formula.

Since $\operatorname{char}(F) = 0$, any S_n-character φ decomposes as $\varphi = \sum_{\lambda \in \operatorname{Par}(n)} a(\lambda)\chi_\lambda$, with $0 \leq a(\lambda) \in \mathbb{Z}$. The corresponding decomposition of $\chi_n(A)$ is denoted by

$$\chi_n(A) = \sum_{\lambda \in \operatorname{Par}(n)} m(\lambda)\chi_\lambda \; .$$

It follows that

$$c_n(A) = \sum_{\lambda \in \operatorname{Par}(n)} m(\lambda) d_\lambda \; .$$

For a review of that approach see [Rgv4].

These sequences of cocharacters and of codimensions satisfy the following three general theorems:

Theorem 1. ([Rgv1], [Lat]) *Given a P.I. algebra A, there exist constants $b, \alpha > 0$ such that $c_n(A) \leq b \cdot \alpha^n$ for all n.*

Theorem 2. ([Amr.Rgv]) *Denote $H(k, \ell; n) = \{(\lambda_1, \lambda_2, \cdots) \vdash n \mid \lambda_{k+1} \leq \ell\}$. Given a P.I. algebra A, there exist k, ℓ such that for all n, $\chi_n(A) = \sum_{\lambda \in H(k,\ell;n)} m_\lambda(A)\chi_\lambda$.*

Theorem 3. ([Brl1], [Brl.Rgv2]) *The multiplicities $m_\lambda(A)$ are polynomially bounded: there exist constants $b, k > 0$ such that for all n and all $\lambda \vdash n$, $m_\lambda(A) \leq b|\lambda|^k$.*

Remarks. (a) Theorem 1 holds in any characteristic, while theorems 2 and 3 do assume that $\operatorname{char}(F) = 0$.

(b) The codimensions $c_n(A)$ were introduced in order to prove that if A and B are two P.I. algebras, then $A \otimes B$ is a P.I. algebra [Reg1].

By now, tight estimates for the asymptotic behaviour of $c_n(A)$ were calculated for infinitely many P.I. algebras. We describe these results in the following sections. The emerging phenomena is that, probably,

$$c_n(A) \underset{n\to\infty}{\simeq} a \cdot n^g \cdot \alpha^n$$

for some constants a, g and α. Furthermore, in all cases computed so far, $1 \le \alpha \in \mathbb{Z}$ and $g \in \frac{1}{2}\mathbb{Z}$.

In fact, we have the following corollary of a result of Kemer [Kem1].

Theorem. (see [Drn.Rgv]) *If $1 \le \alpha \le 2$ then $\alpha = 1$ or $\alpha = 2$. i.e.: either $c_n(A) \le a \cdot n^k$ for some a and k, or $b \cdot 2^n \le c_n(A)$ for some $0 < b$.*

§2. Cocharacters for Additional Structures

Matrix algebras are endowed with the trace function, which gives rise to two types of trace identities — with corresponding cocharacters and codimensions.

We begin with monomials. The two types of trace monomials are

1) the pure:

$$\mathrm{tr}(x_{i_1} \cdots x_{i_a}) \cdot \mathrm{tr}(x_{j_1} \cdots x_{j_b}) \cdots \mathrm{tr}(x_{k_1} \cdots x_{k_c}) \qquad \text{and}$$

2) the mixed

$$x_{i_1}, \cdots x_{i_a} \cdot \mathrm{tr}(x_{j_1} \cdots x_{j_b}) \cdots \mathrm{tr}(x_{k_1} \cdots x_{k_c}) \ .$$

Linear combinations of such monomials yield the corresponding (free) algebras of pure and of mixed trace polynomials. Such polynomials can be evaluated on $M_k(F)$, the $k \times k$ matrices, hence the obvious definitions of pure and of mixed trace identities, $PTI(M_k(F))$ and $MTI(M_k(F))$.

In analogy to the previous, "ordinary" case, restricting to multilinear such polynomials yield the corresponding pure and mixed trace cocharacters and codimensions. The mixed sequences bridge between the pure-trace and the ordinary sequences.

We look now more closely at these invariants. Let $PT_n = PT_n(x_1, \cdots, x_n)$ denote the multilinear pure trace polynomials in $x_1, \cdots, x_n$. Let $\sigma \in S_n$; the correspondence

$$\sigma = (i_1, \cdots, i_a)(j_1, \cdots, j_b) \cdots (k_1, \cdots, k_c) \to$$
$$\to M_\sigma(x_1, \cdots, x_n) = \mathrm{tr}(x_{i_1} \cdots x_{i_a})\,\mathrm{tr}(x_{j_1} \cdots x_{j_b}) \cdots \mathrm{tr}(x_{k_1} \cdots x_{k_c})$$

is a vector space isomorphism $FS_n \cong PT_n$. The natural S_n action $f(x_1, \cdots, x_n) \to f(x_{\tau(1)}, \cdots, x_{\tau(n)})$, $\tau \in S_n$, makes PT_n a left FS_n module, which is thus isomorphic to FS_n over itself — under the S_n conjugation action.

It is well known that $FS_n = \bigoplus_{\lambda \in \mathrm{Par}\,(n)} I_\lambda$, where the I_λ's are the minimal 2-sided ideals. Here we have the important theorem of Procesi and Razmyslov.

Theorem. ([Pro], [Raz1]) *Let $A = M_k(F)$, with $PTI(A)$ its pure trace identities. Under the above isomorphism,*

$$PTI(A) \cap PT_n \cong \bigoplus_{\substack{\lambda \in \mathrm{Par}\,(n) \\ \lambda_{k+1} \neq 0}} I_\lambda \ .$$

Since the S_n character of I_λ, under the conjugation action, is $\chi_\lambda \otimes \chi_\lambda$, it follows that the pure trace cocharacters of $A = M_k(F)$ are

$$PT\chi_n(A) = \chi_{S_n}\big(PT_n/(PT_n \cap PTI(A)\big) = \sum_{\substack{\lambda \in \mathrm{Par}\,(n) \\ \lambda_{k+1} = 0}} \chi_\lambda \otimes \chi_\lambda \ .$$

Here $\otimes$ is the Kronecker product of characters: $(\varphi \otimes \psi)(\sigma) = \varphi(\sigma)\psi(\sigma)$.

Now to the "mixed" case: denote by $MT_n = MT_n(x_1, \cdots, x_n)$ the multilinear mixed trace polynomials in $x_1, \cdots, x_n$.

Any $f \in PT_{n+1}(x_0, \cdots, x_n)$ can be written uniquely as $f = \mathrm{tr}(x_0 \cdot g)$ where $g \in MT_n(x_1, \cdots, x_n)$. Let $B \in M_k(F)$, then $\mathrm{tr}(XB) = 0$ for all $X \in M_k(F)$ if and only if $B = 0$. Thus, in $f = \mathrm{tr}(x_0 g)$, f is a pure trace identity if and only if g is a mixed trace identity, and $g \to \mathrm{tr}(x_0 \cdot g)$ is an identities-preserving vector space isomorphism: $MT_n \cong PT_{n+1}$. Moreover, both spaces are FS_n modules for $S_n = S_n(1, \cdots, n) \subseteq S_{n+1}(0, 1, \cdots, n)$, and the above is an FS_n module isomorphism.

It follows that the n-th mixed trace cocharacter of $M_k(F)$ is

$$MT\chi_n(M_k(F)) = [PT\chi_{n+1}(M_k(F))] \downarrow_{S_n}$$

the restriction of the $n+1$-th pure trace cocharacter to S_n.

It is known that

$$\chi_n(M_k(F)) = \sum_{\substack{\mu \vdash n \\ \mu_{k^2+1} = 0}} m(\mu)\chi_\mu$$

and

$$MT\chi_n(M_k(F)) = \sum_{\substack{\mu \vdash n \\ \mu_{k^2+1} = 0}} \bar{m}(\mu)\chi_\mu \ .$$

Formanek's results in [Frm1], [Frm2] imply

Theorem. ([Formanek]) *In the above, $m(\mu) = \bar{m}(\mu)$ if $\mu_{k^2} \geq 2$ (i.e. for most μ's, $m(\mu) = \bar{m}(\mu)$).*

Finally, we examine involution-cocharacters. Let A be an algebra with an involution $a \to a^*$. Polynomials $f(x_1, x_1^*, x_2, x_2^*, \cdots)$ can be evaluated on A, thus yielding $*$-identities. Such $f(x_1, x_1^*, x_2, x_2^*, \cdots)$ is linear in x_1 if $\deg_{x_1} f + \deg_{x_1^*} f = 1$. The space $V_n^*(x_1, x_1^*, \cdots, x_n, x_n^*)$ of such $*$-multilinear polynomials is $2^n \cdot n!$ dimensional. Here the wreath product $\mathbb{Z}_2 \sim S_n$ acts naturally on V_n^*. $*$-cocharacters are therefore the $\mathbb{Z}_2 \sim S_n$ characters, and their degrees are the $*$-codimensions [Gmb.Reg].

§3. Kemer's Structure Theory

In his solution of the Specht problem, Kemer introduced "Verbally Prime" T ideals.

Definition. ([Kem2]) The T-ideal $I \subseteq F < x >$ is verbally prime if it satisfies: if $f(x_1, \cdots, x_r) \cdot g(x_{r+1}, \cdots, x_s) \in I$ then either $f \in I$ or $g \in I$. A P.I. algebra is verbally prime if its T ideal of identities $I = I(A)$ is verbally prime.

One such algebra is E, the infinite dimensional Grassmann (Exterior) algebra. Kemer then proved

Theorem [Kem2].

(a) *The verbally prime P.I. algebras are: F, $F < x >$, $M_k(F)$ (the $k \times k$ matrices over F), $M_k(E)$ and $M_{k,\ell} \subseteq M_{k+\ell}(E)$. Here $E = E_0 \oplus E_1$, where E_0 is the even and E_1 the odd parts of E, and*

$$M_{k,\ell} = \overset{\quad k \quad\ \ \ell}{\begin{pmatrix} E_0 & E_1 \\ & \\ E_1 & E_0 \end{pmatrix}}\begin{matrix} k \\ \\ \ell \end{matrix}$$

The T-ideals of the above algebras are the verbally prime T-ideals.

(b) *Let $I \subseteq F < x >$ be an arbitrary T-ideal, then there exist verbally prime T-ideals $J_1, \cdots, J_r$ such that*

$$J_{i_1} \cdots J_{i_m} \subseteq I \subseteq J_1 \cap \cdots \cap J_r \ ,$$

$1 \le i_1, \cdots, i_m \le r$.

The above theorem underscores the importance of verbally prime T-ideals, as well as that of intersections and products of such ideals. Estimates of the asymptotics of the codimensions of the corresponding P.I. algebras are discussed in the sequel.

§4. The Codimensions $\mathbf{c_n(M_k(F))}$ and $\mathbf{t_n(M_k(F))}$

Let $t_n(M_k(F))$ denote the (pure) trace codimensions of $M_k(F)$. Its asymptotics, as well as that of the codimensions $c_n(M_k(F))$, is given by:

4.1 Theorem. ([Rgv3]) *As $n \longrightarrow \infty$,*

$$c_{n-1}(M_k(F)) \simeq t_n(M_k(F)) \simeq a \cdot n^g \cdot \alpha^n \ ,$$

where $\alpha = k^2$, $g = \frac{1}{2}(k^2 - 1)$ and

$$a = \left(\frac{1}{\sqrt{2\pi}}\right)^{k-1} \left(\frac{1}{2}\right)^{\frac{1}{2}(k^2-1)} \cdot 1! \cdot 2! \cdots (k-1)! \cdot k^{\frac{1}{2}(k^2+4)} \ .$$

The derivation of the above asymptotics is done by applying first the Procesi–Razmyslov theory of trace identities, and then Formanek's "conductor" theorems, which relate the trace and the ordinary cocharacters:

4.2 Theorem. (a consequence of [Pro], [Raz1]) *The trace cocharacters* $T\chi_n(M_k(F))$ *satisfy:*

$$T\chi_n(M_k(F)) = \sum_{\substack{(\lambda_1,\lambda_2,\cdots)\vdash n \\ \lambda_{k+1}=0}} \chi_\lambda \otimes \chi_\lambda$$

where $\otimes$ *is the Kronecker (i.e. "inner") product. It follows that the trace codimensions satisfy*

$$t_n(M_k(F)) = \sum_{\substack{(\lambda_1,\lambda_2,\cdots)\vdash n \\ \lambda_{k+1}=0}} d_\lambda^2 \ .$$

4.3 Theorem. (a consequence of [Frm1], [Frm2]) *Denote the* n^{th} *cocharacter of* $M_k(F)$ *by*

$$\chi_n(M_k(F)) = \sum_{\substack{(\mu_1,\mu_2,\cdots)\vdash n \\ \mu_{k^2+1}=0}} m_\mu \chi_\mu$$

and let

$$(T\chi_{n+1}(M_k(F))) \downarrow_{S_n} = \sum_{\substack{(\mu_1,\mu_2,\cdots)\vdash n \\ \mu_{k^2+1}=0}} \bar{m}_\mu \chi_\mu$$

where $\downarrow_{S_n}$ *denotes the restriction to* S_n. *If* $\mu_{k^2} \geq 2$ *then* $m_\mu = \bar{m}_\mu$.

The exponential growth of the d_μ's, together with the polynomial growth of both m_μ's and of $\bar{m}_\mu$'s [Brl1], [Rgv3] imply:

4.4 Theorem.

$$c_n(M_k(F)) \simeq t_{n+1}(M_k(F)) = \sum_{\substack{\lambda\vdash n+1 \\ \lambda_{k+1}=0}} d_\lambda^2 \ .$$

The final step now is the computation of the asymptotics of the right-hand-side. This is done in [Rgv2] as follows: First, d_λ of a "generic" λ is simplified by an asymptotic approximation. The corresponding sum is then approximated by $a_1 a_2 n^g \alpha^n$ (g and α as in 4.1), with a_2 explicit and a_1 a Riemann sum. This Riemann sum is then approximated by an integral — the Selberg integral. The evaluation of that integral approximates a_1, hence the approximation of $a_1 a_2$ by a of 4.1.

§5. The Codimensions of the other Verbally Prime Algebras

Parts of the theory of the previous section were generalized to $M_{k,\ell}$. First, Razmyslov [Raz2] applied the trace function of $M_{k,\ell}$ to determine the trace identities of that algebra. Recall that the (super) trace of $A \in M_{k,\ell}$ is the sum of the first k elements minus the sum of the last ℓ elements of the main diagonal of A. In particular Razmyslov's work implies

5.1 Theorem. *Denote* $H(k,\ell;n) = \{(\lambda_1,\lambda_2,\cdots) \vdash n \mid \lambda_{k+1} \leq \ell\}$. *Then the trace codimensions of* $M_{k,\ell}$ *satisfy:*

$$t_n(M_{k,\ell}) = \sum_{\lambda \in H(k,\ell;n)} d_\lambda^2 \ .$$

We note that the precise asymptotics of the right hand side is calculated in [Brl.Rgv1, §7], similar to the computation of the asymptotics mentioned at the end of §4.

Razmyslov [Raz2] also found a central polynomial for $M_{k,\ell}$. Applying it (see [Brl.Rgv3]), Formanek's conductor theorems are partially generalized, thus relating $c_n(M_{k,\ell})$ with $t_{n+1}(M_{k,\ell})$. Together with Theorem 5.1, this yields the following tight sandwich for $c_n(M_{k,\ell})$:

5.2 Theorem. ([Brl.Rgv3]) *There exist* $0 < a_2 \le a_1$ *such that*

$$a_2 n^g \alpha^n \le c_n(M_{k,\ell}) \le a_1 n^g \alpha^n ,$$

where $\alpha = (k+\ell)^2$ *and* $g = -\frac{1}{2}(k^2+\ell^2-1)$.

To sandwich $c_n(M_k(E))$, we apply a special case of a theorem of Kemer [Kmr2] on the tensor products of verbally prime algebras: $M_{k,k}$ and $M_k(E)\otimes E$ satisfy the same identities, hence $c_n(M_{k,k}) = c_n(M_k(E)\otimes E)$. Since in general, $c_n(A\otimes B) \le c_n(A)\cdot c_n(B)$ and since $c_n(E) = 2^{n-1}$ [Kra.Rgv], this implies (see [Brl.Rgv3])

5.3 Theorem. *There exist* $0 < a_2 \le a_1$ *such that*

$$a_2 n^{-\frac{1}{2}(2k^2-1)}(2k^2)^n \le c_n(M_k(E)) \le a_1 n^{-\frac{1}{2}(k^2-1)}(2k^2)^n.$$

We mention two special cases:

5.4 Theorem.

$$\chi_n(E) = \sum_{\lambda\in H(1,1;n)} \chi_\lambda$$

[Ols.Rgv] *and*

$$c_n(E) = 2^{n-1}$$

[Kra.Rgv].

5.5 Theorem. $\chi_n(E\otimes E) = \sum_{\ell=0}^{n}(\sum_{\lambda\in H(1,2;\ell)}\chi_\lambda)\hat{\otimes}\chi_{(n-\ell)}$ [Po], *where* $\hat{\otimes}$ *is the "outer" product. It follows that*

$$c_n(E\otimes E) = c_n(M_{1,1}) \simeq \frac{2}{\sqrt{\pi}}\,\frac{1}{\sqrt{n}}\cdot 4^n \quad \textit{(see (Rgv5])}.$$

5.6. Table. Next, assume that for any verbally prime algebra A there exist constants a, g and α such that $c_n(A) \simeq a n^g \alpha^n$ (for $A = M_{k,\ell}$ or $M_k(E)$, $k \ge 2$, this is still a conjecture). Theorems 4.1, 5.2 and 5.3 can be summarized in the following table

	α	g	a
$M_k(F)$	k^2	$-\frac{1}{2}(k^2-1)$	see 4.1
$M_{k,\ell}$	$k^2+\ell^2$	$-\frac{1}{2}(k^2+\ell^2-1)$	?
$M_k(E)$	$2\cdot k^2$	$-\frac{1}{2}(2k^2-1) \le g \le -\frac{1}{2}(k^2-1)$	?

Finally, we introduce the following notations for the families of verbally prime algebras, according to our present knowledge of the asymptotics of their codimensions:

5.7 Definition.

$$\mathcal{F}_{pr} = \{E, E \otimes E\} \cup \{M_k(F) \mid k = 1, 2, \cdots\}$$

(pr for "precise asymptotics").

$$\mathcal{F}_{ts} = \mathcal{F}_{pr} \cup \{M_{k,\ell} \mid k, \ell \geq 1\} \qquad (ts \text{ for "tight sandwich"}).$$

$$\mathcal{F}_{vp} = \text{All the verbally prime algebras.}$$

Clearly, $\mathcal{F}_{pr} \subseteq \mathcal{F}_{ts} \subseteq \mathcal{F}_{vp}$.

§6. Intersections of Verbally Prime T-ideals

Here we have the following:

6.1 Theorem. ([Brl.Rgv4]) *Let $J_1, \cdots, J_r$ be distinct verbally prime ideals. Then there exists a unique i such that*

$$c_n\left(F < x > /(J_1 \cap \cdots \cap J_r)\right) \simeq c_n(F < x > /J_i).$$

The proof follows from the bounds

$$c_n(F < x > /J_i) \leq c_n\left(F < x > /(J_1 \cap \cdots \cap J_r)\right) \leq \sum_{i=1}^{r} c_n(F < x > /J_i) \qquad (1 \leq i \leq r)$$

and from the fact that Table 5.6 implies that as $n \to \infty$, the upper bound on the right is dominated by a (unique) single summand. A crucial fact here is that if $0 \leq k_1, \ell_1, k_2, \ell_2 \in \mathbb{Z}$ and $k_1^2 + \ell_1^2 = k_2^2 + \ell_2^2$ then either $k_1 = k_2$ (and $\ell_1 = \ell_2$) or $k_1 = \ell_2$ (and $\ell_1 = k_2$).

§7. Products of Verbally Prime T-ideals

We introduce first an extension of 5.7:

7.1 Definition. Define $\bar{\mathcal{F}}_{pr}$ as follows:

$A \in \bar{\mathcal{F}}_{pr}$ if there exist $A_1, \cdots, A_s \in \mathcal{F}_{pr}$ such that $\mathrm{Id}(A) = \mathrm{Id}(A_1) \cdots \mathrm{Id}(A_s)$, where $\mathrm{Id}(A)$ denotes the T-ideal of the identities of A.

Clearly, $\mathcal{F}_{pr} \subseteq \bar{\mathcal{F}}_{pr}$. Define similarly $\mathcal{F}_{ts} \subseteq \bar{\mathcal{F}}_{ts}$ and $\mathcal{F}_{vp} \subseteq \bar{\mathcal{F}}_{vp}$ via products of the corresponding T-ideals. We have

7.2 Theorem. ([Brl.Rgv4])

(1) *Let $A \in \bar{\mathcal{F}}_{pr}$, then $c_n(A) \simeq a \cdot n^g \cdot \alpha^n$ where a, g and α can be calculated explicitly.*

(2) *Let* $A \in \bar{\mathcal{F}}_{ts}$, *then*

$$q_2 n^g \alpha^n \leq c_n(A) \leq q_1 n^g \alpha^n$$

for constants $0 < q_2 \leq q_1$, g *and* α, *and where* g *and* α *can be calculated explicitly.*

(3) *Let* $A \in \bar{\mathcal{F}}_{vp}$, *then*

$$q_2 n^{g_2} \alpha^n \leq c_n(A) \leq q_1 n^{g_1} \alpha^n$$

for constants $0 < q_2 \leq q_1$, g_1, g_2 *and* α, *where* $q_2 \leq q_1$ *and* $g_2 \leq g_1$ *can be bounded and* α *can be calculated explicitly.*

There are two key tools in the proof of 7.2. The first is a consequence of a theorem of Formanek [Frm3] about the Poincaré series of A_1, A_2 and of A:

7.3 Theorem. ([Brl.Rgv4]) *Let* $\mathrm{Id}(A) = \mathrm{Id}(A_1) \cdot \mathrm{Id}(A_2)$, *then*

$$\chi_n(A) = \chi_n(A_1) + \chi_n(A_2) + \chi_{(1)} \hat{\otimes} \sum_{j=0}^{n-1} \chi_j(A_1) \hat{\otimes} \chi_{n-j-1}(A_2) - \sum_{j=0}^{n} \chi_j(A_1) \hat{\otimes} \chi_{n-j}(A_2).$$

Hence, applying "deg", we obtain

$$c_n(A) = c_n(A_1) + c_n(A_2) + n \sum_{j=0}^{n-1} c_j(A_1) \cdot c_{n-j-1}(A_2) - \sum_{j=0}^{n} c_j(A_1) \cdot c_{n-j}(A_2).$$

The second tool is:

7.4 Lemma. ([Bec.Rgv., 1.1]) *Let* $e_1, e_2, \alpha_1, \alpha_2 \in \mathbb{R}$, *and* $\alpha_1, \alpha_2 > 0$. *Then*

$$\sum_{j=0}^{n} \binom{n}{j} j^{e_1} (n-j)^{e_2} \alpha_1^g \alpha_2^{n-j} \underset{n \to \infty}{\simeq} \frac{\alpha_1^{e_1} \alpha_2^{e_2}}{(\alpha_1 + \alpha_2)^{e_1+e_2}} \cdot n^{e_1+e_2} \cdot (\alpha_1 + \alpha_2)^n.$$

Clearly, if we know the asymptotics (resp. tight sandwiches) of $c_n(A_1)$ and $c_n(A_2)$, then lemma 7.4 gives the asymptotics (resp. tight sandwiches) of the 3rd and 4th summands of 7.3. Thus, both summands have exponential growth $(\alpha_1 + \alpha_2)^n$, while for the first two its α_1^n and α_2^n. However, the power of n is $e_1 + e_2 + 1$ in the third summand, and $e_1 + e_2$ in the fourth. It follows that the third summand dominates $c_n(A)$, and this yields the asymptotics (resp. tight sandwich) of $c_n(A)$.

For example we have

7.5 Theorem. ([Brl.Rgv4]) *Let* $c_n(A_j) \simeq a_j \cdot n^{e_j} \alpha_j^n$, $j = 1, \cdots, s$, *and* $\mathrm{Id}(A) = \mathrm{Id}(A_1) \cdots \mathrm{Id}(A_s)$. *Then* $c_n(A) \simeq a n^e \alpha^n$, *where* $\alpha = \alpha_1 + \cdots + \alpha_s$, $e = e_1 + \cdots + e_s + s - 1$ *and where* $a = a_1 \cdots a_s \frac{\alpha_1^{e_1} \cdots \alpha_s^{e_s}}{(\alpha_1 + \cdots + \alpha_s)^e}$.

We take the opportunity to slightly modify Lemma 2.5 of [Brl.Rgv4]:

7.6 Lemma. *Let* $\mathrm{Id}(A) = \mathrm{Id}(A_1) \cdot \mathrm{Id}(A_2)$ *and assume for* $i = 1, 2$

$$q_{2,i} n^{v_i} \alpha_i^n \le c_n(A_i) \le q_{1,i} n^{u_i} \alpha_i^n$$

with $v_i \le u_i$ *and* $0 < q_{1,i}, q_{2,i}$.

Then, for sufficiently large n, there exist $r_1, r_2 \in \mathbb{R}$ such that

$$r_2 n^{v_1+v_2+1} (\alpha_1 + \alpha_2)^n \le c_n(A) \le r_1 n^{u_1+u_2+1} (\alpha_1 + \alpha_2)^n$$

(the upper bound holds for all n).

We sketch the

Proof. For the upper bound, just follow part (a) of [Brl.Rgv4, Lemma 2.5].

To obtain the lower bound, apply 7.5 and sandwich summands 3rd and 4th of 7.3. Summands 1 and 2 of 7.3 can be discarded, and we also see that the 3rd summand dominates the 4th. By slightly decreasing the coefficient of that 3rd summand we obtain a lower bound — if n is sufficiently large.

Induction on s clearly gives the corresponding result when $I = J_1 \cdots J_s$, where $J_1, \cdots, J_s$ are verbally prime ideals. We summarize:

7.7 Theorem.

(1) *Let* $A \in \bar{\mathcal{F}}_{pr}$, *then* $c_n(A) \simeq a \cdot n^g \cdot \alpha^n$, *and the constants* a, g *and* α *can be calculated explicitly. In particular,* $1 \le \alpha \in \mathbb{Z}$ *and* $g \in \frac{1}{2}\mathbb{Z}$.

(2) *Let* $A \in \bar{\mathcal{F}}_{ts}$, *then there exist* $0 < a_2 \le a_1$, g *and* α, *which can be calculated explicitly, such that*

$$a_2 n^g \alpha^n \le c_n(A) \le a_1 n^g \alpha^n$$

if n *is sufficiently large. Again,* $1 \le \alpha \in \mathbb{Z}$ *and* $g \in \frac{1}{2}\mathbb{Z}$.

(3) *Let* $A \in \bar{\mathcal{F}}_{vp}$, *then there exist* $0 < a_2 \le a_1$, $g_2 \le g_1$ *and* α *such that*

$$a_2 n^{g_2} \alpha^n \le c_n(A) \le a_1 n^{g_1} \alpha^n$$

if n *is large. Here, again,* $1 \le \alpha \in \mathbb{Z}$.

§8. Involution Codimensions

Recall. Let A be an algebra with an involution $* : A \to A$. Then polynomials in $\{x_i, x_i^* \mid i = 1, 2, \cdots\}$ can be evaluated on A, giving rise to $*$-polynomial identities, hence to $*$-codimensions. Cocharacters can be defined via the wreath product $B_n = \mathbb{Z}_2 \sim S_n$ [Gmb.Rgv]. If in addition A has a trace, then one also obtains the various $*$-trace-codimensions and cocharacters.

Recall that when $A = M_k(F)$, $*$ is either the transpose or a symplectic involution. Procesi's theory of $*$-trace identities for $M_k(F)$ [Pro] [Lod.Pro] implies:

8.1 Theorem.

1) *$t_n(M_k(F)$, transpose) =*
$= \sum\{d_\lambda \mid \lambda \vdash 2n, \lambda_{k+1} = 0$ *and all* λ_i *are even*$\}$.
2) *$t_n(M_k(F)$, symplectic) =*
$= \sum\{d_\lambda | \lambda \vdash 2n, \lambda_{k+1} = 0$ *and all* λ'_i *are even*$\}$
(here λ' is the conjugate partition of λ).

The asymptotics of the above sums were also calculated in [Rgv2] (see the remarks following 4.4 above), and we obtain

8.2 Corollary.

(1) *$t_n(M_k(F)$, transpose) $\simeq a_t n^{g_t} \alpha_t^n$ where $\alpha_t = k^2$, $g_2 = -\frac{1}{4}k(k-1)$ and*

$$a_t = \left(\frac{k}{2}\right)^{\frac{1}{4}k(k-1)} \cdot \frac{1}{k!}\left[\Gamma\left(\frac{3}{2}\right)\right]^{-k} \cdot \left[\prod_{j=1}^{k} \Gamma(1+\frac{1}{2}j)\right] \cdot \left(\frac{1}{2}\right)^{k-1}$$

(Γ is the Gamma function).

(2) *$t_n(M_k(F)$, symplectic) $\simeq a_s n^{g_s} \alpha_s^n$ where $k = 2N$, $\alpha_s = k^2 = (2N)^2$, $g_s = -\frac{1}{4}k(k+1) = -\frac{1}{2}N(2N+1)$ and*

$$a_s = \left[\left(\frac{1}{\sqrt{2\pi}}\right)^N 2^{(N^2+N+2)/4} \cdot N^{N(7N-1)/4} \cdot \frac{1}{N!} \prod_{j=1}^{N} \Gamma(2j+1)\right] \times \left(\frac{1}{2}\right)^{N(2N+1)/2}$$

As in the case of the ordinary identities, we now prove

8.3 Theorem. ([Brl.Gmb.Rgv]) *Let $* : M_k(F) \to M_k(F)$ be an involution, then $t_n(M_k(F), *) \simeq c_n(M_k(F), *)$.*

The Proof. is obtained by generalizing the many steps in the proof of Theorem 4.1 above (i.e. ordinary P.I.) to the involution case. The ordinary case was based on Formanek's fundamental work on the various Poincaré series of $M_k(F)$, which relates identities with and without trace [Frm1]. All the essential parts of that proof were generalized to the involution case: a large part was generalized by Berele [Brl2] and other parts are generalized in [Brl.Gmb.Rgv].

We remark that the Littlewood-Richardson (L-R) rule plays an important role in the proof: we study the identities with an involution by applying the representation theory of the wreath product $\mathbb{Z}_2 \sim S_n$. Ignoring the involution translates into restriction to S_n. Now, an irreducible $\mathbb{Z}_2 \sim S_n$ character is $\chi_{\lambda,\mu}$, where λ and μ are partitions and $|\lambda| + |\mu| = n$. The restriction to S_n satisfies $\chi_{\lambda,\mu} \downarrow_{S_n} = \chi_\lambda \hat{\otimes} \chi_\mu$, the outer product which is calculated by the L-R rule. Thus, some basic properties of the L-R rule enter our proof, in particular, in lifting the results from the ordinary to the involution case.

§9. Additional Cases

Polynomially Bounded Codimensions

Call $c_n(A)$ "polynomially bounded" if there exist constants a and k such that $c_n(A) \leq an^k$.

Kemer [Kem1] has characterized such P.I. algebras A. Drensky [Drn] has shown that Kemer's characterization implies that for such an A, $c_n(A) \simeq q \cdot n^k$, with $q \in \mathbb{Q}$ and $0 \leq k \in \mathbb{Z}$.

In that case we have:

9.1. Theorem. ([Drn.Rgv]) *Let $c_n(A)$ be polynomially bounded, and let*

$$c_n(A) \simeq q \cdot n^k.$$

1) *Assume $1 \in A$. Then*

$$0 \approx \frac{1}{k!} \leq q \leq \frac{1}{2!} - \frac{1}{3!} \pm \cdots + \frac{(-1)^k}{k!} \approx 1/e.$$

2) *Assume $1 \notin A$. Then for any $0 \leq q \leq \mathbb{Q}$ there exist $1 \leq k \in \mathbb{Z}$ and an algebra A such that $c_n(A) \simeq qn^k$.*

We conclude with the following intriguing theorem of Drensky.

9.2. Theorem. ([Drn1, 5.4]) *Let $T_d = T([x_1, \cdots, x_{d+1}]) \subseteq F < x >$ denote the T ideal generated by $[x_1, \cdots, x_{d+1}]$, where, by induction, $[x_1, \cdots, x_{d+1}] = [[x_1, \cdots, x_d], x_{d+1}]$. Let $U_d = F < x > /T_d$ and let $c_n(U_d)$ denote its codimensions.*

Then there exist polynomials $p_d(n)$ and $q_d(n)$ such that for all n,

$$p_d(n) \cdot 2^n \leq c_n(U_d) \leq q_d(n) \cdot 2^n \ ,$$

and deg $p_d(n) \to \infty$ *as* $d \to \infty$.

Note also that Proposition 6.7 of [Drn1] gives the exact codimensions corresponding to the following identities:

$$[[x_1, x_2], [x_3, x_4]], \qquad c_n = 2^{n-1}(n-1) + 1 - \binom{n}{2} + 2\binom{n}{4},$$

$$[x_1, x_2]^2, \qquad c_n = 2^{n-1}(n-1) + 1 - \binom{n}{2} + 3\binom{n}{4} + 4\binom{n}{5}$$

and

$$[x_2, x_1, x_1, x_1] \qquad c_n = 2^{n-1} + 2\binom{n}{3} + 5\binom{n}{4} + 4\binom{n}{5}.$$

References

[Amr] S.A. Amitsur, The sequence of codimensions of P.I.–algebras, *Israel J. Math.* **47** (1984), 1–22.

[Amr.Rgv] S.A. Amitsur and A. Regev, P.I. algebras and their cocharacters, *J. Algebra* **78** (1982), 248–254.

[Bck.Rgv] W. Beckner and A. Regev, Asymptotic estimates using Probability, *Adv. in Math.*, in press.

[Brl1] A. Berele, Homogeneous polynomial identities, *Israel J. Math.* **42** (1982), 258–272.

[Brl2] A. Berele, Matrices with involutions and invariant theory, *J. Algebra* **135** (1990), 139–164.

[Brl.Gmb.Rgv] A. Berele, A. Giambruno and A. Regev, Involution codimensions and trace codimensions are asymptotically equal, *Israel J. Math.*, in press.

[Brl.Rgv1] A. Berele and A. Regev, Hook Young diagrams with applications to combinatorics and to representations of Lie superalgebras, *Adv. Math.* **64** (1987), 118–175.

[Brl.Rgv2] A. Berele and A. Regev, Applications of hook diagrams to P.I. algebras, *J. Alg.* **82** (1983), 559–567.

[Brl.Rgv3] A. Berele and A. Regev, On the codimensions of the verbally prime P.I. algebras, *Israel J. Math.* **91** (1995), 239–247.

[Brl.Rgv4] A. Berele and A. Regev, Codimensions of products and of intersections of verbally prime T-ideals, *Israel. J. Math.*, in press.

[Drn1] V. Drensky, Codimensions of T-ideals and Hilbert series of relatively free algebras, *J. Algebra* **91** (1984), 1–17.

[Drn2] V. Drensky, Relations for the cocharacter sequences of T ideals, *Proc. International Conf. on Algebra, honoring A. Malcev, Contemp. Math.* **131** (Part 2) (1992), 285–300.

[Drn.Rgv] V. Drensky and A. Regev, Exact asymptotic behaviour of the codimensions of some P.I. algebras, *Israel J. Math.*, in press.

[Frm1] E. Formanek, Invariants of the ring of generic matrices, *J. Algebra* **89** (1984), 178–223.

[Frm2] E. Formanek, A conjecture of Regev on the Capelli polynomial, *J. Algebra* **109** (1987), 93–114.

[Frm3] E. Formanek, Noncommutative Invariant Theory, in Group Actions on Rings ed. S. Montgomery Contemp. Math. Vol.43 AMS, 1985, pp. 87–119.

[Gmb.Rgv] A. Giambruno and A. Regev, Wreath products and P.I. algebras, *J. Pure and Applied Algebra* **35** (1985), 133–150.

[Kem1] A. Kemer, T-ideals with polynomial growth of the codimensions are Specht, *Sib. Math. J.* **19** (1978), 37–48.

[Kem2] A. Kemer Ideals of Identities of associative algebras, *AMS Translations of Mathematical Monographs* Vol. **87** (1988).

[Kra.Rgv] D. Krakowski and A. Regev, The polynomial identities of the Grassmann algebra, *Trns. AMS* **181** (1973), 429–438.

[Lat] V.N. Latyshev, On Regev's theorem on identities in a tensor product of P.I.-algebras, *Usp. Mat. Nauk.* **27** (1972), 213–214. (Russian)

[Lod.Pro] J.L. Loday and C. Procesi, Homology of symplectic and orthogonal algebras, *Adv. in Math.* **69** (1988), 93–108.

[Ols.Rgv] J. Olson and A. Regev, The colength of some T ideals, *J. Algebra* (1976), 100–111.

[Po] A. Popov, Identities of the tensor square of a Grassmann algebra, (English:), *Algebra and Logic* **21** (1982), No.4, 296–316.

[Pro] C. Procesi, The invariant theory of $n \times n$ matrices, *Adv. in Math.* **19** (1976), 306–381.

[Raz1] Yu.P. Razmyslov, Trace identities of full matrix algebras over a field of characteristic zero, *Math. USSR-Isv.* **8** (1974), 727–760.

[Raz2] Yu.P. Razmyslov, Trace identities and central polynomials in the matrix superalgebra $M_{k,\ell}$, *Math USSR-Sbornik* **56** (1987), 187–206.

[Rgv1] A. Regev, Existence of identities in $A \otimes B$, *Israel J. Math.* **11** (1972), 131–152.

[Rg2] A. Regev, Asymptotic values for degrees associated with stripes of Young diagrams, *Adv. Math.* **41** (1981), 115–136.

[Rgv3] A. Regev, Codimensions and trace codimensions of matrices are asymptotically equal, *Israel J. Math.* **47** (1984), 246–250.

[Rgv4] A. Regev, On the codimensions of matrix algebras, in: Some Current Trends in Algebra, Proceedings (Varma 1986), 1988, pp. 162–172, L.N.M 1352, Springer.

[Rgv5] A. Regev, Sign trace identities, *Linear and Multilinear Algebra* **21** (1987), 1–28.

A. Regev
Department of Theoretical Mathematics
The Weizmann Institute of Science
Rehovot 76100
Israel
and
Department of Mathematics
The Pennsylvania State University
University Park, PA 16802
U.S.A.

Canadian Mathematical Society
Conference Proceedings
Volume **22**, 1998

Problems on Group Rings

Klaus W. Roggenkamp

Abstract. We shall report on some recent developments of questions centering around the isomorphism problem and the conjectures of Zassenhaus for integral group rings.

1. The Problems

In what follows $R = \text{alg. int.}(K)$, where K is either a local or a global number field. Given a group G and a commutative ring R, we denote by $RG = \{\sum_{\text{finite}} r_g \cdot g\}$ the GROUP RING. For a normal subgroup N we have the AUGMENTATION SEQUENCE WITH RESPECT TO N

$$0 \longrightarrow I(N) \uparrow G \longrightarrow RG \xrightarrow{\epsilon_N} RG/N \longrightarrow 0\,, \tag{1}$$

which is induced by sending $g \in G$ to $g \cdot N \in G/N$. The kernel of the augmentation map, the AUGMENTATION IDEAL $I(N) \uparrow G$ is the RG-ideal generated by the elements $\{n - 1 : n \in N \setminus \{0\}\}$.

In case $N = 1$, the map $\epsilon_1 =: \epsilon_G$ is called the AUGMENTATION MAP of RG.

By $U(RG)$ we denote the UNITS in RG, and $V(RG)$ are the UNITS OF AUGMENTATION 1. Then obviously $U(RG) = V(RG) \cdot U(R)$. So we do not lose anything if we restrict to units of augmentation one. We also may assume that ring homomorphisms $\alpha: RG \longrightarrow RH$ commute with the augmentation; i.e. are AUGMENTED: Replace $\alpha(g)$ by $\alpha(g) \cdot \epsilon_H(\alpha(g)^{-1})$.

AMS Subject Classification (1991). Primary 16G30.

Keywords and Phrases: Integral group rings, isomorphism problem, Zassenhaus conjectures.

This research was partially supported by the Deutsche Forschungsgemeinschaft.

Definition 1.1.

1. $\mathrm{Aut}(X)$ are the automorphisms of the group — ring — module X and $\mathrm{Out}(X)$ are the outer automorphisms — i.e. the automorphisms modulo inner automorphisms — of the group — ring X.
2. The augmented automorphisms of RG are denoted by $\mathrm{Aut}_n(RG)$, similarly $\mathrm{Out}_n(RG)$ denotes the augmented outer automorphisms — note that inner automorphisms are automatically augmented.

For many years there have been the following OUTSTANDING PROBLEMS which deal with the various ways of how G can be embedded into RG — this is tantamount to questions on the automorphisms of RG.

Problem 1.2.

IP ISOMORPHISM-PROBLEM:
(This was posed in 1939 by G. Higman in his thesis for finite groups): Does $RG \simeq RH$ imply $G \simeq H$?

ZCaut *The* ZASSENHAUS-CONJECTURE FOR AUTOMORPHISMS*:*
Let $\alpha: RG \longrightarrow RG$ be an augmented automorphism. Does this imply that there is a group automorphism $\sigma: G \longrightarrow G$ and a central automorphism[1] *$\gamma: RG \longrightarrow RG$ such that $\alpha = \sigma \cdot \gamma$?*

ZC *The* ZASSENHAUS-CONJECTURE (1976)*:*
Let $RG = RH$ as augmented algebras; does this imply that there is central automorphism $\gamma: RG \longrightarrow RG$ with $\mathrm{Im}(\gamma \downarrow_G) = H$?

ZCsub *The* ZASSENHAUS-CONJECTURE FOR SUBGROUPS*:*
Assume that we have a finite group $U \leq V(RH)$, does this imply that there is a unit $a \in KG$ such that $a \cdot U \cdot a^{-1} \leq G$?

CP *The* CONJUGACY PROBLEM*:*
Let K be a local number field and let G be a finite group with $|G| \cdot R \neq R$. If $RG = RH$ does there exist a unit $u \in RG$ with $u \cdot H \cdot u^{-1} = G$?

CPsub *The* CONJUGACY PROBLEM FOR SUBGROUPS*:*
Let K be a local number field and G a finite group with $|G| \cdot R \neq R$. If there is a finite group $U \leq U(RH)$ does there exist a unit $u \in RG$ with $u \cdot U \cdot u^{-1} \leq G$?

AP *The* AUTOMORPHISM PROBLEM*:*
Is the natural homomorphism

$$\Phi_R(G): \mathrm{Out}(G) \longrightarrow \mathrm{Out}(RG),$$

which is induced from the injection $G \longrightarrow RG$, injective?

Remark 1.3.

1. A less ambitious question than a positive answer to the isomorphism problem would be:
What can be said about the structure of H in terms of G provided $\mathbb{Z}G = \mathbb{Z}H$ as augmented algebras. In [Ro-Sc; 86] and in more generality in [Ki-Ri; 93] a Čech-style cohomology theory has been developed, which allows to describe the finite solvable group H provided $\mathbb{Z}G \simeq \mathbb{Z}H$ in terms of G and certain

[1] This means that the automorphism leaves the centre elementwise fixed.

cocycles. Moreover, this theory also provides the right setup to study the Zassenhaus conjecture.

2. The Zassenhaus conjecture is equivalent to a positive answer to the isomorphism problem and the Zassenhaus conjecture for automorphisms.
3. In case of a finite group, any central automorphism of RG is given by conjugation with unit $a \in KG$. This is the original formulation of the conjectures by Zassenhaus. Zassenhaus has made these conjectures only for finite groups. For infinite groups the formulation I have chosen is weaker than if one would extend directly Zassenhaus' formulation.
4. The Zassenhaus conjecture for group rings makes a very strong statement of how in case $RG = RH$, the group H is embedded into the group ring RG. Even if the Zassenhaus conjecture fails, it would be interesting to know, how the group H is embedded into the group ring RG. Also here the Čhech-style cohomology give an answer for solvable groups (cf. [Ki-Ri; 93]).
5. We shall later also discuss some variations of the Zassenhaus conjectures.

I shall report here on both, some positive and some negative results to these problems. Historically, most of the positive results were found before the counterexamples had been constructed. I am sure that in case we had found the negative results first, we would not have thought that the — very strong — positive results could be possible.

2. Positive Results

Here G is a finite group.

2.1 The Conjugacy Problem

Theorem 2.1. (Roggenkamp–Scott [Ro-Sc; 87]; Weiss [Wei; 88]) *Let K be a local number field and let G be a finite p-group. Then the* CONJUGACY PROBLEM FOR SUBGROUPS *is true.*

This is a p-Sylow theorem for finite p-subgroups of the pro-finite p-group $V(RG)$.

Remark 2.2. For p-groups and K a local number field the CONJUGACY PROBLEM was first proved in [Ro-Sc; 87]. Shortly afterwards, Weiss [Wei; 88] simplified the proof considerably and extended the result to prove the CONJUGACY PROBLEM FOR SUBGROUPS in case R is unramified over $\widehat{\mathbb{Z}}_p$. The final result was obtained in [Rog; 92]; see also [Tho; 94]. G. Thompson extended the techniques of [Ro-Sc; 87] to prove subgroup rigidity, independently of Weiss (loc. cit.).

If one requires a bit less, one obtains a positive answer to the Conjugacy Problem[2]:

Theorem 2.3. (Scott [Sc; 87], 266 and [Sc; 90], p. 259) *Let G be a finite group such that the generalized Fitting subgroup of G is a p-group*[3]. *Assume that R is the*

---

[2] L. L. Scott proved this result while collaborating with the author.

[3] For a solvable group G this means that there exists a prime p so that the group $O_{p'}(G) = 1$. Here $O_{p'}(G)$ is the maximal normal subgroup of G of order prime to p.

ring of integers in a global number field. Then the conjugacy problem has a positive answer.

Remark 2.4. Actually one would expect this result to hold for the ring of algebraic integers in a local number field, in which p is not invertible — for a further discussion we refer to [Sc; 90].

2.2 The Zassenhaus Conjecture

Obviously, both the above theorems imply the Zassenhaus conjecture for global K. Moreover, Theorem 2.3 allows for solvable groups to develop an obstruction theory for the Zassenhaus Conjecture for automorphisms to hold. This gives rise to a Čech style cohomology.

Theorem 2.5. (Weiss [Wei; 91]) *Let G be nilpotent. Then the Zassenhaus Conjecture for subgroups is true.*

Though the Zassenhaus Conjecture fails for solvable groups, there are indications, that it holds for simple groups, Bleher [Ble; 95], Bleher–Hiss–Kimmerle [Bl-Hi-Ki; 96], Peterson [Pet; 76], Bleher [Bl; 96], Bleher–Geck–Kimmerle [Bl-Ge-Ki; 96]:

Proposition 2.6. *The Zassenhaus conjecture holds*

1. *for every finite Coxetergroup (i.e. finite groups generated by reflections),*
2. *for the following list of finite simple groups — p is a rational prime and $m, f \in \mathbb{N}$:*

$$\begin{array}{lllll} SL(2,p^f), & PSL(2,p^f), & {}^2B_2(2^{2m+1}), & {}^2G_2(3^{2m+1}), & {}^2F_4(2^{2m+1}), \\ SL(3,3^m), & SU(3,3^{2m}), & Sp(4,2^m), & G_2(p^m), & {}^3D_4(p^{3m}), \end{array}$$

3. *for every minimal simple group,*
4. *for the Zassenhaus groups,*
5. *for every simple group with an abelian 2-Sylow-subgroup,*
6. *for 15 out of the 26 sporadic simple groups*[4].

2.3 Variations on the Zassenhaus Conjecture

Two variations of the Zassenhaus conjecture are worth mentioning explicitly:

Definition 2.7.

ZC1: Let $u \in V(RG)$ be an element of finite order, then there exists an element $a \in (KG)$ with ${}^a u \in G$.

ZC-P: Let $U \leq V(RG)$ be a finite p-group, then there exists an element $a \in U(KG)$ with ${}^a U \subseteq G$.

To neither of these conjectures there is at present a counterexample, but there is some positive evidence for them.

The following is an interesting observation of Marciniak–Ritter–Sehgal–Weiss. We first have to define some kind of augmentation with respect to conjugacy classes — averaging over the conjugacy classes:

---

[4] This list is increasing steadily.

Definition 2.8. For $a := \sum_{g \in G} r_g \cdot g \in RG$ we define for a conjugacy class C in G the element $a(C) := \sum_{g \in C} r_g$, which we call the PARTIAL AUGMENTATION OF a WITH RESPECT TO C.

Proposition 2.9. ([Ma-Ri-Se-We; 97]) *The Zassenhaus conjecture ZC1 is equivalent to the following statement:*
For every $u \in V(RG)$ of finite order there exists a UNIQUE *conjugacy class C with $u(C) \neq 0$.*

Let us shortly discuss the conceptual setup:

Denote by $\{C_j\}_{1 \leq j \leq h}$ the conjugacy classes of G and by $\{\chi_i\}_{1 \leq i \leq h}$ the characters; moreover we denote by χ_i^j the value of χ_i on an element in C_j.

For an element $x := \sum_{g \in G} x_g \cdot g$, we have the character-value

$$\chi_i(x) = \sum_{g \in g} x_g \cdot \chi(g) = \sum_{j=1}^{h} \left(\sum_{g \in C_j} x_g \right) \chi_i^j = \sum_{j=1}^{h} x(C_j) \cdot \chi_i^j \, .$$

So the partial sums from Definition 2.7 are exactly the coefficients which occur in the character-values.

As for the SECOND MODIFICATION, we have the following general result for finite solvable groups, which is only ZC-P for Sylow-subgroups.

Theorem 2.10. ([Ki-Ro; 91]) *Let G be a finite solvable group, and let H be a group basis of $\mathbb{Z}G$*[5]. *Let P be Sylow p-subgroup of H. Then there exists a unit $a \in \mathbb{Q}G$ such that $a \cdot P \cdot a^{-1}$ is a Sylow p-subgroup of G.*

The most complete result for ZC-P is given by an application of Weiss' result [Wei; 88]:

Theorem 2.11. (41.12 from [Seh; 93]) *The Zassenhaus conjecture* ZC-P *(cf. Definition* 2.7*) holds for $\mathbb{Z}G$ provided G has a normal Sylow p-subgroup (i.e. every p-subgroup of $V(\mathbb{Z}G)$ is rationally conjugate to a subgroup of P).*

Another class of groups for which this holds is the following:

Proposition 2.12. ([Do-Ju; 96]) *Assume that G is a finite nilpotent-by-nilpotent group, then* ZC-P *(cf. Definition* 2.7*) holds for $\mathbb{Z}G$.*

Proof. We use induction and we have to distinguish two cases: If Proposition 2.15 below can not be applied, then G must have a normal Sylow p-subgroup and the result follows from Theorem 2.11. □

In a similar spirit one proves:

Proposition 2.13. *Let G be a finite Sylow-tower group, then* ZC-P *holds for $\mathbb{Z}G$.*

Frobenius groups[6] are treated in

---

[5] I.e. $\mathbb{Z}G = \mathbb{Z}H$ as augmented algebras.

[6] See Theorem 2.21.

Proposition 2.14. ([Do-Ju-Mi; 96]) *Let G be a finite Frobenius group. Then*

1. *G satisfies* ZC-P *for $p > 2$.*
2. *G satisfies* ZC-2 *if the symmetric group S_5 on 5 letters is not a homomorphic image of G*

The next result is an observation of Dokuchaev and Juriaans [Do-Ju; 96]. The proof we give here extends his result considerably.

Proposition 2.15. ([Do-Ju; 96]) *Let $N \triangleleft G$ and assume that $U \leq V(RG)$ is a finite subgroup with $(|N|, |U|) = 1$. Denote by $\overline{U}$ the image of U in $V(RG/N)$. Then the Zassenhaus conjecture holds for U if and only if it holds for $\overline{U}$.*

A crucial ingredient in the proof is the following:

Claim 2.16. *Let K be a local number field and let $U \leq V(RG)$ be a finite subgroup with $(p, |U|) = 1$. Then ${}^aU \leq G$ for some $a \in U(KG)$ if and only if ${}^uU \leq G$ for some $u \in U(RG)$. In particular, if ${}^aU \leq G$ for some $a \in U(KG)$, then U is part of a group basis.*

Proof. We only need to describe the argument in one direction. Let $\phi^{-1}: U \longrightarrow H \leq G$ be the isomorphism induced by conjugation with $a \in U(KG)$. We now consider two $(H \times G)$-R-bi-modules — i.e. left $R(H \times G^{op})$-modules. The first is ${}_1M_1$, which is RG considered in the natural way as $(H \times G)$-bi-module. The other bi-module is ${}_\phi M_1$, which is RG considered as $(H \times G)$-bi-module, where however the left action is twisted by ϕ; the right action is still the old one.

Since $(p, |H|) = 1$, the Sylow p-subgroup of $H \times G$ is still the Sylow p-subgroup of G, and hence both modules — note that the right action is in both cases the regular action on RG — are projective as $H \times G$-modules, since their restriction to a Sylow p-subgroup is.

The hypothesis that the Zassenhaus conjecture is true for U now amounts to the statement, that the $H \times G$ modules $K \otimes_R {}_1M_1$ and $K \otimes_R {}_\phi M_1$ are isomorphic. However, for projective modules this implies that already the modules ${}_1M_1$ and ${}_\phi M_1$ are isomorphic.

But this implies that U and H are conjugate in $V(RG)$. □

The next observation is some kind of a PRO-FINITE VERSION OF THE THEOREM OF SCHUR–ZASSENHAUS.

Claim 2.17. *Let K be a global number field with semi-local ring of integers R such that for every $p \notin \pi(N)$*[7] *we have $p \cdot R = R$ for a normal subgroup $N \triangleleft G$. Assume that $U \leq V(RG)$ is a finite subgroup with $(|N|, |U|) = 1$. Denote by $\overline{U}$ the image of U in $V(RG/N)$. Assume that $H \leq G$ with $\overline{U} = \overline{H}$. Then U and H are conjugate in $V(RG)$.*

Proof. The hypotheses show that U and H are isomorphic, hence we have to show with the notation of the proof of Claim 2.16 that the $R(H \times G)$-bi-modules $B_1 := {}_1M_1$ and $B_2 := {}_\phi M_1$ are isomorphic. Since R is semi-local, this is true if and only if for $i = 1$ and 2 the modules $\overline{B_i}^n := B_i/(\mathrm{rad}(R)^n \cdot B_i)$ are isomorphic provided n is sufficiently large.

---

[7] $\pi(N)$ denotes the set of prime divisors of $|N|$.

Hence it suffices to show that the groups U and H are conjugate in

$$V((R/\operatorname{rad}(R)^n)G).$$

We have the split exact sequence

$$0 \longrightarrow 1 + I_{R/\operatorname{rad}(R)^n}(N) \uparrow^G \longrightarrow \widetilde{E} \longrightarrow \overline{H} \longrightarrow 0$$

for which both, H and U are complements. Since $(|1+I_{R/\operatorname{rad}(R)^n}(N) \uparrow^G|, |H|) = 1$, we conclude that both complements are conjugate in $\widetilde{E}$. □

Remark 2.18. The proof — an analysis of J. M. Maranda's arguments [Mar; 55] — actually shows, that H and U are conjugate in E, is the inverse image of $\overline{H}$ under the natural map $V(RG) \longrightarrow V(R(G/N))$.

We are now in the position to prove Proposition 2.15:

Proof of [Proposition 2.15]. We only need to prove one direction. So assume that the Zassenhaus conjecture holds for $\overline{U}$. We look at the short exact sequence

$$0 \longrightarrow I(N) \uparrow^G \longrightarrow RG \longrightarrow RG/N \longrightarrow 0.$$

We have to distinguish two cases:

Case 1: Let $\wp \in \max(R)$ be a prime relatively prime to $|N|$, then the above sequence is split and so the rational conjugation on RG/N extends to a rational conjugation in $R_\wp G$, which has the effect, that now H and U have the same image in $R_\wp G/N$.

Case 2: Let $\wp \in \max(R)$ be a prime dividing $|N|$, then by Claim 2.16 the groups $\overline{H}$ and $\overline{H}$ are conjugate in the units of $R_\wp G/N$; this conjugating unit can be lifted to a unit in $R_\wp G$, and after conjugating with this unit, we may assume that now H and U have the same image in $R_\wp G/N$.

In this latter case we can now apply Claim 2.17 to obtain the desired result. □

A careful analysis of the proof shows, that we actually have proved the following:

Proposition 2.19. *Let K be a global number field and with semi-local ring of integers R such that for every $\wp \notin \pi(N)$ we have $\wp \cdot R = R$ for a normal subgroup $N \triangleleft G$. Assume that $U \leq V(RG)$ is a finite subgroup with $(|N|, |U|) = 1$. Denote by $\overline{U}$ the image of U in $V(RG/N)$. Assume that $H \leq G$ with ${}^{\overline{a}}\overline{U} = \overline{H}$ for some $\overline{a} \in U(KG/N)$. Then U and H are conjugate in $V(RG)$.*

2.4 The Isomorphism Problem

Though the isomorphism Problem is open for finite solvable groups, the simple groups seem to be much more rigid:

Theorem 2.20. (Kimmerle–Lyons–Sandling–Teague [Ki-Ly-Sa-Te; 90]) *The finite simple groups are determined by their integral group rings.*

More generally, even the chief factors of a finite group are determined by its integral groups ring.

We also have several classes of solvable groups, for which the Isomorphism-Problem has a positive solution, Kimmerle [Ki; 91], Kimmerle–Roggenkamp [Ki-Ri; 93], Roggenkamp–Scott [Ro-Sc; 86], Roggenkamp–Zimmermann [Ro-Zi; 92].

Theorem 2.21. *The Isomorphism Problem has a positive solution for the following class of finite groups:*

1. *If G is nilpotent-by abelian,*
2. *if G is abelian by nilpotent,*
3. *if G is a Frobenius group*[8]*,*
4. *if G is a two-Frobenius group*[9]*.*

3. Negative Results

3.1 The Isomorphism Problem

Theorem 3.1. (Roggenkamp–Zimmermann [Ro-Zi; 95 1], [Ro-Zi; 95 2]) *There exist two non-isomorphic finite-by-cyclic groups $\mathcal{G}$ and $\mathcal{H}$, and a suitable algebraic number field K such that*

$$R\mathcal{G} \simeq R\mathcal{H}.$$

The FIRST INGREDIENT to prove this result is a observation of M. Mazur, who gave a construction of groups, which allows to decide, when the group rings are isomorphic. This construction depends on an automorphism of a group, which is not inner, which becomes inner however in the group rings (cf. the Automorphism Problem). The construction of a finite group which has such an automorphism is the SECOND INGREDIENT. It was produced in joint work A. Zimmermann.

Mazur's Observation:

M. Mazur elaborated on the following construction of semi-direct products.

Let H be a group and let β be an automorphism of H. Then we define the group $H \rtimes_\beta C_\infty$ as follows.

As a set $H \rtimes_\beta C_\infty$ is the product $H \times C_\infty$, where C_∞ is the infinite cyclic group. The multiplication in it is defined for

$$g, g' \in G\,,\; z, z' \in C_\infty \quad \text{as} \quad (g, z) \cdot (g', z') = \left(g \cdot \beta^z(g'), z \cdot z'\right).$$

For the interpretation of the symbol β^z, we identify C_∞ with $\mathbb{Z}$.

A direct consequence of the construction is:

Lemma 3.2. (Mazur [Maz; 95])

1. *$H \rtimes_\beta C_\infty \simeq H \times C_\infty$ for a finite group H if and only if β is inner.*
2. *Moreover for an integral domain R of characteristic zero, we have for the group rings $R(H \rtimes_\beta C_\infty) \simeq R(H \times C_\infty)$ provided β becomes inner in RH.*

Together with A. Zimmermann we have constructed a finite group G and an automorphism β of G, which is not inner, but which lies in the kernel of the natural homomorphism

$$\mathrm{Out}(G) \to \mathrm{Out}(\mathbb{Z}_\pi G)\,;$$

---

[8] Recall, that a Frobenius group G is one which has a subgroup H, such that $H \cap {}^gH = 1$ for $g \in G \setminus H$. Then there exists a normal subgroup N, the Frobenius kernel with complement H.

[9] A 2-Frobenius group G has two normal subgroups $N \leq T \leq G$ such that T is a Frobenius group with kernel N and G/N is a Frobenius group with kernel T/N.

here $\mathbb{Z}_\pi$ is the semi-localization of $\mathbb{Z}$ at a finite set of rational primes, which contains all the prime divisors of $|G|$.

Remark 3.3. One might be tempted to ask, why this construction does not work for finite groups; i.e. why can't we replace C_∞ by C_n, a cyclic group of order n. An analysis of the above arguments shows, that then u[10] must have order dividing n. On the other hand, one knows, that for finite groups F_1 and F_2, an isomorphism $\mathbb{Z}F_1 \simeq \mathbb{Z}F_2$ implies that F_1 is a direct product of two groups if and only if F_2 is a direct product of two groups of the same order. A consequence of these considerations is that the conjugating element u cannot have finite order.

THE ISOMORPHISM PROBLEM FOR FINITE GROUPS IS STILL WIDE OPEN[11].

3.2 The Automorphism Problem

It follows from a result of D. Coleman [Col; 64] that the natural map

$$\Phi : \mathrm{Out}(G) \to \mathrm{Out}(RG)$$

is injective for p-groups. Here R is the ring of integers in a global or a local number field K, in which p is not invertible.

Moreover, J. Krempa and Jacowski–Marciniak [Ja-Ma; 87] have shown, that for a group G with normal 2-Sylow-subgroups, the map Φ is injective; they also showed that the kernel is an elementary abelian 2-group.

However, there was not known an example of a finite group G, where this map is not injective for a global field K.

If one wants to construct a homomorphism $\alpha \in \mathrm{Aut}(G)$ with $\Phi(\alpha)$ inner on RG, then

1. α must be the identity on the conjugacy classes of G,
2. α must be inner on the Sylow p-subgroups, as follows from a careful analysis of Coleman's proof.

To construct such an α is is a purely group theoretical problem, which is easily solved. However, in order to show that α becomes inner on RG we have to use the FOLLOWING INGREDIENTS FROM INTEGRAL REPRESENTATION THEORY:

1. We show that α is inner on $\mathbb{Z}G$ semi-locally. From this one can not automatically conclude that α is inner on $\mathbb{Z}G$. The obstruction is an element in the class group of $\mathbb{Z}G$. We use class field theory — this is where K enters — to kill this obstruction.
2. The passage from the local to the semi-local situation becomes possible by interpreting automorphisms as invertible bi-modules and using Fröhlich's exact sequence of Picard groups [Fro; 73].
3. Finally we are in the local resp. complete situation. Here we develop some type of Clifford theory[12] [Ro-Ty; 92] to show, that α acts as inner automorphism on the inertia groups, after having applied the theorem of

---

[10] Recall the β is conjugation by u.

[11] Cf. the remark at the end of this paper.

[12] The author has learnt the way of using Clifford theory integrally for modules from Leonard Scott.

Noether–Deuring, to pass to a splitting field. The main point in our construction is, to involve quaternion groups in order to keep the inertia groups small.

3.3 Zassenhaus Conjecture

Theorem 3.4. (Roggenkamp–Scott [Ro-Sc; 86]) *There exists a finite meta-cyclic group G and an augmented automorphism $\alpha: \mathbb{Z}G \longrightarrow \mathbb{Z}G$, which is a counterexample for the Zassenhaus Conjecture for automorphisms and thus also for the Zassenhaus Conjecture.*

Remark 3.5. The example constructed above was of theoretical nature: We used theoretical arguments — Clifford theory, invertible bi-modules and K-theory — to show the existence of the automorphism, though explicit groups were given. We showed that a group of order $2^6 \cdot 3^2 \cdot 5$ provided a counterexample over $\mathbb{Z}$, and mentioned an infinite family of similarly constructed groups providing semi-local counterexamples.

Later, Klingler [Kl; 91], using one of these and a direct construction of our automorphism, gave an example of order $2^6 \cdot 3 \cdot 5 \cdot 7$.

3.4 Local-Global Principle for Automorphisms

For a group ring RG we denote by $\text{Outcent}(RG)$ the group of outer central automorphisms of RG, and by $\text{Picent}(RG)$ the group of isomorphism classes of invertible RG-RG-bi-modules, where the centre of RG acts in the same way from the right and from the left.

Then there is Fröhlich's localization sequence for the ring of integers R in an algebraic number field with localizations $R_\wp$ for $\wp \in \max(R)$. We denote by $Z(RG)$ the centre of RG, and for a commutative rings S we write $CL(S)$ for the class group of S. $CL_{RG}(Z(RG))$ stands for those ideal classes $(\mathfrak{a})$ of $Cl(Z(RG))$ such that $\mathfrak{a} \cdot RG$ is a principal ideal.

Theorem 3.6. (Fröhlich [Fro; 73]) *The following is a commutative diagram with exact rows:*

$$\begin{array}{ccccccccc} 0 & \longrightarrow & CL(Z(RG)) & \longrightarrow & \text{Picent}(RG) & \xrightarrow{\lambda_M} & \prod_{\wp \in \max(R)} \text{Picent}\, R_\wp G & \longrightarrow & 0 \\ & & \uparrow & & \uparrow & & \uparrow & & \\ 0 & \longrightarrow & CL_{RG}(Z(RG)) & \longrightarrow & \text{Outcent}(RG) & \xrightarrow{\lambda_A} & \prod_{\wp \in \max(R)} \text{Outcent}(R_\wp G). & & \end{array}$$

The theory of genera shows, that the map λ_M is surjective. In addition we have $\text{Outcent}(R_\wp G) \simeq \text{Picent}(R_\wp G)$. Hence the above commutative diagram shows: Given a family $(\alpha_\wp) \in \text{Autcent}(R_\wp G)$, there exists an invertible bi-module B which localizes to modules which correspond to $(\alpha_\wp)$; it is not clear though that there exist a CENTRAL automorphism α of RG which localizes to $(\alpha_\wp) \in \text{Autcent}(R_\wp G)$.

Problem 3.7. *Is the map λ_A surjective? With other words, can we always find in the genus of an invertible bi-module a representative, which is free on one side?*

In all the examples computed by us, the answer was positive.

3.5 Completing Subgroups to Group Bases

In what follows let G be a finite p-group, though the problems are also of importance of nilpotent groups. Let R be the ring of algebraic integers in a global number field K, and let $\widehat{R}$ be the completion of R at a prime above p. Then we know, that every finite subgroup $U \leq V(\widehat{R}G)$ can be extended to a GROUP BASIS; i.e. a subgroup $U \leq U_0$ with $U_0 \simeq G$.

So looking at the group ring RG in the global situation, there are two immediate problems:

Problem 3.8. *Let H be a finite subgroup of G. We denote by $CL_H(RG)$ the subgroups U of RG such that there exists $a \in U(KG)$ with $a \cdot U \cdot a^{-1} = H$ modulo conjugation in RG. Describe the group $CL_H(RG)$!*

Some progress has been made here in the thesis of A. Zimmermann [Zi; 93]:

If for example one wants to describe the involutions in $\mathbb{Z}G$, this amounts to finding the integral points $\mathbb{P}$ on an integral quadric Q: Let

$$x = \sum_{g \in G} z_g \cdot g \in \mathbb{Z}G\,.$$

The element x is an involution if and only if its coefficients $\{z_g\}_{g\in G}$ lie on the intersection of the quadrics defined by

$$q_G = \sum_{k \in G} \left(\sum_{g \cdot h = k} z_g \cdot z_h \right) \cdot k - 1\,.$$

Note that q_G is actually a family of quadrics:

$$\sum_{g \cdot h = 1} z_g \cdot z_h - 1\,,$$

$$\sum_{g \cdot h = k} z_g \cdot z_h \quad \mathit{for} \quad k \neq 1\,.$$

On the integral variety $V_G^2(\mathbb{Z})$ (i.e. the integers satisfying q_G) the unit group $V(\mathbb{Z}G)$ acts by conjugation, and the orbit space $V_G^2(\mathbb{Z})/V(\mathbb{Z}G)$ then describes the conjugacy classes of involutions in $\mathbb{Z}G$.

We can interpret Weiss' result as follows: The orbits of the 2-adic points $V_G^2(\widehat{\mathbb{Z}})$ under the 2-adic unit group are parameterized by the conjugacy classes of involutions in the 2-group G.

Integrally the situation is quite different.

Theorem 3.9. (A. Zimmermann [Zi; 95]) *Let D_n be the dihedral group of order 2^{n+1}, $n > 1$, then the number of orbits in $V_{D_n}^2(\mathbb{Z})/V(\mathbb{Z}D_n)$ is given by*

$$|V_{D_n}^2(\mathbb{Z})/V(\mathbb{Z}D_n| = 1 + 2 \cdot 2^{n-1} \cdot \prod_{k=1}^{n} h_{2^k}^{+}\,,$$

where $h_{2^k}^+$ is the class number of the maximal real subfield of the 2^k-th cyclotomic field.

The other part of Weiss' result can be interpreted as follows: Each orbit of the 2-adic unit group on $V_{D_n}^2(\widehat{\mathbb{Z}})$ meets the group in exactly one conjugacy class of involutions in D_n ; i.e. the involutions in the group lie transversally to the orbits.

One can not expect such a result globally, since not all group bases are conjugate; however, an analogy would be:

Problem 3.10. *Let $U \leq V(RG)$ be a finite group. Can we find $a \in N_{U(KG)}(RG)$ such that $a \cdot U \cdot a^{-1} \leq G$? This implies in particular, that U is part of a group basis.*

THIS PROBLEM IS STILL OPEN.

For the Dihedral groups though we have a result:

Theorem 3.11. (Zimmermann [Zi; 95]) *The number of elements in the orbit space $V_{D_n}^2(\mathbb{Z})/V(\mathbb{Z}D_n)$ which are part of a group basis is $1 + 2 \cdot 2^{n-1}$.*

Thus for the dihedral groups of order 2^n a positive answer is EQUIVALENT TO AN OPEN CONJECTURE OF H. COHN [An-Ch-Ha; 65] on the class number of the maximal real subfield of the 2^k-th cyclotomic field. He has conjectured $h_{2^k}^+ = 1$.

Acknowledgement

I would like to thank the referee for some clarifying comments.

Remark made in proof reading

In October 1996, M. Hertweck, Univ. Stuttgart, has constructed two non-isomorphic finite groups G and H with $\mathbb{Z}G \simeq \mathbb{Z}H$.

References

[An-Ch-Ha; 65] N.C. Ankency, S. Chowla and H. Hasse, On the class number of the maximal real subfield of a cyclotomic field, *J. Reine Angew. Math.* **217** (1995), 217–220.

[Ble; 95] F. M. Bleher, Tensor products and a conjecture of Zassenhaus, *Arch. Math.* **64** (1995), 289–298.

[Bl;96] F. M. Bleher, Automorphismen von Gruppenringen und Blocktheorie, Dissertation, Universität Stuttgart, 1996.

[Bl-Ge-Ki;96] F. M. Bleher, M. Geck and W. Kimmerle, Automorphisms of generic Iwahori-Hecke algebras and integral group rings of finite Coxeter groups, Preprint 1996.

[Bl-Hi-Ki; 96] F. M. Bleher, W. Kimmerle and G. Hiss, Auto-equivalences of Blocks and a Conjecture of Zassenhaus, *J. Pure and Applied Alg* (1996).

[Col; 64] D. Coleman, On the modular group ring of a p-group, *Proc. AMS* **15** (1964), 511–514.

[Dok; 96] M. A. Dokuchaev, Finite Subgroups of Units in Integral Group Rings, *IME-USP* **2** (1996) (to appear).

[Do-Ju; 96] M. A. Dokuchaev and S. O. Juriaans, Finite subgroups of integral group rings, *Can. J. Math.* (to appear).

[Do-Ju-Mi; 96] M. A. Dokuchaev, S. O. Juriaans and C. Polcino Milies, Integral group rings of Frobenius Groups and the Conjectures of H. J. Zassenhaus, preprint.

[Fro; 73] A. Fröhlich, The Picard groups of non-commutative rings, in particular of orders, *Trans. Amer. Math. Soc.* **180** (1973), 1–46.

[Hig; 39] G. Higman, Units in group rings, D. phil. theses, Oxford Univ. (1939).

[Ja-Ma; 87] S. Jackowski and Z. Marciniak, Group automorphisms inducing the identity map on cohomology, *J. of Pure and Appl. Algebra* **44** (1987), 241–250.

[Ki;91] W. Kimmerle, Beiträge zur ganzzahligen Darstellungtheorie endlicher Gruppen, *Bayr. Math. Schriften* Heft **36** (1991), 1–139.

[Ki-Ly-Sa-Te; 90] W. Kimmerle, R. Lyons, R. Sandling and D. Teague, Composition factors from the group ring and Artin's theorem on orders of simple groups, *Proceedings LMS* (3) **60** (1990), 89–122.

[Ki-Ro; 91] W. Kimmerle and K. Roggenkamp, A Sylow-like Theorem for Integral Group Rings of Finite Solvable Groups, *Arch. Math.* **60** (1993), 1–6.

[Ki-Ri; 93] W. Kimmerle and K. Roggenkamp, Projective Limits of Groups Rings, *J. Pure Appl. Algebra* **88** (1993), 119–142.

[Kl; 91] L. Klingler, Construction of a counterexample to a conjecture of Zassenhaus, *Comm. Alg.* **19 (8)** (1991), 2303–2330.

[Mar; 55] J. M. Maranda, On P-adic integral representations of finite groups, *Can. J. Math.* **5** (1953), 344–355.

[Ma-Ri-Se-We; 97] Z. Marciniak, J. Ritter, S. K. Sehgal and A. Weiss, Torsion units in integral group rings of some met-abelian groups, II, *Journal of Number Theory* **25** (1987), 340–352.

[Maz; 95] M. Mazur, On the isomorphism problem for integral group rings of infinite groups, *Expo. Math.* **13,** No. 5 (1995), 433–445.

[Pet; 76] G. Peterson, Automorphisms of the integral group rings of S_n, *Proceedings AMS* Vol. **59,** No.1 (1976), 14–18.

[Rog; 92] K. W. Roggenkamp, Subgroup rigidity of p-adic group rings (Weiss' arguments revisited), *J. Lon. Math. Soc.* **46** (1992), 432–448.

[Ro-Sc; 86] K. W. Roggenkamp and L. L. Scott, The Isomorphism Theorem for Integral Group Rings of Nilpotent by Abelian Groups, manuscript (1986).

[Ro-Sc; 87] K. W. Roggenkamp K. W. and L. L. Scott, Isomorphisms of p-adic group rings, *Annals of Mathematics* **126** (1987), 593–647.

[RoSc; 86] K. W. Roggenkamp K. W. and L. L. Scott, A counterexample to a conjecture of H. Zassenhaus, manuscript (1986).

[Ro-Ty; 92] K. W. Roggenkamp and M. V. Taylor, *Groups Rings and Class groups*, DMV-Seminar 18, Birkhäuser, Basel, 1992.

[RoZi; 92] K. W. Roggenkamp and A. Zimmermann, On the isomorphism problem for integral group rings of finite groups, *Arch. Math.* **59** (1992), 534–544.

[RoZi; 95 1] K. W. Roggenkamp and A. Zimmermann, Outer automorphisms may become inner in their integral group rings, *J. Pure Appl. Algebra* **103** (1995), 91–99.

[RoZi; 95 2] K. W. Roggenkamp and A. Zimmermann, A counterexample for the isomorphism-problem of polycyclic groups, *J. Pure Appl. Algebra* **103** (1995), 101–103.

[Sc; 87] L. L. Scott, Recent progress on the isomorphism problem, *Proceedings of Symposia in Pure Mathematics* **47** (1987), 269–273.

[Sc; 90] L. L. Scott, *Defect groups and the isomorphism problem, Représentations linéaires des groupes finis*, Proc. Colloq. Luminy, France, 1988; *Astérisque* (1990), 181–182.

[Seh; 83] S. K. Sehgal, *Torsion units in integral group rings*, Proc. Nato Institute on Methods in Ring Theory, Antwerp, D. Riedel, Dordrecht, 1983, pp. 497–504.

[Seh; 93] S. K. Sehgal, *Units of Integral Group Rings*, Longman's, Essex, 1993.

[Tho; 94] G. Thompson, Subgroup rigidity in finite dimensional group algebras over p-groups, *Trans. Am. Math. Soc.* **341** (1994), 423–447.

[Wei; 88] A. Weiss, p–adic rigidity of p–torsion, *Annals of Mathematics* (1988), 317–332.

[Wei; 91] A. Weiss, Torsion units in integral groups rings, *J. Reine Angew. Math.* **415** (1991), 175–187.

[Zi; 90] A. Zimmermann, *Das Isomorphieproblem ganzzahliger Gruppenringe für Gruppen mit abelschem Normalteiler und Quotienten, der eine Vermutung von Hans Zassenhaus erfüllt*, Diplomarbeit, Universität Stuttgart, 1990.

[Zi; 93] A. Zimmermann, *Endliche Untergruppen der Einheitengruppe ganzzahliger Gruppenringe*, Dissertation, Universität Stuttgart, 1992.

[Zi; 95] A. Zimmermann, On the Torsion Units in Integral Group Rings of Dihedral 2-Groups, *Journal of Algebra* **175** (1995), 122–136.

K. W. Roggenkamp
Mathematisches Institut B, Universität Stuttgart
Pfaffenwaldring 57
D–70550 Stuttgart
Germany
e-mail: kwr@mathematik.uni-stuttgart.de

Canadian Mathematical Society
Conference Proceedings
Volume **22**, 1998

Tame Module Categories of Finite Dimensional Algebras

Andrzej Skowroński

0. Introduction

The present notes are an extended version of the talk given during the Ring Theory Conference held at Miskolc in July 1996.

The representation theory of associative finite dimensional algebras has left its traces in many recent developments of algebra and other areas of Mathematics, especially Lie theory, algebraic geometry, singularity theory, deformation theory, invariant theory, finite group representation theory and mathematical physics. We refer only to the Proceedings [9], [33], [67] for some articles showing an interaction between the representation theory of finite dimensional algebras and related fields. The main aim of these notes is to show an internal beauty and richness of the modern representation theory of finite dimensional algebras. We shall present some recent results on tame module categories of finite dimensional algebras over an algebraically closed field. From Drozd's Tame and Wild Theorem, the class of finite dimensional algebras over an algebraically closed field may be divided into two disjoint classes. One class consists of algebras for which the indecomposable modules occur, in each dimension d, in a finite number of discrete and a finite number of one-parameter families. The second class is formed by the wild algebras whose representation theory is as complicated as the study of finite dimensional vector spaces with two noncommuting endomorphisms, for which the classification is a well-known unsolved problem. Hence, we can realistically hope to describe modules only for tame algebras. For a finite dimensional associative K-algebra A

AMS Subject Classification (1991). 14L30, 15A63, 16G20, 16G60, 16G70, 16S90, 18G20.

Research partially supported by the Polish Scientific Grant KBN No. 2P03A 020 08.

This paper is in a final form and no version of it will be submitted for publication elsewhere.

with an identity over an algebraically closed field K, we shall discuss connections between: the representation type of A, arithmetic properties of the Euler and Tits quadratic forms on the Grothendieck group $K_0(A)$ of A, properties of discrete and one-parameter families of indecomposable finite dimensional A-modules, geometric and homological properties of indecomposable finite dimensional A-modules, degenerations of modules and algebras, and the behaviour of connected components of the Auslander–Reiten quiver Γ_A of A in the category $\operatorname{mod} A$ of all finite dimensional A-modules. For general background we refer the reader to [6], [44], [78] and [81].

We divide the notes into the following parts:

1. Preliminaries on module categories.
2. Tame and wild algebras.
3. Quivers and their representations.
4. Auslander–Reiten quiver.
5. Component quiver.
6. Affine varieties of modules.
7. Degenerations of algebras.
8. Integral quadratic forms.
9. Tame quasitilded algebras.
10. Tame simply connected algebras.

1. Preliminaries on Module Categories

Throughout this article A will denote a fixed finite dimensional associative K-algebra with an identity over an algebraically closed field K. We denote by $\operatorname{mod} A$ the category of finite dimensional (over K) right A-modules and by $\operatorname{ind} A$ its full subcategory consisting of indecomposable modules. The term module is used for an object of $\operatorname{mod} A$ if not specified otherwise. We shall denote by $\operatorname{rad}(\operatorname{mod} A)$ the **Jacobson radical** of $\operatorname{mod} A$, that is, the ideal of $\operatorname{mod} A$ generated by all noninvertible morphisms in $\operatorname{ind} A$. The **infinite radical** $\operatorname{rad}^\infty(\operatorname{mod} A)$ of $\operatorname{mod} A$ is the intersection of all powers $\operatorname{rad}^i(\operatorname{mod} A)$, $i \geq 1$, of $\operatorname{rad}(\operatorname{mod} A)$. We shall denote by $D : \operatorname{mod} A \to \operatorname{mod} A^{\text{op}}$ the standard duality $\operatorname{Hom}_K(-, K)$, where A^{op} is the opposite algebra to A.

The main problem of the representation theory of algebras is to describe the structure of the category $\operatorname{mod} A$. It is well-known that any module M from $\operatorname{mod} A$ can be written as a (finite) direct sum $M = M_1 \oplus \ldots \oplus M_r$ of indecomposable modules, and the theorem of Krull–Schmidt asserts that such a decomposition is unique (up to isomorphism). Hence, for many purposes is sufficient to deal only with the category $\operatorname{ind} A$. Without loss of generality we may assume that A is connected, that is, has no non-trivial central idempotents. Moreover, up to Morita equivalence, we may assume that A is basic, that is, A modulo its radical $\operatorname{rad} A$ is a product of copies of K.

Let $1 = e_1 + \ldots + e_n$ be a decomposition of the identity of A into a sum of primitive orthogonal idempotents. Then it is known that:

- $P_1 = e_1 A, \ldots, P_n = e_n A$ is a complete set of pairwise nonisomorphic indecomposable projective A-modules;
- $S_1 = e_1 A / e_1 \operatorname{rad} A, \ldots, S_n = e_n A / e_n \operatorname{rad} A$ is a complete set of pairwise nonisomorphic simple A-modules;

- $I_1 = D(Ae_1), \dots, I_n = D(Ae_n)$ is a complete set of pairwise nonisomorphic indecomposable injective A-modules.

Clearly, for any $1 \le i \le n$, P_i is a projective cover of S_i and I_i is an injective envelope of S_i. We deal with the problem how to built all objects of $\operatorname{ind} A$ from P_i, S_i, I_i, $1 \le i \le n$.

An important role in the study of indecomposable A-modules is played by the **Auslander–Reiten translations** $D\operatorname{Tr}$ and $\operatorname{Tr} D$. Let X be a module from $\operatorname{mod} A$ and $P' \xrightarrow{f} P \longrightarrow X \longrightarrow 0$ its minimal projective presentation. Then the transpose $\operatorname{Tr} X$ of X is the cokernel of the map $\operatorname{Hom}(f, A)$ in $\operatorname{mod} A^{\mathrm{op}}$. Hence, $D\operatorname{Tr} X$ and $\operatorname{Tr} DX$ are modules in $\operatorname{mod} A$. It is known that if X is indecomposable nonprojective (respectively, indecomposable noninjective) then $D\operatorname{Tr} X$ is indecomposable noninjective (respectively, $\operatorname{Tr} DX$ is indecomposable nonprojective). Clearly, $D\operatorname{Tr} P = 0$ for any projective A-module P and $D\operatorname{Tr} I = 0$ for any injective A-module I. A morphism $f : X \to Y$ in $\operatorname{mod} A$ is called **irreducible** if f is neither a split epimorphism nor a split monomorphism and, if there is a factorization $f = gh$, then either g is a split monomorphism or h is a split epimorphism. It is known that a morphism $f : X \to Y$ with X and Y from $\operatorname{ind} A$ is irreducible if and only if f belongs to $\operatorname{rad}(X, Y) \setminus \operatorname{rad}^2(X, Y)$. Let Z be a module from $\operatorname{mod} A$. Then a **minimal right almost split morphism** for Z is a morphism $g : E \to Z$ in $\operatorname{mod} A$ such that:

(1) g is not a split epimorphism,
(2) if $g = hg$ then h is an automorphism of E, and
(3) if $h : E' \to Z$ is not a split epimorphism,

then there exists $h' : E' \to E$ such that $h = h'g$. A minimal left almost split morphism for Z is defined dually. An **Auslander–Reiten sequence** (**almost split sequence**) in $\operatorname{mod} A$ is a short exact sequence

$$0 \longrightarrow X \xrightarrow{f} E \xrightarrow{g} Y \longrightarrow 0$$

such that f is a minimal left almost split morphism for X and g is a minimal right almost split morphism for Y. Observe that for such an exact sequence, the modules X and Y are indecomposable. The following theorem proved by Auslander and Reiten in [5] plays a crucial role in the representation theory of finite dimensional algebras.

Theorem 1.1.
(i) *For every indecomposable nonprojective A-module Y there exists a unique (up to isomorphism) Auslander–Reiten sequence $0 \to X \to E \to Y \to 0$. Moreover, then X is isomorphic to $D\operatorname{Tr} Y$.*
(ii) *For every indecomposable noninjective A-module X there exists a unique (up to isomorphism) Auslander–Reiten sequence $0 \to X \to E \to Y \to 0$. Moreover, then Y is isomorphic to $\operatorname{Tr} DX$.*
(iii) *If P is an indecomposable projective A-module then the inclusion $\operatorname{rad} P \hookrightarrow P$ is a minimal right almost split morphism for P.*
(iv) *If I is an indecomposable injective A-module then the projection $I \to I/\operatorname{soc} I$ is a minimal left almost split morphism for I.*

Given Y from $\operatorname{ind} A$ and a minimal right almost split morphism $g : E \to Y$ for Y the irreducible morphisms ending at Y are described as follows: a morphism $u : Z \to Y$ in $\operatorname{mod} A$ is irreducible if and only if $u = vg$ for some split monomorphism

$v : Z \to E$. Dually, if X is from $\operatorname{ind} A$ and $f : X \to E$ is a minimal left almost split morphism for X, then a morphism $h : X \to Z$ in $\operatorname{mod} A$ is irreducible if and only if $h = fp$ for some split epimorphism $p : E \to Z$. Therefore, we may recover the irreducible morphisms ending (respectively, starting) at the indecomposable A-modules from the minimal right (respectively, left) almost split morphisms for indecomposable A-modules described by the above theorem.

We note also that the Auslander–Reiten translations $D\operatorname{Tr}$ and $\operatorname{Tr} D$ reduce the computation of Ext-groups to that of Hom-groups. Namely, for modules X and Y from $\operatorname{mod} A$, let $\underline{\operatorname{Hom}}_A(X,Y)$ be the factor group of $\operatorname{Hom}_A(X,Y)$ modulo the subgroup of all maps $X \to Y$ in $\operatorname{mod} A$ which factor through a projective module. Similarly, let $\overline{\operatorname{Hom}}_A(X,Y)$ be the factor group of $\operatorname{Hom}_A(X,Y)$ modulo the subgroup of all maps $X \to Y$ in $\operatorname{mod} A$ which factor through an injective module. Then we have the following Auslander–Reiten formulas [5].

Theorem 1.2. *Let X and Y be A-modules. Then there are isomorphisms*

$$D\overline{\operatorname{Hom}}_A(Y, D\operatorname{Tr} X) \simeq \operatorname{Ext}^1_A(X,Y) \simeq D\underline{\operatorname{Hom}}(\operatorname{Tr} DY, X).$$

Moreover, the following theorem characterizes the A-modules of projective (respectively, injective) dimension at most one [10].

Theorem 1.3. *Let X be an A-module. Then*

(i) $\operatorname{pd}_A X \leq 1$ *if and only if* $\operatorname{Hom}_A(D({}_AA), D\operatorname{Tr} X) = 0$.
(ii) $\operatorname{id}_A X \leq 1$ *if and only if* $\operatorname{Hom}_A(\operatorname{Tr} DX, A_A) = 0$.

It is well-known that any module X from $\operatorname{mod} A$ has a finite chain of submodules

$$0 = X_0 \subseteq X_1 \subseteq \ldots \subseteq X_l = X$$

with X_i/X_{i-1} being simple for all $1 \leq i \leq l$, called a **composition series** of X. A useful point of view concerning the composition series of a module from $\operatorname{mod} A$ is to study the Grothendieck group $K_0(A)$ of the category $\operatorname{mod} A$. The group $K_0(A)$ is defined as the factor group of the free abelian group generated by the isomorphism classes $[M]$ of all modules M in $\operatorname{mod} A$ modulo the subgroup generated by the expressions $[X] + [Z] - [Y]$ for all exact sequences $0 \to X \to Y \to Z \to 0$ in $\operatorname{mod} A$. It follows from the Jordan–Hölder theorem that the images of the isomorphism classes $[S_1], \ldots, [S_n]$ of the simple A-modules form a $\mathbb{Z}$-basis of the group $K_0(A)$, and hence we may identify $K_0(A)$ with $\mathbb{Z}^n$. We may then assign to each module M from $\operatorname{mod} A$ its dimension-vector $\underline{\dim} M \in K_0(A) = \mathbb{Z}^n$, being the collection of the multiplicities of simple modules $S_1, \ldots, S_n$ in the composition series of M. It is known that

$$\big(\dim_K \operatorname{Hom}_A(P_i, M)\big)_{1 \leq i \leq n} = \underline{\dim} M = \big(\dim_K \operatorname{Hom}_A(M, I_i)\big)_{1 \leq i \leq n}.$$

One of the main problems of the representation theory is to determine the set of dimension-vectors of indecomposable modules. Next we may try to classify the indecomposable modules having a fixed dimension-vector. In particular, it would be interesting to find sufficient conditions for an indecomposable module M to be uniquely determined (up to isomorphism) by its dimension-vector. The following theorem proved by Reiten, Skowroński and Smalø [74] is presently the most general fact in this direction.

Theorem 1.4. *Let M be an indecomposable A-module which does not lie on a short cycle $M \to N \to M$ of nonzero nonisomorphism in* $\operatorname{ind} A$. *Then M is uniquely determined by its dimension-vector.*

It is known that if an indecomposable A-module M does not lie on a short cycle in $\operatorname{ind} A$ then $\operatorname{Ext}_A^1(M, M) = 0$. The problem whether an indecomposable A-module M with $\operatorname{Ext}_A^1(M, M) = 0$ is uniquely determined by $\underline{\dim} M$ remains still open.

2. Tame and Wild Algebras

An intuitive notion of wild algebras was built on investigations of Corner and Brenner who showed that there are algebras A such that for any finite dimensional algebra B there is a full exact embedding of $\operatorname{mod} B$ into $\operatorname{mod} A$. Denote by $F_m = K\langle x_1, x_2, \dots, x_m\rangle$, $m \geq 2$, the free associative K-algebra in n noncommuting generators $x_1, \dots, x_n$. Then the category $\operatorname{mod} F_n$ of finite dimensional right F_n-modules consists of $(m+1)$-tuples $(V, \varphi_1, \dots, \varphi_n)$ where V is a finite dimensional K-vector space and φ_i, $1 \leq i \leq n$, are K-linear endomorphisms of V. Any finite dimensional K-algebra B can be written as a factor algebra $K\langle x_1, \dots, x_m\rangle / I$ for some m, and $\operatorname{mod} B$ is the full subcategory of $\operatorname{mod} F_m$ consisting of all modules annihilated by I. Then there is a full exact embedding $\operatorname{mod} B \to \operatorname{mod} F_2$ which assigns to each B-module $(V, \varphi, \dots, \varphi_m)$ the F_2-module (V^{m+2}, α, β) where

$$\alpha = \begin{bmatrix} 0 & 1 & & & & \\ & 0 & 1 & & \text{\Large 0} & \\ & & \ddots & \ddots & & \\ & & & \ddots & \ddots & \\ & \text{\Large 0} & & & 0 & 1 \\ & & & & & 0 \end{bmatrix} \quad \text{and} \quad \beta = \begin{bmatrix} 0 & & & & & \\ 1 & 0 & & & \text{\Large 0} & \\ \varphi_1 & 1 & 0 & & & \\ & \ddots & \ddots & \ddots & & \\ & & \ddots & \ddots & \ddots & \\ & \text{\Large 0} & & \varphi_m & 1 & 0 \end{bmatrix}.$$

Following Drozd [36] an algebra A is said to be **wild** if there is an F_2–A-bimodule M such that ${}_{F_2}M$ is free of finite rank and the functor

$$- \otimes_{F_2} M : \ \operatorname{mod} F_2 \to \operatorname{mod} A$$

preserves indecomposability and isomorphism classes. Hence, if an algebra A is wild, the study of indecomposable A-modules is as complicated as that for any finite dimensional algebra B, an impossible task! An example of wild algebra is provided by the following algebra of dimension 5

$$\Lambda = \begin{bmatrix} K & K^3 \\ 0 & K \end{bmatrix} = \left\{ \begin{bmatrix} a & (u, v, w) \\ 0 & b \end{bmatrix} ; \ a, b, u, v, w \in K \right\}.$$

In this case, the F_2–Λ-bimodule M defining the required functor $- \otimes_{F_2} M : \operatorname{mod} F_2 \to \operatorname{mod} \Lambda$ is defined as follows: ${}_{F_2}M = F_2 \oplus F_2$ and the right Λ-module structure on M is given by the formula

$$(f, g) * \begin{bmatrix} a & (u, v, w) \\ 0 & b \end{bmatrix} = (af, ufx_1 + vf + wfx_2 + bg)$$

for $f, g \in F_2 = K\langle x_1, x_2\rangle$, $a, b, u, v, w \in K$.

Following Drozd [36], an algebra A is said to be **tame** if, for any dimension d, there exists a finite number of $K[x]$–A-bimodules M_i, $1 \leq i \leq n_d$, which are free finite rank left modules over the polynomial algebra $K[x]$ in one variable, and all but finitely many isoclasses of indecomposable A-modules of dimension d are of the form $K[x]/(x-\lambda) \otimes_{K[x]} M_i$ for some $\lambda \in K$ and some i. Denote by $\mu_A(d)$ the least number of $K[x]$–A-bimodules satisfying the above conditions for d. Hence, if an algebra A is tame, then the indecomposable A-modules occur, in each dimension d, in a finite number of discrete and a finite number of one-parameter families. The following algebra, considered already by Weierstrass and Kronecker,

$$H = \begin{bmatrix} K & K^2 \\ 0 & K \end{bmatrix} = \left\{ \begin{bmatrix} a & (u,v) \\ 0 & b \end{bmatrix} ;\ a, b, u, v \in K \right\}$$

is tame, and $\mu_A(d) \leq 1$ for any $d \geq 1$.

The following Tame and Wild Theorem proved in 1979 by Drozd [36] is remarkable for the modern representation theory of algebras.

Theorem 2.1. *Every algebra A is either tame or wild, and not both.*

Among the tame algebras we may distinguish the class of **representation-finite algebras**, having only finitely many isomorphism classes of indecomposable modules. The representation theory of representation-finite algebras is presently rather well understood. One of the remarkable results on this class of algebras is the following validity of the second Brauer–Thrall conjecture.

Theorem 2.2. *An algebra A is representation-finite if and only if $\mu_A(d) = 0$ for all $d \geq 1$.*

A proof of the above theorem was announced by Nazarova and Roiter in 1973. For fields of the characteristic different from 2, the first complete proof of the second Brauer–Thrall conjecture has been given only in 1985 by Bautista [7]. The assumption on the characteristic of the base field has later been removed by Bongartz [14]. Moreover, modifications of the proof have been published by Fichbacher [38] and Bretscher–Todorov [22]. All these proofs rely on the following result proved by Bautista, Gabriel, Roiter and Salmeron [8].

Theorem 2.3. *Every representation-finite algebra A admits a multiplicative basis.*

By a multiplicative basis of an algebra A we mean a basis of A as a K-vector space such that the multiplication of any two basis vectors is either a basis vector or zero. It follows from the above theorem that there are only finitely many isomorphism classes of representation-finite algebras of any fixed dimension.

The representation theory of tame representation-infinite algebras is only emerging. Presently the most accessible are (tame) algebras of polynomial growth. Following [83], [84], an algebra A is said to be of **polynomial growth** if there is a positive integer m such that $\mu_A(d) \leq d^m$ for all $d \geq 1$. We shall discuss the representation theory of important classes of polynomial growth algebras later. We refer to [23], [26], [27], [35], [37], [49], [90], [98] for some important classes (string algebras, special biserial algebras, clannish algebras, tame blocks of group algebras, ...) of tame nonpolynomial growth algebras.

We end this section with some remarks on a different approach to tame algebras due to Crawley–Boevey [28], [29]. In order to avoid the difficulties of dealing with families of modules, Crawley–Boevey has proposed to consider instead generic modules. A **generic module** over an algebra A is by definition an indecomposable infinite dimensional right A-module M such that M considered as a left $\mathrm{End}_A(M)$-module is of finite length (called the endolength of M). The endomorphism ring of a generic module always is a local ring [28].

We have the following fact proved in [28].

Theorem 2.4. *An algebra A is representation-infinite if and only if there is a generic A-module.*

In [76] Ringel has shown that, for any representation-infinite tame algebra H of global dimension one (hereditary algebra) there exists a unique generic module, and whose endomorphism ring is a divison algebra.

Following [28] an algebra A is said to be **generically tame** if, for each $d \in \mathbb{N}$, there are only finitely many isomorphism classes of generic A-modules of endolength d. The following theorem proved by Crawley–Boevey [28] gives another view on tame algebras.

Theorem 2.5. *Let A be an algebra. The following conditions are equivalent:*

(i) *A is tame.*
(ii) *A is generically tame.*
(iii) $\mathrm{End}_A(G)/\operatorname{rad}\mathrm{End}_A(G) \simeq K(x)$ *for any generic A-module G.*

We mention also the following relationship between the generic modules and one-parameter families of finite dimensional indecomposable modules over tame algebras established in [28]. Let G be a generic module over a tame algebra A, and R a (commutative) pricipal ideal domain which is finitely generated over K. Then a **realization** of G over R is a finitely generated R–A-bimodule M such that if L is the quotient field of R, then $G \simeq L \otimes_R M$ and $\dim_L(L \otimes_K M)$ is equal to the length of G over $\mathrm{End}_A(G)$. It is shown in [28] that if G has a realization over R then R is isomorphic to a finitely generated localization $K[x, f^{-1}(x)]$ of $K[x]$. Conversely, every generic A-module G has a realization which is free over any such R. We have the following fact proved in [28].

Theorem 2.6. *Let A be a tame algebra. For each generic A-module G choose a realization M_G, say over R_G. Then, for each $d \in \mathbb{N}$, all but finitely many isomorphism classes of indecomposable A-modules of dimension d are of the form $R_G/(p)^n \otimes_{R_G} M_G$ for some generic A-modules G, some $n \geq 1$, and some prime $p \in R_G$.*

3. Quivers and Their Representations

The aim of this section is to present the category $\operatorname{mod} A$ as the category of K-linear representations of a quiver with relations. This way we get a concrete description of the modules in terms of vector spaces together with linear transformations.

Following Gabriel [39], [40], by a **quiver** $Q = (Q_0, Q_1)$ we mean an oriented graph Q with the set Q_0 of vertices and the set Q_1 of arrows between vertices. We denote by $s : Q_1 \to Q_0$ and $e : Q_1 \to Q_0$ the maps which assign to each arrow $\alpha : x \to y$ in Q_1 its source $x = s(\alpha)$ and its end $y = e(\alpha)$, respectively. A path in the quiver Q is an ordered sequence of arrows $p = \alpha_1 \dots \alpha_n$ with $e(\alpha_t) = s(\alpha_{t+1})$ for $1 \leq t < n$, $n \geq 1$. If moreover $s(\alpha_1) = e(\alpha_n)$, p is said to be an oriented cycle. To each vertex $i \in Q_0$ we attach a trivial path e_i such that $s(e_i) = i = e(e_i)$. The **path algebra** KQ of Q is the K-vector space with the paths of Q as basis and the multiplication defined as follows. For nontrivial paths $p = \alpha_1 \dots \alpha_r$ and $q = \beta_1 \dots \beta_s$ we set $pq = \alpha_1 \dots \alpha_r \beta_1 \dots \beta_s$ if $e(\alpha_r) = s(\beta_1)$ and $pq = 0$ otherwise. For a nontrivial path $p = \alpha_1 \dots \alpha_r$ and $i \in Q_0$ we set $e_i p = p$ if $i = s(\alpha_1)$, $pe_i = p$ if $e(\alpha_r) = i$, and $e_i p = 0$, $pe_i = 0$ otherwise. Finally, we put $e_i e_i = e_i$ and $e_i e_j = 0$ for $i \neq j$. It is clear that KQ is an associative K-algebra and e_i, $i \in Q_0$, are pairwise orthogonal idempotents in KQ. If Q_0 is finite then $\sum_{i \in Q_0} e_i$ is the identity of KQ. If Q_0 is infinite then KQ has no identity. Observe also that KQ is finite dimensional if and only if Q is finite (the sets Q_0 and Q_1 are both finite) and Q has no oriented cycles. For each $m \geq 1$ we denote by $(KQ)_m$ the subspace of KQ generated by all paths in Q of length $\geq m$. An ideal I of KQ such that $(KQ)_m \subseteq I \subseteq (KQ)_2$ for some $m \geq 2$ is called **admissible**. It is easy to see that any admissible ideal I of KQ is generated by a system of linear forms $w = w_1\lambda_1 + \dots + w_t\lambda_t$ (called **K-linear relations** in Q), where $\lambda_1, \dots, \lambda_t$ are elements of K and $w_1, \dots, w_t$ are paths of length ≥ 2 in Q having a common source and a common end. Conversely, any system of K-linear relations in Q generates an admissible ideal of KQ. By a **bound quiver** we mean a pair (Q, I) consisting of a quiver Q and an admissible ideal I of KQ. Then the factor algebra KQ/I is called the **bound quiver algebra** of (Q, I).

Let $Q = (Q_0, Q_1)$ be a quiver. A **finite dimensional representation** of Q over K is a system $V = (V_i, \varphi_\alpha)_{i \in Q_0, \alpha \in Q_1}$ where V_i, $i \in Q_0$, are K-vector spaces such that $\sum_{i \in Q_0} \dim_K V_i < \infty$, and $\varphi_\alpha : V_{s(\alpha)} \to V_{e(\alpha)}$, $\alpha \in Q_1$, are K-linear maps. A morphism $f : V \to V'$ of two representations $V = (V_i, \varphi_\alpha)$ and $V' = (V'_i, \varphi'_\alpha)$ of Q is a system $f = (f_i)_{i \in Q_0}$ of K-linear maps $f_i : V_i \to V'_i$, $i \in Q_0$, such that $f_{s(\alpha)}\varphi'_\alpha = \varphi_\alpha f_{e(\alpha)}$ for all arrows $\alpha \in Q_1$. We then get the category $\operatorname{rep}_K Q$ of finite dimensional representations of the quiver Q over K. For each admissible ideal I of KQ we denote by $\operatorname{rep}_K(Q, I)$ the full subcategory of $\operatorname{rep}_K Q$ consisting of all representations $V = (V_i, \varphi_\alpha)$ of Q such that $\sum_{1 \leq j \leq t} \varphi_{\alpha_{1,j}} \cdots \varphi_{\alpha_{m_j,j}} \lambda_j = 0$ for any K-linear relation $\sum_{1 \leq j \leq t} \alpha_{1,j} \dots \alpha_{m,j} \lambda_j \in I$.

The direct sum $V \oplus V'$ of two representations $V = (V_i, \varphi_\alpha)$ and $V' = (V'_i, \varphi'_\alpha)$ in $\operatorname{rep}_K(Q)$ (respectively, $\operatorname{rep}_K(Q, I)$) is defined by $(V \oplus V')_i = V_i \oplus V'_i$ for $i \in Q_0$, and

$$\begin{bmatrix} \varphi_\alpha & 0 \\ 0 & \varphi'_\alpha \end{bmatrix} : V_{s(\alpha)} \oplus V'_{s(\alpha)} \longrightarrow V_{e(\alpha)} \oplus V'_{e(\alpha)}$$

for $\alpha \in Q_1$. A representation V is called indecomposable if $V \neq 0$ and if $V = V' \oplus V''$ implies $V' = 0$ or $V'' = 0$. As a consequence of Theorem 3.1, the Theorem of Krull–Schmidt implies that any finite dimensional representation of (Q, I) can be written as a direct sum of indecomposable representations, and this decomposition is unique up to isomorphism.

We have the following important result proved by Gabriel in [42].

Theorem 3.1. *Let Q be a quiver and I be an admissible ideal of KQ. Then* *there is an equivalence of categories*

$$\mathrm{mod}(KQ/I) \simeq \mathrm{rep}_K(Q, I).$$

Another important observation due to Gabriel [40], [42] is the following theorem.

Theorem 3.2. *Let A be an arbitrary (finite dimensional, basic, connected) algebra. Then there is a finite connected quiver $Q = Q_A$ and an admissible ideal I in KQ such that A is isomorphic to the bound quiver algebra KQ/I.*

The quiver $Q = Q_A$, called the **Gabriel quiver** of A, is uniquely determined by A. The set Q_0 of vertices of Q is a complete set $e_1, \dots, e_n$ of primitive orthogonal idempotents of A, and the number of arrows from e_i to e_j is equal to $\dim_K e_i(\mathrm{rad}\, A)e_j/e_i(\mathrm{rad}^2\, A)e_j$. One can also prove that

$$\dim_K e_i(\mathrm{rad}\, A)e_j/e_i(\mathrm{rad}^2\, A)e_j = \dim_K \mathrm{Ext}^1_A(S_i, S_j).$$

Observe, that there is a bijection between the set of vertices of Q_A and the set $\{[S_1], \dots, [S_n]\}$ of isoclasses of simple A-modules.

The above two theorems reduce the study of the category $\mathrm{mod}\, A$ of finite dimensional modules over an algebra A to the study of the category $\mathrm{rep}_K(Q, I)$ of finite dimensional representations of an associated bound quiver (Q, I). In order to illustrate this idea consider two matrix algebras

$$A = \begin{bmatrix} K & K & K & K^2 \\ 0 & K & 0 & K \\ 0 & 0 & K & K \\ 0 & 0 & 0 & K \end{bmatrix} \quad \text{and} \quad B = \begin{bmatrix} K & K & K & K \\ 0 & K & 0 & K \\ 0 & 0 & K & K \\ 0 & 0 & 0 & K \end{bmatrix}$$

and the quiver

$$Q: \quad \begin{array}{ccccc} & & 1 & & \\ & \alpha\swarrow & & \searrow\gamma & \\ 2 & & & & 3 \\ & \beta\searrow & & \swarrow\sigma & \\ & & 4 & & \end{array}$$

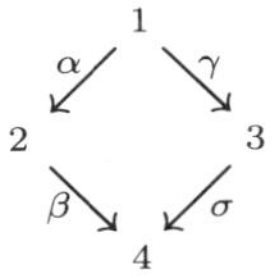

It is not hard to see that $Q_A = Q = Q_B$, $A = KQ$ and $B = KQ/I$, $I = (\alpha\beta - \gamma\sigma)$. Hence, $\mathrm{mod}\, A$ is equivalent to the category $\mathrm{rep}_K(Q)$ of K-linear representations

$$\begin{array}{ccccc} & & V_1 & & \\ & \varphi_\alpha\swarrow & & \searrow\varphi_\gamma & \\ V_2 & & & & V_3 \\ & \varphi_\beta\searrow & & \swarrow\varphi_\sigma & \\ & & V_4 & & \end{array}$$

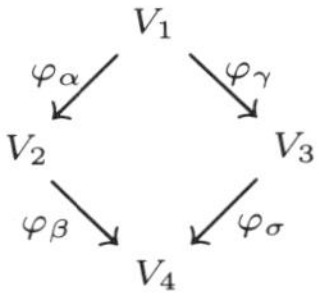

with V_1, V_2, V_3, V_4 finite dimensional K-vector spaces, and $\mathrm{mod}\, B$ is equivalent to the full subcategory $\mathrm{rep}_K(Q, I)$ of K-linear representations

$$(V_1, V_2, V_3, V_4, \varphi_\alpha, \varphi_\beta, \varphi_\gamma, \varphi_\sigma)$$

satisfying the condition $\varphi_\alpha\varphi_\beta = \varphi_\gamma\varphi_\sigma$. A direct checking shows that

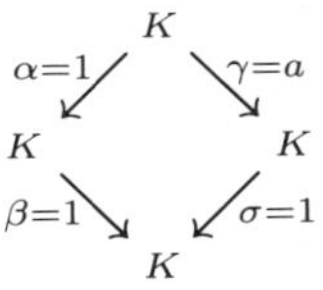

$$\begin{array}{ccccc} & & K & & \\ & \alpha=1\swarrow & & \searrow\gamma=a & \\ K & & & & K \\ & \beta=1\searrow & & \swarrow\sigma=1 & \\ & & K & & \end{array} \qquad a \in K$$

is an infinite family of pairwise nonisomorphic indecomposable objects of $\operatorname{rep}_K Q$, and hence A is representation-infinite. On the other hand, simple linear algebra considerations show that following representations

$$I_1 = S_1 = \begin{smallmatrix} & K & \\ 0 & & 0 \\ & 0 & \end{smallmatrix}, \qquad S_2 = \begin{smallmatrix} & 0 & \\ K & & 0 \\ & 0 & \end{smallmatrix}, \qquad S_3 = \begin{smallmatrix} & 0 & \\ 0 & & K \\ & 0 & \end{smallmatrix},$$

$$P_4 = S_4 = \begin{smallmatrix} & 0 & \\ 0 & & 0 \\ & K & \end{smallmatrix}, \qquad I_2 = \begin{smallmatrix} & K & \\ K & & 0 \\ & 0 & \end{smallmatrix}, \qquad I_3 = \begin{smallmatrix} & K & \\ 0 & & K \\ & 0 & \end{smallmatrix},$$

$$P_2 = \begin{smallmatrix} & 0 & \\ K & & 0 \\ & K & \end{smallmatrix}, \qquad P_3 = \begin{smallmatrix} & 0 & \\ 0 & & K \\ & K & \end{smallmatrix}, \qquad P_1 = I_4 = \begin{smallmatrix} & K & \\ K & & K \\ & K & \end{smallmatrix},$$

$$P_1/S_4 = \begin{smallmatrix} & K & \\ K & & K \\ & 0 & \end{smallmatrix}, \qquad \operatorname{rad} P_1 = \begin{smallmatrix} & 0 & \\ K & & K \\ & K & \end{smallmatrix},$$

form a complete set of isomorphism classes of indecomposable objects in $\operatorname{rep}_K(Q, I)$. In particular, B is a representation-finite algebra having only 11 isoclasses of indecomposable modules.

We also note that an algebra A is hereditary ($\operatorname{gl.dim} A \leq 1$) if and only if Q_A has no oriented cycles and $A \simeq KQ_A$. The representation type of hereditary algebras is known. The following theorem due to Gabriel [39] describes all representation-finite hereditary algebras.

Theorem 3.3. *Let A be a hereditary algebra. Then A is representation-finite if and only if A is isomorphic to the path algebra of one of the Dynkin quivers*

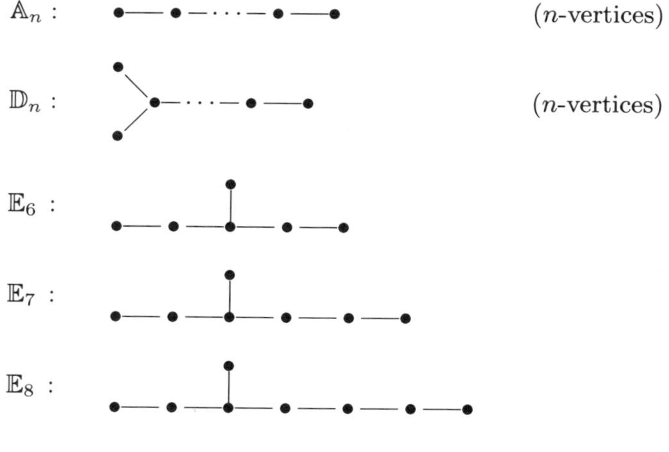

where $\bullet\!\!-\!\!\!-\!\!\bullet$ *means* $\bullet \longrightarrow \bullet$ *or* $\bullet \longleftarrow \bullet$.

The following theorem due to Nazarova [68] and Donovan–Freislich [34] describes all representation-infinite tame hereditary algebras.

Theorem 3.4. *Let A be a hereditary algebra. Then A is representation-infinite tame if and only if A is the path algebra of one of the extended Dynkin quivers*

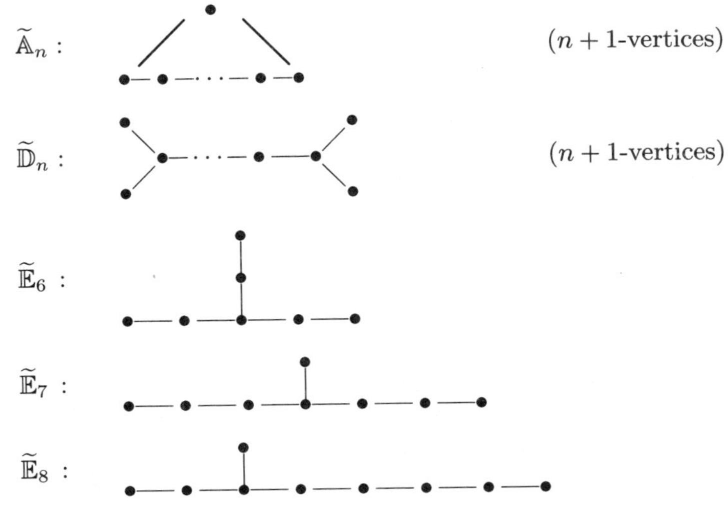

where $\bullet\!\!-\!\!\!-\!\!\bullet$ *means* $\bullet \longrightarrow \bullet$ *or* $\bullet \longleftarrow \bullet$.

We refer also to [32] for a complete description of indecomposable finite dimensional modules over tame hereditary algebras.

We shall also exhibit the class of canonical algebras introduced by Ringel [78], playing an important role in the representation theory of algebras. Let n be a positive integer $n \geq 2$, $p = (p_0, p_1, \dots, p_n)$ be an $(n+1)$-tuple of integers $p_i \geq 2$, and $\lambda = (\lambda_0, \lambda_1, \dots, \lambda_n)$ be an $(n+1)$-tuple of pairwise different elements of $\mathbb{P}_1(K) = K \cup \{\infty\}$, normalized such that $\lambda_0 = \infty$, $\lambda_1 = 0$, $\lambda_2 = 1$. Consider the quiver $\Delta(p_0, \dots, p_n)$:

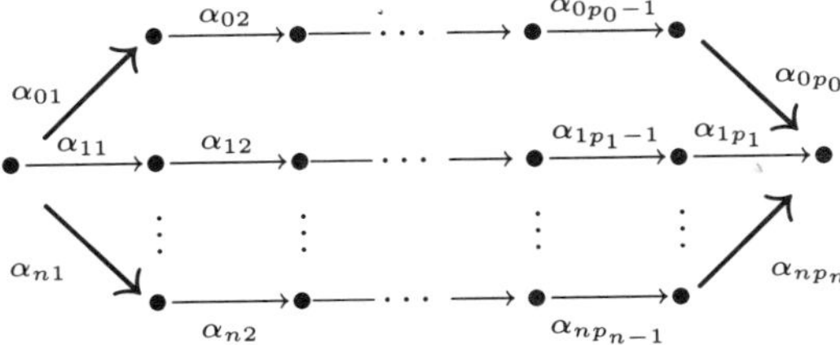

and the admissible ideal $I(\lambda_0, \dots, \lambda_n)$ in $K\Delta(p_0, \dots, p_n)$ generated by

$$\alpha_{i1}\alpha_{i2}\dots\alpha_{ip_i} + \alpha_{01}\alpha_{02}\dots\alpha_{0p_0} + \lambda_i\alpha_{11}\alpha_{12}\dots\alpha_{1p_1}, \qquad \text{for } i = 2, \dots, n.$$

Then

$$C(p, \lambda) = K\Delta(p_0, \dots, p_n)/I(\lambda_0, \dots, \lambda_n)$$

is called the **canonical algebra** and $p = (p_0, p_1, \dots, p_n)$ its weight sequence. It is easy to check that $C(p, \lambda)$ is representation-infinite and $\operatorname{gl.dim} C(p, \lambda) = 2$. We note also that there is a close relation between the category $\operatorname{mod} C(p, \lambda)$ and the category of coherent sheaves over a corresponding weighted projective line (see [45]). The tame canonical algebras and their module categories has been described by Ringel in [78]. In particular, we have the following theorem.

Theorem 3.5. *Let $C = C(p, \lambda)$ be a canonical algebra with a weight sequence $p = (p_0, p_1, \dots, p_n)$, $n \geq 2$. Then C is tame if and only if $\sum_{0 \leq i \leq n} \frac{1}{p_i} \leq 1$.*

Observe that $\sum_{0 \leq i \leq n} \frac{1}{p_i} \leq 1$ if and only if $p = (p_0, p_1, \dots, p_n)$ is one of the sequences $(2, 2, n-2)$, $(2, 3, 3)$, $(2, 3, 4)$, $(2, 3, 5)$, $(2, 3, 6)$, $(2, 4, 4)$, $(3, 3, 3)$ and $(2, 2, 2, 2)$. The representation theory of canonical algebras of types $(2, 2, n-2)$, $(2, 3, 3)$, $(2, 3, 4)$, $(2, 3, 5)$ is closely related (see [78]) to the representation theory of hereditary algebras of extended Dynkin quivers $\widetilde{\mathbb{D}}_n$, $\widetilde{\mathbb{E}}_6$, $\widetilde{\mathbb{E}}_7$, $\widetilde{\mathbb{E}}_8$, respectively. The canonical algebras of types $(2, 3, 6)$, $(2, 4, 4)$, $(3, 3, 3)$ and $(2, 2, 2, 2)$ are called tubular, and for their representation theory we refer to [78].

We shall devote the final part of this section to an another application of bound quiver presentation of an algebra A. Namely, with the help of covering techniques [20], [35], [43], we may frequently reduce the study of $\operatorname{mod} A$ to that for much simpler simply connected subalgebras of a universal cover of A. Let $A = KQ/I$, for a finite connected bound quiver (Q, I). A K-linear relation $\rho = \sum_{1 \leq j \leq m} w_j \lambda_j \in I$ is said to be minimal if $m \geq 2$ and for each nonempty proper subset J of $\{1, \dots, m\}$, $\sum_{j \in J} w_j \lambda_j \notin I$ holds. Denote by $m(I)$ the set of minimal relations of the ideal I, and by $\pi_1(Q, x_0)$ the fundamental group of Q at a fixed vertex $x_0 \in Q_0$. Let $N(Q, m(I), x_0)$ be the normal subgroup of $\pi_1(Q, x_0)$ generated by all elements of the form $[wvu^{-1}w^{-1}]$ where w is a walk from x_0 to x and u, v are paths from x to y such that there is an element $\sum_{1 \leq j \leq m} w_j \lambda_j \in m(I)$ with $u = w_j$, $v = w_r$ for some $1 \leq j, r \leq m$. Following [50] and [65] the **fundamental group** $\pi_1(Q, I)$ of (Q, I) is defined to be the group

$$\pi_1(Q, I) = \pi_1(Q, x_0)/N(Q, m(I), x_0).$$

We may now construct a Galois covering

$$F : K\widetilde{Q}/\widetilde{I} \to KQ/I = A$$

with group $\pi_1(Q, I)$ called the **universal Galois covering** of KQ/I. Let W be the topological universal cover of Q with base point at x_0, that is, the vertices of W are the homotopy classes of walks starting at x_0 and for two such vertices y_1, y_2 we set $y_1 \to y_2$ in W if and only if there exist representatives w_1, w_2 of the classes y_1, y_2, respectively, and an arrow α in Q with $w_2 = w_1 \alpha$. There is a natural (left) action of $\pi_1(Q, x_0)$ on W. Then the normal subgroup $N(Q, m(I), x_0)$ acts on W and we denote by $\widetilde{Q}$ the orbit quiver $W/N(Q, m(I), x_0)$. Moreover, the induced action of $\pi_1(Q, I)$ on $\widetilde{Q}$ gives a map $p : \widetilde{Q} \to Q$ of quivers. Thus we have a homomorphism $p : K\widetilde{Q} \to KQ$ of K-algebras. Finally, we denote by $\widetilde{I}$ the ideal in $K\widetilde{Q}$ generated by all liftings, through p, of generators of I. Consequently, we obtain the required Galois covering $F : K\widetilde{Q}/\widetilde{I} \to KQ/I$ with group $\pi_1(Q, I)$. Moreover, there is (see [20]) an associated push-down functor

$$F_\lambda : \operatorname{mod} K\widetilde{Q}/\widetilde{I} \to \operatorname{mod} KQ/I = \operatorname{mod} A.$$

If $\pi_1(Q, I)$ is torsion-free then F_λ induces an injection from the set of $\pi_1(Q, I)$-orbits of isoclasses of finite dimensional indecomposable $K\widetilde{Q}/\widetilde{I}$-modules into the set of isoclasses of finite dimensional indecomposable KQ/I-modules (A-modules).

In fact, in many cases it is a bijection (see [20], [21], [35], [43], [90]). For example, it is the case if $A = KQ/I$ is representation-finite. The following example illustrates the above considerations.

For $n \geq 2$, consider the triangular matrix algebra

$$\Lambda_n = \begin{bmatrix} K[x]/(x^n) & K[x]/(x^n) \\ 0 & K[x]/(x^n) \end{bmatrix}.$$

Then $\Lambda_n = KQ/I_n$ where Q is the quiver

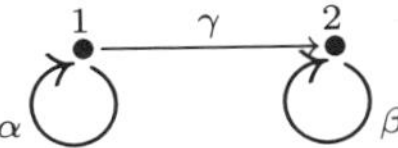

and I_n is the admissible ideal of KQ generated by α^n, β^n and $\alpha\gamma - \gamma\beta$. Therefore, the classification of indecomposable finite dimensional Λ_n-modules is equivalent to the classification of indecomposable representations

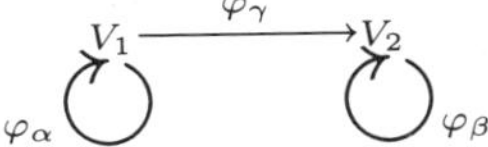

of Q over K with V_1, V_2 finite dimensional and $\varphi_\alpha^n = 0$, $\varphi_\beta^n = 0$, $\varphi_\alpha\varphi_\gamma = \varphi_\gamma\varphi_\beta$. It is rather difficult to describe such indecomposable representations, and hence decide when Λ_n is tame. The main problem is created by the presence of endomorphisms φ_α and φ_β. On the other hand, $\pi_1(Q, I_n)$ is an infinite cyclic group, and we have a Galois covering $F : \widetilde{\Lambda}_n = K\widetilde{Q}/\widetilde{I}_n \to KQ/I_n = \Lambda_n$ with group $\pi_1(Q, I_n)$, where $\widetilde{Q}$ is the infinite quiver

$$\begin{array}{ccccccccccc}
\cdots \leftarrow & (1,i-1) & \overset{\alpha_{i-1}}{\longleftarrow} & (1,i) & \overset{\alpha_i}{\longleftarrow} & (1,i+1) & \overset{\alpha_{i+1}}{\longleftarrow} & (1,i+2) & \leftarrow \cdots \\
 & \big\downarrow{\scriptstyle \gamma_{i-1}} & & \big\downarrow{\scriptstyle \gamma_i} & & \big\downarrow{\scriptstyle \gamma_{i+1}} & & \big\downarrow{\scriptstyle \gamma_{i+2}} & \\
\cdots \leftarrow & (2,i-1) & \overset{\beta_{i-1}}{\longleftarrow} & (2,i) & \overset{\beta_i}{\longleftarrow} & (2,i+1) & \overset{\beta_{i+1}}{\longleftarrow} & (2,i+2) & \leftarrow \cdots
\end{array}$$

and $\widetilde{I}_n$ is generated by all elements $\alpha_{i+n} \cdots \alpha_{i+1}$, $\beta_{i+n} \cdots \beta_{i+1}$ and $\alpha_i\gamma_i = \gamma_{i+1}\beta_i$, for all $i \in \mathbb{Z}$. Due to the fact that $\widetilde{Q}$ has no oriented cycles we may easily find some wild finite bound subquivers of $(\widetilde{Q}, \widetilde{I}_n)$ in the case when $n \geq 5$. For $n \leq 4$, using the one-point extension process of known tame finite bound subquivers of $(\widetilde{Q}, \widetilde{I}_n)$, we may deduce (see [82]) that the classification of indecomposable objects in $\operatorname{rep}_K(\widetilde{Q}, \widetilde{I}_n)$, and hence $\operatorname{rep}_K(Q, I_n)$ is possible. In this case, the push-down functor $F_\lambda : \widetilde{\Lambda}_n \to \Lambda_n$ is dense, that is, any module in $\operatorname{mod} \Lambda_n$ is the image of a module from $\operatorname{mod} \widetilde{\Lambda}_n$ via F_λ. Hence we deduce that Λ_n is representation-finite for $n \leq 3$ [20], Λ_4 is tame representation-infinite, and Λ_n is wild for all $n \geq 5$ [82].

4. Auslander–Reiten Quiver

The vertices of the Gabriel quiver Q_A of an algebra A correspond to the isoclasses of simple A-modules. In order to deal with all indecomposable A-modules, it is convenient to consider a more complicated quiver Γ_A, called the **Auslander–Reiten quiver** of A. The vertices of Γ_A are the isoclasses $[X]$ of objects of $\operatorname{ind} A$.

For two vertices $[X]$ and $[Y]$ of Γ_A the number of arrows from $[X]$ to $[Y]$ is equal to $\dim_K \operatorname{rad}(X, Y)/\operatorname{rad}^2(X, Y)$. Thus the number of arrows from $[X]$ to $[Y]$ is the number of linearly independent irreducible morphisms starting at X and ending at Y. Moreover, we have the translations τ_A and τ_A^- induced by $D\operatorname{Tr}$ and $\operatorname{Tr} D$. In fact, the set of vertices of Γ_A can be considered as the set of vertices of a 2-dimensional simplicial complex (geometric realization of Γ_A [20]). Observe also that Γ_A is locally finite, that is, each vertex of Γ_A is the source (respectively, end) of at most finitely many arrows. Clearly, A is representation-finite if and only if Γ_A is finite. If A is representation-infinite (and connected) then Γ_A decomposes into a union of countable (infinite) connected components. It is known (see [3]) that two vertices $[X]$ and $[Y]$ of Γ_A belong to the same connected component of Γ_A if and only if there exists a sequence of indecomposable A-modules $X = Z_0, Z_1, \ldots, Z_s = Y$, $s \geq 1$, such that, for each $1 \leq i \leq s$, either $\operatorname{Hom}_A(Z_{i-1}, Z_i)/\operatorname{rad}^\infty(Z_{i-1}, Z_i) \neq 0$ or $\operatorname{Hom}_A(Z_i, Z_{i-1})/\operatorname{rad}^\infty(Z_i, Z_{i-1}) \neq 0$. Moreover, any indecomposable A-module M determines a countable set of indecomposable A-modules, namely those whose isomorphism classes belong to the same component as $[M]$.

Sometimes it is easier to describe a full component (shape) given by an indecomposable module that an individual module. Therefore, the shapes of connected components in Γ_A are the first basic invariants of the category $\operatorname{mod} A$. We know that $\operatorname{ind} A$ admits only finitely many isoclasses of projective (respectively, injective) objects. Hence, all but finitely many connected components of Γ_A do not contain vertices given by the projective or injective modules. Such components of Γ_A are said to be **regular**. Observe that, if $\mathcal{C}$ is a regular connected component of Γ_A, then the translations τ_A and τ_A^- are mutually inverse isomorphisms of the quiver $\mathcal{C}$.

Let $\Delta = (\Delta_0, \Delta_1)$ be a locally finite quiver without oriented cycles. We denote by $\mathbb{Z}\Delta$ the translation quiver defined as follows: $\mathbb{Z} \times \Delta_0$ is the set $(Z\Delta)_0$ of vertices of $Z\Delta$, given an arrow $\alpha : x \to y$ from Δ_1, there are arrows $(i, \alpha) : (i, x) \to (i, y)$ and $(i, \alpha)' : (i, y) \to (i + 1, x)$, $i \in \mathbb{Z}$, and the translations τ and τ^- of $\mathbb{Z}\Delta$ are defined by $\tau(i, x) = (i - 1, x)$, $\tau^-(i, x) = (i + 1, x)$ for $i \in \mathbb{Z}$, $x \in \Delta_0$. In particular, if Δ is the quiver

$$\mathbb{A}_\infty : \ 0 \to 1 \to 2 \to \cdots$$

then $\mathbb{Z}\mathbb{A}_\infty$ is of the form

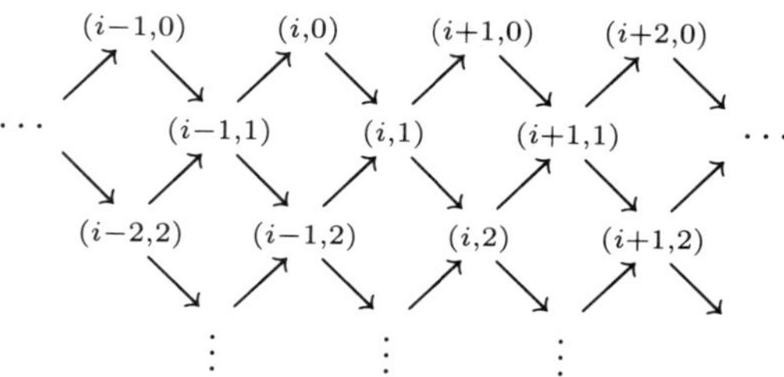

with $\tau(i, j) = (i - 1, j)$ and $\tau^-(i, j) = (i + 1, j)$ for $i \in \mathbb{Z}$, $j \in \mathbb{N}$. For $r \geq 1$, denote by $\mathbb{Z}\mathbb{A}_\infty/(\tau^r)$ the translation quiver obtained from $\mathbb{Z}\mathbb{A}_\infty$ by identifying each vertex (i, j) of $\mathbb{Z}\mathbb{A}_\infty$ with the vertex $\tau^r(i, j)$ and each arrow $x \to y$ in $\mathbb{Z}\mathbb{A}_\infty$ with the arrow $\tau^r x \to \tau^r y$, and call it the **stable tube of rank** r. A stable tube of rank 1 is said to be **homogeneous**. Observe that any vertex in a stable tube of rank r is τ-periodic with period r.

The following two theorems proved by Happel, Preisner and Ringel [52] and Zhang [99], respectively, describe the general shape of regular components of an Auslander–Reiten quiver.

Theorem 4.1. *A regular connected component of an Auslander–Reiten quiver Γ_A contains a τ_A-periodic vertex if and only if $\mathcal{C}$ is a stable tube.*

Theorem 4.2. *Let $\mathcal{C}$ be a regular connected component of an Auslander–Reiten quiver Γ_A containing no τ_A-periodic module. Then $\mathcal{C}$ is of the form $\mathbb{Z}\Delta$ for some locally finite quiver Δ without oriented cycles.*

If H is a representation-infinite tame hereditary algebra then the regular components of Γ_A form a $\mathbb{P}_1(K)$-family of stable tubes, and at most 3 of them are nonhomogeneous [32]. For H wild hereditary all regular components of Γ_A are of the form $\mathbb{Z}\mathbb{A}_\infty$, and there are infinitely many of them [77].

It is not clear which translation quivers $\mathbb{Z}\Delta$ occur as regular components of Γ_A. We know from [80] that, if Δ is a connected finite quiver with at least three vertices and no oriented cycle, then $\mathbb{Z}\Delta$ occurs as a regular component of some Γ_A if and only if Δ is neither Dynkin nor an extended Dynkin quiver. Moreover, it has been proved by Crawley–Boevey and Ringel [31] that if Δ is a locally finite, connected quiver without oriented cycles and such that after deleting finitely many vertices and arrows we obtain a disjoint union of quivers of type $\mathbb{A}_\infty$, then $\mathbb{Z}\Delta$ occurs as a regular component of some Γ_A.

It has been proved by Hoshino [56] that if A is a representation-infinite (connected) algebra and X an indecomposable A-module such that $X \simeq D\operatorname{Tr} X$, then $[X]$ lies in a homogeneous tube of Γ_A. Moreover, we have the following important result on tame algebras proved by Crawley–Boevey [25].

Theorem 4.3. *Let A be a tame algebra. Then, for each dimension d, almost all indecomposable A-modules X of dimension d are $D\operatorname{Tr}$-invariant, and hence lie in homogeneous tubes of Γ_A. In particular, all but countably many regular components of Γ_A are homogeneous tubes.*

It is expected (see [79]) that any regular connected component of the Auslander–Reiten quiver of a tame algebra A is either a stable tube or one of two forms $\mathbb{Z}_\infty\mathbb{A}_\infty$, $\mathbb{Z}\mathbb{D}_\infty$, where

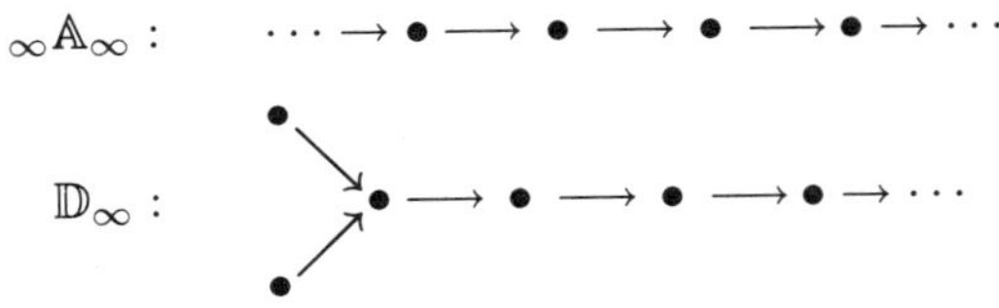

We note also the following consequence of Theorem 4.3: if A is a tame algebra then each connected component of Γ_A admits the isoclasses of at most finitely many modules of any fixed dimension. In general it is not the case as an example due to Liu and Schultz [64] shows.

We end this section with some examples of Auslander–Reiten quivers.

Example 4.4. Let $A = KQ/I$ where Q is the quiver

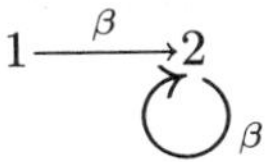

and the ideal I of KQ is generated by β^2. Then it is rather easy to see that there are only 7 indecomposable pairwise nonisomorphic A-modules and Γ_A is of the form

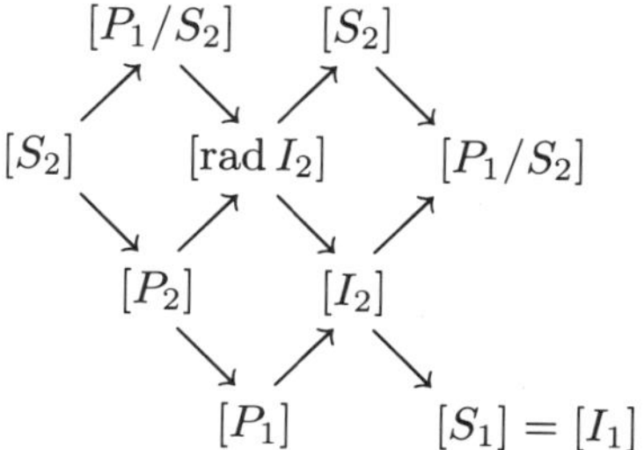

Observe that $[S_2]$, $[P_1/S_2]$, $[\operatorname{rad} I_2]$ are unique τ_A-periodic vertices of Γ_A.

Example 4.5. Let $A = KQ/I$ where Q is the quiver

$$1 \underset{\gamma}{\overset{\alpha}{\leftrightarrows}} 3 \underset{\sigma}{\overset{\beta}{\rightleftarrows}} 2$$

and the ideal I of KQ is generated by $\gamma\beta$, $\sigma\alpha$ and $\alpha\gamma - \beta\sigma$. Then A is a representation-finite selfinjective (projective modules are injective) and Γ_A is the following "cylinder"

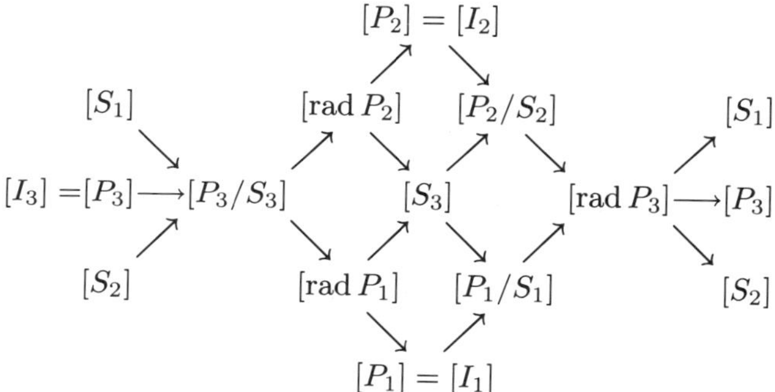

Observe that all nonprojective vertices of Γ_A are τ_A-periodic and with the same period equal 3.

Example 4.6. Let $A = KQ/I$ where Q is the quiver

$$1 \underset{\gamma}{\overset{\alpha}{\leftrightarrows}} 3 \underset{\sigma}{\overset{\beta}{\rightleftarrows}} 2$$

(as above) and I is the ideal of KQ generated by $\gamma\alpha$, $\sigma\beta$ and $\alpha\gamma - \beta\sigma$. Then A is a representation-finite selfinjective algebra and Γ_A is the following "Möbius band"

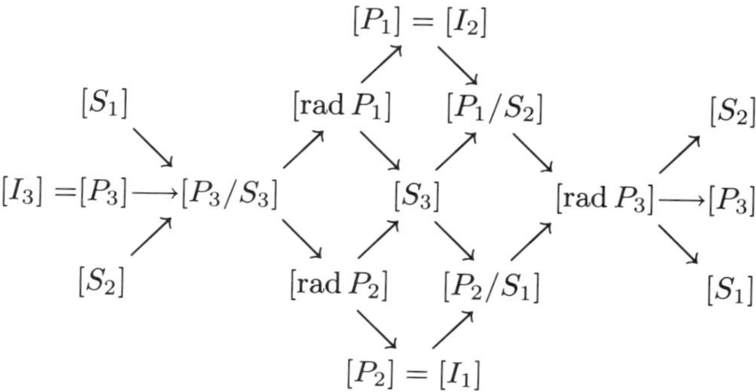

Observe that all nonprojective vertices of Γ_A are τ_A-periodic: $[S_3]$, $[P_3/S_3]$, $[\operatorname{rad} P_3]$ have period 3 but $[S_1]$, $[P_2/S_1]$, $[\operatorname{rad} P_2]$, $[S_2]$, $[P_1/S_2]$, $[\operatorname{rad} P_2]$ have period 6.

Example 4.7. Let A be the path algebra KQ of the following Dynkin quiver Q of type $\mathbb{D}_6$

$$\begin{array}{ccccccccc} & & 1 & & & & & & \\ & & \uparrow & & & & & & \\ 2 & \longrightarrow & 3 & \longleftarrow & 4 & \longrightarrow & 5 & \longrightarrow & 6 \end{array}$$

Then A has $30 = 6 \cdot 5$ isomorphism classes of indecomposable A-modules and Γ_A is a directed quiver of the form

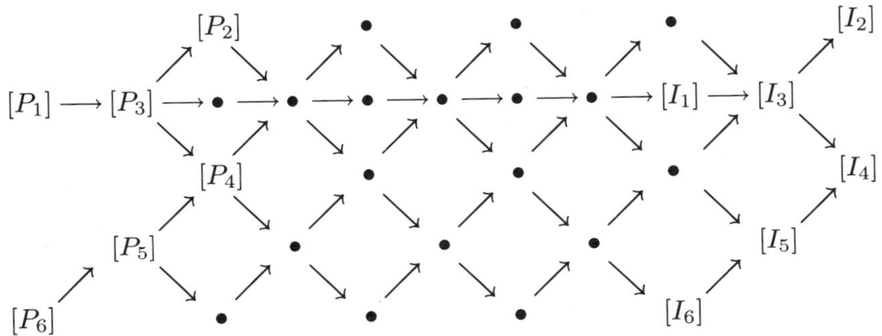

Example 4.8. Let A be the path algebra KQ of the following extended Dynkin quiver Q of type $\widetilde{\mathbb{E}}_7$

$$\begin{array}{ccccccccccccc} & & & & & & 1 & & & & & & \\ & & & & & & \downarrow & & & & & & \\ 5 & \longrightarrow & 4 & \longrightarrow & 3 & \longrightarrow & 2 & \longleftarrow & 6 & \longleftarrow & 7 & \longleftarrow & 8 \end{array}$$

Then Γ_A consists of a preprojective component $\mathcal{P}$ of the form

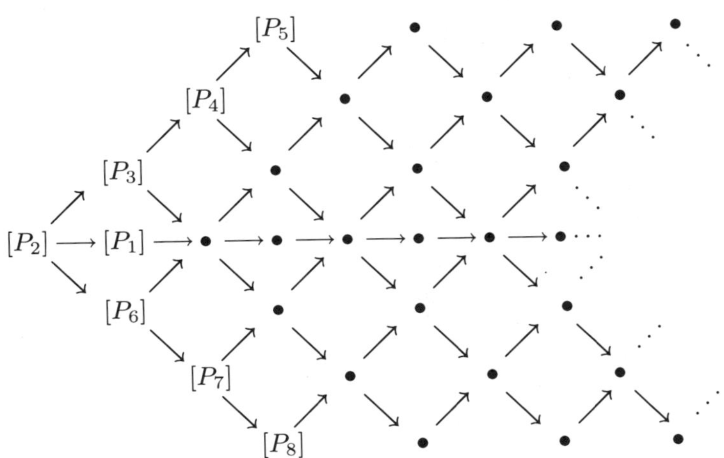

a preinjective component $\mathcal{Q}$ of the form

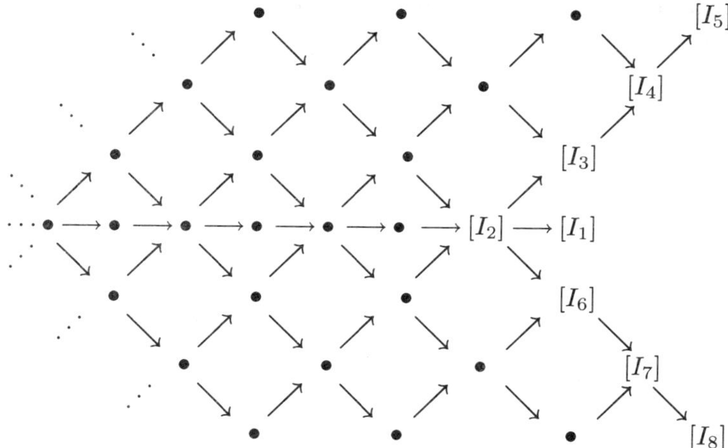

and a $\mathbb{P}_1(K)$-family $\mathcal{T} = (\mathcal{T}_\lambda)_{\lambda \in \mathbb{P}_1(K)}$ of stable tubes. Moreover, in $\mathcal{T}$ one tube has rank 2, one tube has rank 3, one tube has rank 4, and the remaining tubes are homogeneous.

5. Component Quiver

The Auslander–Reiten quiver Γ_A of an algebra A describes mainly the quotient category $\operatorname{mod} A/\operatorname{rad}^\infty(\operatorname{mod} A)$. If A is representation-finite then $\operatorname{rad}^\infty(\operatorname{mod} A) = 0$ and we may recover the morphisms in $\operatorname{mod} A$ from the quiver Γ_A. On the other hand, if A is representation-infinite and X, Y are indecomposable A-modules whose isomorphism classes lie in different connected components of Γ_A then $\operatorname{Hom}_A(X,Y) = \operatorname{rad}^\infty(X,Y)$. Hence $\operatorname{rad}^\infty(\operatorname{mod} A)$ contains lot of information on the representation theory of A. For example, it has been proved by Kerner and the author [60] that if $\operatorname{rad}^\infty(\operatorname{mod} A)$ is nilpotent then A is tame. Further, by a result of Coelho, Marcos, Merklen and the author [24], A is representation-infinite if and only if $(\operatorname{rad}^\infty(\operatorname{mod} A))^2 \neq 0$. It would be also interesting to know how to distinguish the representation types of A (finite type, polynomial growth, tame, wild) by the behaviour of the connected components of Γ_A in the category $\operatorname{mod} A$. In order to study the above questions, the author introduced in [86] the **component quiver** Σ_A of A. The vertices of Σ_A are the connected components of Γ_A. Two connected components $\mathcal{C}$ and $\mathcal{D}$ of Γ_A are connected in Σ_A by an arrow $\mathcal{C} \to \mathcal{D}$ if $\operatorname{rad}^\infty(X,Y) \neq 0$ for some indecomposable A-modules X, Y with $[X] \in \mathcal{C}$ and $[Y] \in \mathcal{D}$. It follows from [78] that the component quiver of a tame hereditary algebra (respectively, tame canonical algebra) is directed, that is, has no oriented cycles. Clearly, if A is a representation-finite algebra then Σ_A consists of exactly one vertex, namely the quivery Γ_A.

Example 5.1. Let A be the path algebra KQ of an extended Dynkin quiver Q. We know [78] that Γ_A consists of a preprojective component $\mathcal{P}$, a preinjective component $\mathcal{Q}$ and a family $\mathcal{T}_\lambda$, $\lambda \in \mathbb{P}_1(K)$, of stable tubes. Then the component quiver Σ_A of A has $\mathcal{P}$, $\mathcal{Q}$, $\mathcal{T}_\lambda$, $\lambda \in \mathbb{P}_1(K)$, as the vertices. Moreover, it follows from [78] that the arrows in Σ_A are of the form: $\mathcal{P} \to \mathcal{Q}$, $\mathcal{P} \to \mathcal{T}_\lambda$ and $\mathcal{T}_\lambda \to \mathcal{Q}$, for all $\lambda \in \mathbb{P}_1(K)$.

The following facts proved by the author in [87] and [88] show that the shape of Σ_A carries important information on the Auslander–Reiten components and the representation theory of A.

Theorem 5.2. *Let A be an algebra, $\mathcal{C}$ a connected component of Γ_A, and assume that Σ_A has no loop at $\mathcal{C}$. Then*

(i) *All but finitely many τ_A-orbits of $\mathcal{C}$ are periodic.*
(ii) *$\mathcal{C}$ admits the isoclasses of at most finitely many modules of any fixed dimension.*
(iii) *If $\mathcal{C}$ is regular then $\mathcal{C}$ is either a stable tube or of the form $\mathbb{Z}\Delta$ for some finite quiver Δ with at least three vertices and without oriented cycles.*

Theorem 5.3. *Let A be an algebra such that Σ_A is directed. Then*

(i) *A is of polynomial growth.*
(ii) *All but finitely many τ_A-orbits in Γ_A are periodic.*
(iii) *All regular connected components of Γ_A are stable tubes.*

6. Affine Varieties of Modules

In this section we assign to A a family of affine algebraic varieties of modules. This allows to study finite dimensional A-modules using geometric methods, in particular methods from algebraic transformation groups and invariant theory (see [61], [62]).

Let $A = KQ/I$ for $Q = Q_A$ and an admissible ideal I of KQ. Fix a vector $z \in K_0(A) = \mathbb{Z}^n$ ($n = |Q_0|$) with nonnegative coordinates. Denote by $\mathrm{mod}_A(z)$ the set of all representations V in $\mathrm{rep}_K(Q, I)$ with $V_i = K^{z_i}$ for all $i \in Q_0$. A representation V in $\mathrm{mod}_A(z)$ is given by $z_{e(\alpha)} \times z_{s(\alpha)}$-matrices $V(\alpha)$ determining the maps $\varphi_\alpha : K^{s(\alpha)} \to K^{e(\alpha)}$, $\alpha \in Q_1$, in the canonical bases of K^{z_i}, $i \in Q_0$. Moreover, the matrices $V(\alpha)$, $\alpha \in Q_1$, satisfy the relations

$$\sum_{1 \le j \le t} V(\alpha_{1,j}) \dots V(\alpha_{m_j,j})\lambda_j = 0$$

for all K-linear relations $\sum_{1 \le j \le t} \alpha_{1,j} \dots \alpha_{m_j,j}\lambda_j \in I$. Therefore, $\mathrm{mod}_A(z)$ is a closed subset of $\prod_{\alpha \in Q_1} K^{z_{e(\alpha)} z_{s(\alpha)}}$ in the Zariski topology, and so $\mathrm{mod}_A(z)$ is an affine variety. We call $\mathrm{mod}_A(z)$ the **affine variety of A-modules of dimension-vector z**. We note that $\mathrm{mod}_A(z)$ is not necessarily irreducible. Clearly, it is the case when $I = 0$. The affine (reductive) algebraic group $G(z) = \prod_{i \in Q_0} \mathrm{Gl}_{z_i}(K)$ acts on the variety $\mathrm{mod}_A(z)$ by conjugation:

$$(g \cdot V)(\alpha) = g_{s(\alpha)} V(\alpha) g_{e(\alpha)}^{-1}$$

for $g = (g_i) \in G(z)$, $V \in \mathrm{mod}_A(z)$, $\alpha \in Q_1$.

We shall identify $\mathrm{mod}\, A$ with $\mathrm{rep}_K(Q, I)$ and an A-module M of dimension-vector z with the corresponding point of the variety $\mathrm{mod}_A(z)$. Then it is easy to see that two A-modules M and N in $\mathrm{mod}_A(z)$ are isomorphic if and only if M and N belong to the same $G(z)$-orbit. In this setting a number of interesting questions arises very naturally: what are the irreducible components, orbit closures and singularities in the varieties $\operatorname{mod}_A(z)$. It is known that all varieties $\operatorname{mod}_A(z)$, $z \in K_0(A)$, are nonsingular if and only if $A = KQ$ (that is, $I = 0$)[16]. Moreover, the following known facts show a connection between homological and geometric properties of modules (see [62], [47], [48], [72]).

Theorem 6.1. *Let M be a module in $\operatorname{mod}_A(z)$. Then*

(i) *If $\operatorname{Ext}_A^1(M,M) = 0$ then the $G(z)$-orbit $G(z)M$ of M is an open subset of $\operatorname{mod}_A(z)$ and its closure $\overline{G(z)M}$ is an irreducible component of $\operatorname{mod}_A(z)$.*
(ii) *If $\operatorname{Ext}_A^2(M,M) = 0$ then M is a nonsingular point of $\operatorname{mod}_A(z)$.*

One of the important problems of the geometric classification of finite dimensional modules is to study the degenerations of modules. For $M, N \in \operatorname{mod}_A(z)$, we say that N is a **degeneration** of M if N belongs to the closure of $G(z)M$, and we denote this fact by $M \leq_{\deg} N$, and not by $N \leq_{\deg} M$ as one might expect. Then $\leq_{\deg}$ is a partial order on the set of isomorphism classes (orbits) of modules of a given dimension-vector. It is not clear how to characterize the partial order $\leq_{\deg}$ in terms of representation theory. However, there are two other partial orders $\leq_{\text{ext}}$ and $\leq$ on the isomorphism classes of A-modules defined in terms of extensions and homomorphisms as follows [1], [19], [75]:

- $M \leq_{\text{ext}} N \Leftrightarrow$ there are modules M_i, U_i, V_i and exact sequences $0 \to U_i \to M_i \to V_i \to 0$ in $\operatorname{mod} A$ such that $M = M_1$, $M_{i+1} = U_i \oplus V_i$ and $N = M_{s+1}$ for some natural number s.
- $M \leq N \Leftrightarrow \dim_K \operatorname{Hom}_A(M,X) \leq \dim_K \operatorname{Hom}_A(N,X)$ holds for all modules X from $\operatorname{mod} A$.

We note that $\leq$ is a partial order by a result of Auslander [4] (see also [15]). Furthermore, $M \leq N$ is equivalent to the inequalities

$$\dim_K \operatorname{Hom}_A(X,M) \leq \dim_K \operatorname{Hom}_A(X,N)$$

for all modules from $\operatorname{mod} A$. We have the following relation between the above partial orders.

Theorem 6.2. *Let M and N be modules in a variety $\operatorname{mod}_A(z)$. Then the following implications hold:*

$$M \leq_{\text{ext}} N \Rightarrow M \leq_{\deg} N \Rightarrow M \leq N.$$

Unfortunately, the reverse implications are not true in general, and it is interesting to find out when they are. But for tame hereditary algebras we have the following facts proved by Bongartz in [19] (see also [17]) and [18], respectively.

Theorem 6.3. *Let A be the path algebra of a Dynkin quiver. Then the partial orders $\leq_{\text{ext}}$, $\leq_{\deg}$ and $\leq$ coincide for all A-modules of any fixed dimension-vector.*

Theorem 6.4. *Let A be the path algebra of an extended Dynkin quiver. Then the partial orders $\leq_{\deg}$ and $\leq$ coincide for all A-modules of any fixed dimension-vector.*

Recently (November 1996) my student G. Zwara proved the following fact which completes the above theorem (see [100]).

Theorem 6.5. *Let A be the path algebra of an extended Dynkin quiver. Then the partial orders $\leq_{\text{ext}}$ and $\leq_{\text{deg}}$ coincide for all A-modules of any fixed dimension-vector.*

We note that, if $M <_{\text{ext}} N$ for some A-modules M and N, then N is decomposable. The following example shows that there are very simple (representation-finite) algebras for which there exist indecomposable modules M and N such that $M <_{\text{deg}} N$, and hence $M \not\leq_{\text{ext}} N$. Let A be the bound quiver algebra KQ/I given by the quiver

$$Q: \quad \bullet \xrightarrow{\alpha} \bullet \circlearrowleft_{\beta}$$

and the ideal I of KQ generated by β^2. Then it is easy to see that for the indecomposable representations

$$M: \quad K \xrightarrow{\left[\begin{smallmatrix}1\\0\end{smallmatrix}\right]} K^2 \circlearrowleft \left[\begin{smallmatrix}0&0\\1&0\end{smallmatrix}\right]$$

and

$$N: \quad K \xrightarrow{\left[\begin{smallmatrix}0\\1\end{smallmatrix}\right]} K^2 \circlearrowleft \left[\begin{smallmatrix}0&0\\1&0\end{smallmatrix}\right]$$

of A, M degenerates to N.

The following theorem has been proved by the author and Zwara in [96].

Theorem 6.6. *Let A be an algebra and assume that there is an integer m such that for any sequence*

$$M_r <_{\text{deg}} \cdots <_{\text{deg}} M_2 <_{\text{deg}} M_1$$

of degenerations of indecomposable A-modules $M_1, \ldots, M_r$, the inequality $r \leq m$ holds. Then A is tame. In particular, A is tame if, for any degeneration $M <_{\text{deg}} N$ of A-modules, the module N is decomposable.

Let $a_1 = 1, a_2, \ldots, a_m$ be a K-basis of A. Then we have the associated structure constants a_{kji}, $1 \leq i, j, k \leq m$, defined by

$$a_j a_i = \sum_{k=1}^{m} a_{kji} a_k.$$

For a positive integer d, we may consider the affine variety $\operatorname{mod}_d A$ of d-dimensional (right) A-modules, consisting of m-tuples $M = (M_1, \ldots, M_m) \in M_{d\times d}(K^m)$ of $d \times d$ matrices with coefficients in K such that M_1 is the identity matrix and

$$M_j M_i = \sum_{k=1}^{m} M_k a_{kji}$$

for all $1 \leq i, j \leq m$. Observe that $\operatorname{mod}_d A$ is a closed subset of K^{md^2} in the Zariski topology. A d-dimensional A-module M can be regarded as a K-algebra homomorphism

$$M : A \to \operatorname{End}_K(K^d) = M_{d\times d}(K)$$

and hence we may identify M with the m-tuple $(M_1, \dots, M_m) \in \mathrm{mod}_d A$ given by $M_i = M(a_i)$ for any $1 \le i \le m$. The general linear group $\mathrm{Gl}_d(K)$ acts on $\mathrm{mod}_d A$ by conjugation:

$$g * M = (gM_1g^{-1}, \dots, gM_mg^{-1})$$

for all $g \in \mathrm{Gl}_d(K)$ and $m = (M_1, \dots, M_m) \in \mathrm{mod}_d A$. Observe also that two d-dimensional A-modules M and N are isomorphic if and only if M and N belong to the same $\mathrm{Gl}_d(K)$-orbit in $\mathrm{mod}_d A$. For a vector $z \in K_0(A) = \mathbb{Z}^n$ and $|z| = z_1 + \dots + z_n$, we may consider the variety $\mathrm{mod}_A(z)$ as the affine subvariety

$$\{M \in \mathrm{mod}_{|z|} A \mid M(e_i) = E_i \text{ for all } 1 \le i \le n\}$$

of $\mathrm{mod}_{|z|} A$ where, for each $1 \le i \le n$, E_i is the matrix of the projection $K = \bigoplus_{i=1}^n K^{z_i} \to K^{z_i}$. One can prove that, if M, N are two d-dimensional A-modules and N belongs to the Zariski closure of the $\mathrm{Gl}_d(K)$-orbit of M in $\mathrm{mod}_d A$ (N is a degeneration of M in $\mathrm{mod}_d A$) then $\underline{\dim} M = \underline{\dim} N$. Moreover, if $z \in \mathbb{N}^n$, $|z| = z_1 + \dots + z_n$, and $M, N \in \mathrm{mod}_A(z)$ then $N \in \overline{G(z) \cdot M}$ in $\mathrm{mod}_A(z)$ if and only if $N \in \overline{\mathrm{Gl}_{|z|}(K) * M}$ in $\mathrm{mod}_{|z|} A$. Therefore, the study of degenerations in $\mathrm{mod}_d A$ is equivalent to that in the corresponding subvarieties $\mathrm{mod}_A(z)$, where z ranges all dimension-vectors $z \in K_0(A)$ with $|z| = d$.

For all $d \in \mathbb{N}$ consider the following closed and $\mathrm{Gl}_d(K)$-invariant subsets

$$\mathrm{mod}_{d,t} A = \{M \in \mathrm{mod}_d A \mid \dim_K \mathrm{End}_A(M) \ge t\}, \qquad t = 1, \dots, d^2,$$

of $\mathrm{mod}_d A$. We then get the following characterizations of tame (respectively, representation-finite) algebras proved by Geiss in [46].

Theorem 6.7. *An algebra A is tame if and only if we have the inequalities* $\dim \mathrm{mod}_{d,t} A \le d + (d^2 - t)$ *for all $d \in \mathbb{N}$, $t \in \{1, \dots, d^2\}$.*

Theorem 6.8. *An algebra A is representation-finite if and only if we have the inequalities* $\dim \mathrm{mod}_{d,t} A \le d^2 - t$ *for all $d \in \mathbb{N}$, $t \in \{1, \dots, d^2\}$.*

7. Degenerations of Algebras

For $d \in \mathbb{N}$ and $V = K^d$, the K-bilinear maps $V \times V \to V$ form a vector space $\mathrm{Hom}_K(V \otimes V, V)$ of dimension d^3, and hence an affine variety. The associative maps form a closed subset ass_d of $\mathrm{Hom}_K(V \otimes V, V)$ in the Zariski topology. Further, the associative K-algebra structures with 1 on $V = K^d$ form an open subset alg_d of ass_d. Hence, alg_d is an affine variety, called the affine variety of d-dimensional K-algebras. On alg_d operates the general linear group $\mathrm{Gl}_d(K)$ by transport of structure (see [41]). Clearly, two algebras A and B in alg_d are isomorphic if and only if A and B belong to the same $\mathrm{Gl}_d(K)$-orbit in alg_d. For $A, B \in \mathrm{alg}_d$, we say that B is a **degeneration** of A if B belongs to the closure of $\mathrm{Gl}_d(K)A$ in the Zariski topology of alg_d, and we denote this fact by $A \le_{\mathrm{deg}} B$. It is easy to see that in alg_2 we have the degeneration $K \times K \le_{\mathrm{deg}} K[x]/(x^2)$, and alg_2 is the closure of the $\mathrm{Gl}_2(K)$-orbit of $K \times K$. In alg_3 we have the degenerations

$$K \times K \times K \le_{\mathrm{deg}} K \times K[x]/(x^2) \le_{\mathrm{deg}} K[x]/(x^3) \le_{\mathrm{deg}} K[x, y]/(x^2, xy, y^2)$$

and

$$\begin{bmatrix} K & K \\ 0 & K \end{bmatrix} \le_{\mathrm{deg}} K[x, y]/(x^2, xy, y^2).$$

In [41] Gabriel proved the following remarkable result.

Theorem 7.1. *For each $d \in \mathbb{N}$, the representation-finite algebras in* alg_d *form an open subset.*

As a direct consequence we get the following fact.

Corollary 7.2. *Let A and B be two algebras in* alg_d *such that $A \leq_{\deg} B$ and B is representation-finite. Then A is also representation-finite.*

It is not clear whether the tame algebras in alg_d form an open subset. But, applying Theorem 6.7, Geiss was able to prove in [46] the following important result.

Theorem 7.3. *Let A and B be two algebras in* alg_d *such that $A \leq_{\deg} B$ and B is tame. Then A is also tame.*

It is still an open question whether and algebra A is of polynomial growth if so is a degeneration B of A. It follows from the above results that a convenient way to decide whether a given algebra A is tame (respectively, representation-finite) is to find a suitable tame (respectively, representation-finite) degeneration of A. We illustrate this idea by the following example (see [46]). Let Q be the quiver

$$\alpha \circlearrowleft \bullet \circlearrowright \beta$$

and I the ideal in KQ generated by the elements β^2, $(\beta\alpha)^2$, $(\alpha\beta)^2$ and $\alpha^2 - \beta\alpha\beta$. Moreover, let J be the ideal in KQ generated by the elements β^2, $(\beta\alpha)^2$, $(\alpha\beta)^2$ and α^2. Further, put $A = KQ/I$ and $B = KQ/J$. Then $A, B \in \mathrm{alg}_5$ and it is not hard to see that $A \leq_{\deg} B$. The algebra B is special biserial (in the sense of [92]), and hence is tame (see [98], [35], [23]). Therefore, A is also tame. In fact, both A and B are not of polynomial growth (see [83]). Recall that an algebra A is called biserial if every indecomposable nonuniserial projective left or right A-module P contains two uniserial submodules whose sum is the unique maximal submodule of P and whose intersection is either zero or simple. A module is uniserial if it has a unique composition series. Special biserial algebras are biserial [92]. Applying a modification of Theorem 7.3 and the fact that special biserial algebras are tame, Crawley–Boevey proved in [30] the following result.

Theorem 7.4. *All biserial algebras are tame.*

8. Integral Quadratic Forms

A polynomial $q = q(x_1, \dots, x_n)$ in n variables with integral coefficients is said to be an **integral (unit) quadratic form** provided it is of the form

$$q(x_1, \dots, x_n) = \sum_{1 \leq i \leq n} x_i^2 + \sum_{1 \leq i < j \leq n} q_{ij} x_i x_j$$

where $q_{ij} \in \mathbb{Z}$. We endow $\mathbb{Z}^n$ with a partial order defined componentwise: $z = (z_1, \dots, z_n) \in \mathbb{Z}^n$ is said to be positive, written $z > 0$, provided $z \neq 0$ and $z_i \geq 0$ for all i, $1 \leq i \leq n$. The integral quadratic form q in n variables is said to be **weakly positive** provided $q(z) > 0$ for all $z > 0$ in $\mathbb{Z}^n$, and **weakly nonnegative** provided $q(z) \geq 0$ for all $z \geq 0$ in $\mathbb{Z}^n$. Moreover, an element $z \in \mathbb{Z}^n$ satisfying $q(z) = m$ is said to be an m-**root** of q. Usually, 1-roots are called **roots**.

In the representation theory of algebras an important role is played by two integral quadratic forms: the Euler form and the Tits form. Let $A = KQ/I$ and assume that the quiver Q is connected and without oriented cycles. Moreover, let $Q_0 = \{1, \dots, n\}$. Since A is a triangular algebra ($Q_A = Q$ has no oriented cycles), we infer that $\operatorname{gl.dim} A < \infty$. Consider the **Cartan matrix**

$$C_A = (\dim_K \operatorname{Hom}_A(P_i, P_j))_{1 \leq i,j \leq n}$$

of A. By our assumption on A, C_A is invertible over $\mathbb{Z}$, and we get an integral quadratic form $\chi_A : \mathbb{Z}^n \to \mathbb{Z}^n$ on $K_0(A) = \mathbb{Z}^n$ defined by

$$\chi(x) = x C_A^{-t} x^t \qquad \text{for } x \in \mathbb{Z}^n.$$

It has been proved by Ringel [78] that if X is an A-module then

$$\chi_A(\underline{\dim} X) = \sum_{i \geq 0} (-1)^i \dim_K \operatorname{Ext}_A^i(X, X).$$

Hence, χ_A is called the **Euler form** of A.

The **Tits form** q_A of A is the integral quadratic form $q_A : \mathbb{Z}^n \to \mathbb{Z}$ defined by

$$q_A(x) = \sum_{i \in Q_0} x_i^2 - \sum_{\alpha \in Q_1} x_{s(\alpha)} x_{e(\alpha)} + \sum_{i,j \in Q_0} r_{ij} x_i x_j$$

for $x \in \mathbb{Z}^n$, where r_{ij} is the number of K-linear relations with source at i and end at j, for a minimal (finite) set $\mathcal{R}$ of K-linear relations generating the ideal I. It has been proved by Bongartz [11] that $r_{ij} = \dim_K \operatorname{Ext}_A^2(S_i, S_j)$, and hence r_{ij} does not depend on the choice of $\mathcal{R}$. We know that the number of arrows in Q with source at i and end at j is equal to $\dim_K \operatorname{Ext}_A^1(S_i, S_j)$. Therefore,

$$q_A(x) = \sum_{i \in Q_0} x_i^2 - \sum_{i,j \in Q_0} x_i x_j \dim_K \operatorname{Ext}_A^1(S_i, S_j) + \sum_{i,j \in Q_0} x_i x_j \dim_K \operatorname{Ext}_A^2(S_i, S_j).$$

Then we get the following important result of Bongartz [11].

Theorem 8.1. *If* $\operatorname{gl.dim} A \leq 2$ *then* $\chi_A = q_A$.

Observe that if A is hereditary ($\operatorname{gl.dim} A \leq 1$) then

$$\chi_A(x) = q_A(x) = \sum_{i \in Q_0} x_i^2 - \sum_{\alpha \in Q_1} x_{s(\alpha)} x_{e(\alpha)}.$$

The following theorem proved by Gabriel [39] characterizes the representation-finite hereditary algebras in terms of the Tits (Euler) form.

Theorem 8.2. *Let A be a hereditary algebra. The following conditions are equivalent:*

(i) *A is representation-finite.*
(ii) *q_A is positive definite.*
(iii) *q_A is weakly positive.*

Moreover, if A is representation-finite, then $\underline{\dim}$ *induces a bijection between the set of isomorphism classes of indecomposable A-modules and the set of positive roots of q_A.*

Observe that Theorems 3.3 and 8.2 establish a remarkable connection between the indecomposable representations of quivers of finite type (Dynkin quivers) and the positive roots of the finite dimensional simple Lie algebras.

We have also the following characterization of tame hereditary algebras due to Donovan–Freislich [34] and Nazarova [68] (see also [32], [78]).

Theorem 8.3. *Let A be a hereditary algebra. Then the following conditions are equivalent:*

(i) *A is tame.*
(ii) *q_A is positive semidefinite.*
(iii) *q_A is weakly nonnegative.*

Moreover, if A is tame then $\operatorname{ind} A$ *is controlled by q_A.*

We say that A is **controlled** by q_A if:

- $q_A(\underline{\dim}X) \in \{0,1\}$ for any $X \in \operatorname{ind} A$.
- For any connected vector $x \in \mathbb{N}^n$ with $q_A(x) = 1$ there is precisely one isomorphism class of X in $\operatorname{ind} A$ such that $\underline{\dim}X = x$.
- For any connected vector $x \in \mathbb{N}^n$ with $q_A(x) = 0$ there is an infinite family $(X_\lambda)_\lambda$ of pairwise nonisomorphic indecomposable A-modules with $\underline{\dim}X_\lambda = x$.

If A is representation-infinite tame hereditary then $\{x \in \mathbb{Z}^n; q_A(x) = 0\} = \mathbb{Z}\delta$ for a positive 0-root δ. Hence, combining Theorems 3.4 and 8.3 we get that the dimension-vectors of indecomposable representations of extended Dynkin quivers correspond to the positive roots (real and imaginary) of Kac–Moody Lie algebras of finite Gelfand–Kirillov dimension (see [57]).

For A of arbitrary (finite) global dimension the forms χ_A and q_A may have completely different values on the dimension-vectors of A-modules. But still we have the following inequalities established by de la Peña and the author [72]:

$$q_A(z) \geq \dim G(z) - \dim_X \operatorname{mod}_A(z) \geq \dim_K \operatorname{End}_A(X) - \dim_K \operatorname{Ext}^1_A(X,X),$$

for any module X in $\operatorname{mod}_A(z)$, $z \in K_0(A)$, where $\dim_X \operatorname{mod}_A(z)$ denotes the local dimension of $\operatorname{mod}_A(z)$ at the point X. The following theorem follows by an argument due to Tits.

Theorem 8.4. *Let A be a representation-finite algebra. Then for any positive vector $z \in K_0(A) = \mathbb{Z}^n$ the inequality* $\dim G(z) - \dim \operatorname{mod}_A(z) > 0$ *holds. In particular, q_A is weakly positive.*

The following fact is due to de la Peña [71].

Theorem 8.5. *Let A be a tame algebra. Then for any positive vector $z \in K_0(A) \in \mathbb{Z}^n$ the inequality* $\dim G(z) - \dim \operatorname{mod}_A(z) \geq 0$ *holds. In particular, q_A is weakly nonnegative.*

Unfortunately, the weak positivity (respectively, weak nonnegativity) of q_A does not imply in general that A is representation-finite (respectively, tame) [11]. We shall show in the next sections that if A satisfies some good homological or topological conditions, then still q_A controls the representation type of A.

9. Tame Quasitilted Algebras

Following Happel–Reiten–Smalø [53] an algebra A is said to be **quasitilted** if $\operatorname{gl.dim} A \leq 2$ and for each module X from $\operatorname{ind} A$ we have $\operatorname{pd}_A X \leq 1$ or $\operatorname{id}_A X \leq 1$. It is shown in [53] that an algebra A is quasitilted if and only if A is of the form $A = \operatorname{End}_{\mathcal{H}}(T)$ where T is a tilting object in a locally finite hereditary abelian K-category $\mathcal{H}$. Therefore, the derived category $D^b(\operatorname{mod} A)$ of bounded complexes of finite dimensional A-modules is equivalent, as a triangular category, to the derived category $D^b(\mathcal{H})$ of $\mathcal{H}$ [51]. It is known [53] that every quasitilted algebra A is isomorphic to a triangular matrix algebra

$$\begin{bmatrix} R & {}_RM_S \\ 0 & S \end{bmatrix}$$

where R and S are hereditary, and hence the Gabriel quiver Q_A of A has no oriented cycles. One wide class of quasitilted algebras is formed by the tilted algebras. Recall that a **tilted algebra** [54] is an algebra of the form $\operatorname{End}_H(T)$, where H is a hereditary algebra and $T \in \operatorname{mod} H$ is a tilting H-module, that is, $\operatorname{Ext}^1_H(T,T) = 0$ and T is a direct sum of n (= rank of $K_0(A)$) pairwise nonisomorphic indecomposable H-modules. It is known that the class of tilted algebras coincides with the class of quasitilted algebras A whose derived category $D^b(\operatorname{mod} A)$ is equivalent to $D^b(\operatorname{mod} H)$ for a hereditary algebra H. Moreover, it has been proved in [53] that every representation-finite quasitilted algebra is tilted. We refer to [54], [58], [59], [78] for the representation theory of tilted algebras. The second known class of quasitilted algebras is formed those algebras having the derived category equivalent to the derived category $D^b(\operatorname{mod} C(p,\lambda))$ for some canonical algebra $C(p,\lambda)$, and we call them **quasitilted algebras of canonical type**. The representation theory of quasitilted algebras of canonical type has been established by Lenzing–Skowroński [63] and Meltzer [66]. In particular, it is shown in [63] that every quasitilted algebra of canonical type is a semiregular branch enlargement of a tilt of a canonical algebra by a tilting module of positive rank. There is an interesting open problem whether any quasitilted algebra is tilted or of canonical type? Recently the author proved in [91] the following characterization of tame quasitilted algebras which in particular solves the above problem in the tame case.

Theorem 9.1. *Let A be a quasitilted algebra. The following conditions are equivalent:*

(i) *A is tame.*
(ii) *A is of polynomial growth.*
(iii) *$q_A(= \chi_A)$ is weakly nonnegative.*
(iv) *$\operatorname{ind} A$ is controlled by q_A.*
(v) *$\dim_K \operatorname{Ext}^1_A(X,X) \leq \dim_K \operatorname{End}_A(X)$ for any module X in $\operatorname{ind} A$.*
(vi) *$\operatorname{rad}^\infty(X,X) = 0$ for any module X in $\operatorname{ind} A$.*
(vii) *Σ_A is directed.*
(viii) *A is tame tilted or tame of canonical type.*

We have also the following characterization of representation-finite quasitilted algebras [53], [78].

Theorem 9.2. *Let A be a quasitilted algebra. The following conditions are equivalent:*

(i) *A is representation-finite.*
(ii) *$q_A(=\chi_A)$ is weakly positive.*
(iii) *Γ_A is finite and directed.*
(iv) *Σ_A is finite.*
(v) *A is representation-finite tilted.*
(vi) *$\mathrm{Ext}^1_A(X,X)=0$ and $\mathrm{End}_A(X)\simeq K$ for any module X in $\operatorname{ind} A$.*

Moreover, if A is representation-finite, then $\underline{\dim}$ induces a bijection between the set of isoclasses of indecomposable A-modules and the set of positive roots of q_A.

The following geometric characterization of tame quasitilted algebras has been proved recently by the author and Zwara [96] (using Theorem 9.1 and [93]).

Theorem 9.3. *Let A be a quasitilted algebra. Then A is tame if and only if for any (proper) degeneration $M <_{\deg} N$ of A-modules, the module N is decomposable.*

We mention that there exist tame quasitilted algebras for which the partial orders $\leq_{\mathrm{ext}}$ and $\leq_{\deg}$ do not coincide.

10. Tame Simply Connected Algebras

In the representation theory of finite dimensional algebras an important role is played by simply connected algebras. The importance of simply connected algebras follows from the fact that often we may reduce, with the help of coverings, study of the module category of an algebra to that for the corresponding simply connected algebras. This is known to be the case for all representation-finite algebras (see [21]) and many important classes of tame algebras. Recall that following [2] an algebra is called **simply connected** if A is triangular (the Gabriel quiver Q_A of A has no oriented cycles) and for any presentation $A\simeq KQ/I$ of A as a bound quiver algebra, the fundamental group $\pi_1(Q,I)$ is trivial. It was shown in [84] that a triangular algebra A is simply connected if and only if A does not admit a proper Galois covering. Following [85] an algebra A is called **strongly simply connected** if every full convex subalgebra of A is simply connected. It was shown in [85] that a triangular algebra A is strongly simply connected if and only if, for every convex subcategory C of A, the Hochschild cohomology group $H^1(C,C)$ vanishes (that is, any derivation $\delta : C\to C$ is an inner derivation). Clearly, if the quiver Q_A is a tree then A is strongly simply connected. Moreover, it is known that every representation-finite simply connected algebra is strongly simply connected [21]. We also note that (strongly) simply connected algebras may be of arbitrary large finite global dimension.

In [55] the class of critical algebras (preprojective tilts of hereditary algebras of extended Dynkin types $\widetilde{\mathbb{D}}_n$, $n\geq 4$, $\widetilde{\mathbb{E}}_p$, $6\leq p\leq 8$) has been classified by quivers and relations. They are strongly simply connected and minimal representation-finite, that is, they are representation-infinite but all proper convex subcategories are representation-finite. The following characterization of representation-finite (strongly) simply connected algebras is due to Bongartz [11], [12], [13].

Theorem 10.1. *Let A be a strongly simply connected algebra. The following conditions are equivalent:*

(i) *A is representation-finite.*
(ii) *A does not contain a convex subcategory which is critical.*
(iii) *q_A is weakly positive.*
(iv) *Γ_A is finite and directed.*
(v) *Σ_A is finite.*
(vi) $\mathrm{Ext}_A^1(X,X)=0$ *and* $\mathrm{End}_A(X)\simeq K$ *for any module X in* $\mathrm{ind}\,A$.

Moreover, if A is representation-finite, then $\underline{\dim}$ *induces a bijection between the set of isomorphism classes of indecomposable A-modules and the set of positive roots of q_A.*

In [97] (respectively, [69]) a class of minimal wild (respectively, tame minimal nonpolynomial growth) strongly simply connected algebras, called **hypercritical algebras** (respectively, ***pg*-critical algebras**) has been classified by quivers and relations. An algebra A is called minimal wild (respectively, minimal nonpolynomial growth) if A is wild (respectively, is not of polynomial growth) but every proper convex subcategory of A is not wild (respectively, is of polynomial growth). It is known by a result of de la Peña [70] that the Tits form q_A of a strongly simply connected algebra A is weakly nonnegative if and only if A does not contain a hypercritical convex subcategory.

The following characterization of polynomial growth strongly simply connected algebras has been proved by the author in [89].

Theorem 10.2. *Let A be a strongly simply connected algebra. The following conditions are equivalent:*

(i) *A is of polynomial growth.*
(ii) *A does not contain a convex subcategory which is pg-critical or hypercritical.*
(iii) $\mathrm{rad}^\infty(X,X)=0$ *for any module X in* $\mathrm{ind}\,A$.
(iv) *Σ_A is directed.*

In fact we established in [89] a rather complete representation theory of strongly simply connected algebras of polynomial growth. It is applied in the joint work by de la Peña and the author [72] to prove the following geometric and homological characterizations of polynomial growth strongly simply connected algebras.

Theorem 10.3. *Let A be a strongly simply connected algebra. The following conditions are equivalent:*

(i) *A is of polynomial growth.*
(ii) *For any module X in* $\mathrm{ind}\,A$ *and* $z=\underline{\dim}X$ *we have*

$$\chi_A(z)=\dim G(z)-\dim_X \mathrm{mod}_A(z)\geq 0.$$

(iii) *q_A is weakly nonnegative and* $\mathrm{Ext}_A^2(X,X)=0$ *for any module X in* $\mathrm{ind}\,A$.
(iv) $\dim_K \mathrm{Ext}_A^1(X,X)\leq \dim_K \mathrm{End}_A(X)$ *and* $\mathrm{Ext}_A^r(X,X)=0$ *for any module X in* $\mathrm{ind}\,A$ *and $r\geq 2$.*

Observe that by Theorems 6.2 and 10.3 we get that all indecomposable finite dimensional modules over polynomial growth strongly simply connected algebras are nonsingular. We have also the following fact on the values of the Tits and Euler forms on the dimension-vectors of indecomposable modules [73].

Theorem 10.4. *Let A be a strongly simply connected algebra of polynomial growth. Then there is a natural number m such that*

$$0 \leq \chi_A(\underline{\dim} X) \leq q_A(\underline{\dim} X) \leq m$$

for any module X in $\operatorname{ind} A$.

It follows from [89] that, if A is a polynomial growth strongly simply connected algebra, X is a module in $\operatorname{ind} A$ and $q_A(\underline{\dim} X) = 0$ (respectively, $\chi_A(\underline{\dim} X) = 0$) then there are infinitely many pairwise nonisomorphic indecomposable A-modules Y with $\underline{\dim} Y = \underline{\dim} X$. We note also that for each $r \geq 2$, there exists a polynomial growth strongly simply connected algebra A of global dimension 3 and a module X in $\operatorname{ind} A$ such that $q_A(\underline{\dim} X) = r + 1 > 2 = \chi_A(\underline{\dim} X)$. On the other hand, for arbitrary positive integers s and r there exists a polynomial growth strongly simply connected algebra A of global dimension 2 such that: for each $d \in \{1, \ldots, r\}$ there exist pairwise nonisomorphic modules $X_1^{(d)}, \ldots, X_s^{(d)}$ in $\operatorname{ind} A$ with $\underline{\dim} X_1^{(d)} = \ldots = \underline{\dim} X_s^{(d)} = x^{(d)}$ and $q_A(x^{(d)} = \chi_A(x^{(d)}) = d$ (see [73] for details). Then the following question arises naturally. Do the dimension-vectors of indecomposable modules over a polynomial growth strongly simply connected algebra A form a root system, extending the one considered by Kac [57]?

We shall also mention that for nonpolynomial growth tame strongly simply connected algebras the behaviour of the Tits and Euler form is even more interesting. Namely, there exists a tame strongly simply connected algebra A of global dimension 3 which admits a family X_n, $n \geq 1$, of modules in $\operatorname{ind} A$ such that $q_A(\underline{\dim} X_n) = 1 + 2n$ and $\chi_A(\underline{\dim} X_n) = 1 - 3n$ (negative!).

The following two theorems proved by the author and Zwara in [94] and [95], respectively, give addition information on the behaviour of discrete indecomposable modules over polynomial growth strongly simply connected algebras.

Theorem 10.5. *Let A be a strongly simply connected algebra. The following conditions are equivalent:*

(i) *A is of polynomial growth.*
(ii) *A is tame and there exists $m \in \mathbb{N}$ such that, for each $x \in \mathbb{N}^m$, there are at most m pairwise nonisomorphic indecomposable A-modules X with $\underline{\dim} X = x$ and $X \not\simeq D \operatorname{Tr} X$.*
(iii) *q_A is weakly nonnegative and there exists $m \in \mathbb{N}$ such that, for each $x \in \mathbb{N}^m$ with $q_A \neq 0$, there are at most m pairwise nonisomorphic indecomposable A-modules X with $\underline{\dim} X = x$.*

Theorem 10.6. *Let A be a strongly simply connected algebra of polynomial growth. Then*

(i) *Every module X from* $\operatorname{ind} A$ *with $\operatorname{Ext}_A^1(X, X) = 0$ is uniquely determined (up to isomorphism) by $\underline{\dim} X$.*
(ii) *The number of isomorphism classes of modules X in* $\operatorname{ind} A$ *with $\operatorname{Ext}_A^1(X, X) = 0$ but $\operatorname{End}_A(X) \not\simeq K$ is finite.*

We end the paper with the following characterization of polynomial growth strongly simply connected algebras in terms of degenerations of indecomposable modules, proved by the author and Zwara in [96].

Theorem 10.7. *Let A be a strongly simply connected algebra. The following conditions are equivalent:*

(i) *A is of polynomial growth.*

(ii) *For A-modules M, M', N such that $M <_{\deg} N$, $M' <_{\deg} N$ and N is indecomposable, $M \simeq M'$ and is indecomposable.*

(iii) *There is an integer m such that for any sequence*

$$M_r <_{\deg} \dots <_{\deg} M_2 <_{\deg} M_1$$

of degenerations of indecomposable A-modules, the inequality $r \leq m$ holds.

References

[1] S. Abeasis and A. del Fra, Degenerations for representations of quiver of type $\mathbb{A}_m$, *J. Algebra* **93** (1985), 376–412.

[2] I. Assem and A. Skowroński, On some classes of simply connected algebras, *Proc. London Math. Soc.* **56** (1988), 417–450.

[3] M. Auslander, Applications of morphisms determined by modules, Representation Theory of Algebras, Lecture Notes in Pure Applied Math., vol. 37, Marcel Dekker, 1976, pp. 245–327.

[4] M. Auslander, Representation theory of finite dimensional algebras, *Contemp. Math.* **13** (1982), 27–39.

[5] M. Auslander and I. Reiten, Representation theory of artin algebras III, *Comm. Algebra* **3** (1975), 239–294.

[6] M. Auslander, I. Reiten and S.O. Smalø, *Representation theory of Artin Algebras*, Cambridge Studies in Adv. Math. 36, (Cambridge University Press, 1995.

[7] R. Bautista, On algebras of strongly unbounded representation type, *Comment. Math. Helv.* **60** (1985), 392–399.

[8] R. Bautista, P. Gabriel A. V. Roiter and L. Salmeron, Representation-finite algebras and multiplicative bases, *Invent. Math.* **81** (1985), 217–285.

[9] R. Bautista, R. Martinez-Villa and J.A. de la Peña (editors), *Representation Theory of Algebras and Related Topics*, CMS Conference Proceedings 19 (AMS 1996).

[10] K. Bongartz, Tilted algebras, In: Representations of Algebras, Lecture Notes in Math. 903, Springer Verlag, 1981, pp. 26–38.

[11] K. Bongartz, *J. London Math. Soc.* **28** (1985), Algebras and quadratic forms, 461–469.

[12] K. Bongartz, Critical simply connected algebras, *Manuscr. Math.* **46** (1984), 117–136.

[13] K. Bongartz, A criterion for finite representation type, *Math. Annalen* **269** (1984), 1–12.

[14] K. Bongartz, Indecomposables are standard, *Comment. Math. Helv.* **60** (1985), 400–410.

[15] K. Bongartz, A generalization of theorem of M. Auslander, *Bull. London Math. Soc.* **21** (1989), 255–256.

[16] K. Bongartz, A geometric version of the Morita equivalence, *J. Algebra* **139** (1991), 159–171.

[17] K. Bongartz, Minimal singularities for representations of Dynkin quivers, *Comment. Math. Helv.* **69** (1994), 575–611.

[18] K. Bongartz, Degenerations for representations of tame quivers, *Ann. Sci. École Norm. Sup.* **28** (1995), 647–668.

[19] K. Bongartz, On degenerations and extensions of finite dimensional modules, *Advances Math.* **121** (1996), 245–287.

[20] K. Bongartz and P. Gabriel, Covering spaces in representation theory, *Invent. Math.* **65** (1981), 331–378.

[21] O. Bretscher and P. Gabriel, The standard form of a representation-finite algebra, *Bull. Soc. Math. France* **111** (1983), 21–40.

[22] O. Bretscher and G. Todorov, On a theorem of Nazarova and Roiter, Representation Theory I, Lecture Notes in Math. 1177, Springer Verlag, 1986, pp. 50–54.

[23] M. C. R. Butler and M. C. Ringel, Auslander-Reiten sequences with few middle terms and applications to string algebras, *Comm. Algebra* **15** (1987), 145-179.

[24] F. U. Coelho, E. N. Marcos, H. A. Merklen and A. Skowroński, Module categories with infinite radical square zero are of finite type, *Comm. Algebra* **22** (1994), 4511–4517.

[25] W.W. Crawley-Boevey, On tame algebras and bocses, *Proc. London Math. Soc.* **56** (1988), 451–483.

[26] W.W. Crawley-Boevey, Functorial filtrations II: Clans and the Gelfand problem, *J. London Math. Soc.* **40** (1989), 9–30.

[27] W.W. Crawley-Boevey, Functorial filtrations III: Semi-dihedral algebras, *J. London Math. Soc.* **40** (1989), 31–39.

[28] W.W. Crawley-Boevey, Tame algebras and generic modules, *Proc. London Math. Soc.* **63** (1991), 241–265.

[29] W. W. Crawley-Boevey, *Modules of finite length over their endomorphism rings* in: Representations of Algebras and Related Topics, London Math. Soc. Lecture Note Series 168, Cambridge University Press, 1992, pp. 127–184.

[30] W.W. Crawley-Boevey, Tameness of biserial algebras, *Arch. Math.* **65** (1995), 399–407.

[31] W.W. Crawley-Boevey and C.M. Ringel, Algebras whose Auslander-Reiten quivers have large regular components, *J. Algebra* **153** (1992), 494–516.

[32] V. Dlab and C.M. Ringel, *Indecomposable representations of graphs and algebras*, Memoirs Amer. Math. Soc. 173, 1976.

[33] V. Dlab and L.L. Scott (editors), *Finite Dimensional Algebras and Related Topics*, NATO ASI Series, Series C: Mathematical and Physical Sciences 424, Kluwer Acad. Publ., 1994.

[34] P. Donovan and M.R. Freislich, *The representation theory of finite graphs and associated algebras*, Carleton Lecture Notes 5, Ottawa, 1975.

[35] P. Dowbor and A. Skowroński, Galois coverings of representation-infinite algebras, *Comment Math. Helv.* **62** (1987), 311–337.

[36] Ju.A. Drozd, Tame and wild matrix problems, Representation Theory II, Lecture Notes in Math. 832, Springer Verlag, 1980, pp. 242–258.

[37] K. Erdmann, *Blocks of tame representation type and related algebras*, Lecture Notes in Math. 1428, Springer Verlag, 1990.

[38] U. Fischbacher, *Une nuvelle preuve d'un théorème de Nazarova et Roiter*, C. R. Acad. Sci. Paris, Sér. I 300, 1985, pp. 259–262.

[39] P. Gabriel, Unzerlegbare Darstellungen I, *Manuscr. Math.* **6** (1972), 71–103.

[40] P. Gabriel, Indecomposable representations II, *Symp. Math.* **11** (1973), 81–104.

[41] P. Gabriel, Finite representation type is open, Representations of Algebras, Lecture Notes in Math. 488, Springer Verlag, 1975, pp. 132–155.

[42] P. Gabriel, Auslander-Reiten sequences and representation-finite algebras, Representation Theory I, Lecture Notes in Math. 831, Springer Verlag, 1980, pp. 1–71.

[43] P. Gabriel, The universal cover of a representation-finite algebra, Representations of Algebras, Lecture Notes in Math. 903, Springer Verlag, 1981, pp. 68–105.

[44] P. Gabriel and A. V. Roiter, *Algebra VIII, Representations of Finite Dimensional Algebras*, Encyclopedia of Math. Sciences 73, Springer Verlag, 1992.

[45] W. Geigle and H. Lenzing, A class of weighted projective curves arising in the representation theory of finite dimensional algebras, Singularities, Representations of Algebras, and Vector Boundles, Lecture Notes in Math. 1273, Springer Verlag, 1987, pp. 265–297.

[46] C. Geiss, On degenerations of tame and wild algebras, *Arch. Math.* **64** (1995), 11–16.

[47] C. Geiss, Geometric methods in representation theory of finite dimensional algebras, Representation Theory of Algebras and Related Topics, CMS Conference Proc. 14 (AMS), 1996, pp. 53–63.

[48] C. Geiss and J.A. de la Peña, On the deformation theory of finite dimensional algebras, *Manuscr. Math.* **88** (1995), 191–208.

[49] I.M. Gelfand and V.A. Ponomarev, Indecomposable representations of the Lorentz group, *Uspekhi Math. Nauk* **23** (1968), 3–60.

[50] E.L. Green, Graphs with relations, coverings and groups-graded algebras, *Trans. Amer. Math. Soc.* **279** (1983), 297–310.

[51] D. Happel, On the derived category of a finite dimensional algebra, *Comment. Math. Helv.* **62** (1987), 339–389.

[52] D. Happel, U. Preiser and C. M. Ringel, Vingberg's characterization of Dynkin diagrams using subadditive functions with application to $D\,\mathrm{Tr}$-periodic modules, Representation Theory II, Lecture Notes in Math. 832, Springer Verlag, 1980, pp. 280–294.

[53] D. Happel, I. Reiten and S.O. Smalø, Tilting in abelian categories and quasitilted algebras, *Memoirs Amer. Math. Soc.* **575** (1996).
[54] D. Happel and C.M. Ringel, Tilted Algebras, *Trans. Amer. Math. Soc.* **274** (1982), 399–443.
[55] D. Happel and D. Vossieck, Minimal algebras of infinite representation type with preprojective component, *Manuscr. Math.* **42** (1983), 221–243.
[56] M. Hoshino, $D\,\mathrm{Tr}$-invariant modules, *Tsukuba J. Math.* **7** (1983), 205–214.
[57] V.G. Kac, Infinite root systems, representations of graphs and invariant theory, *Invent. Math.* **56** (1980), 57–92.
[58] O. Kerner, Tilting wild algebras, *J. London Math. Soc.* **39** (1989), 29–47.
[59] O. Kerner, Stable components of wild tilted algebras, *J. Algebra* **142** (1991), 35–57.
[60] O. Kerner and A. Skowroński, On module categories with nilpotent infinite radical, *Compositio Math.* **77** (1991), 313–333.
[61] H. Kraft, Geometric methods in representation theory, Representations of Algebras, Lecture Notes in Math. 944, Springer Verlag, 1982, pp. 180–258.
[62] H. Kraft, *Geometrische Methoden in der Invariantheorie*, Vieweg Verlag, 1984.
[63] H. Lenzing and A. Skowroński, Quasi-tilted algebras of canonical type, *Colloq. Math.* **71** (1996), 161–181.
[64] S. Liu and R. Schultz, The existence of bound infinite $D\,\mathrm{Tr}$-orbits, *Proc. Amer. Math. Soc.* **122** (1994), 1003–1005.
[65] R. Martinez and J.A. de la Peña, The universal cover of a quiver with relations, *J. Pure Applied Algebra* **30** (1983), 277–292.
[66] H. Meltzer, Auslander-Reiten components for concealed canonical algebras, *Colloq. Math.* **71** (1996), 183–202.
[67] G. Michler and C.M. Ringel (editors), *Representation Theory of Finite Groups and Finite-Dimensional Algebras*, Birkhäuser Verlag, 1991.
[68] L. Nazarova, Representations of quivers of infinite type, *Izv. Akad. Nauk SSSR, Ser. Mat.* **37** (1973), 752–791.
[69] R. Nörenberg and A. Skowroński, Tame minimal non-polynomial growth simply connected algebras, *Colloq. Math.*, in press.
[70] J.A. de la Peña, Algebras with hypercritical Tits form, Topics in Algebra, Banach Center Publications Vol. 26, Part 1, PWN, Warszawa, 1983, pp. 353–369.
[71] J.A. de la Peña, On the dimension of the module-varieties of tame and wild algebras, *Comm. Algebra* **19** (1991), 1795–1807.
[72] J.A. de la Peña and A. Skowroński, Geometric and homological characterizations of polynomial growth strongly simply connected algebras, *Invent. Math.* **126** (1996), 287–296.
[73] J. A. de la Peña and A. Skowroński, The Tits and Euler forms of a tame algebra, Preprint, Mexico, 1997.
[74] I. Reiten, A. Skowroński and S.O. Smalø, Short chains and short cycles of modules, *Proc. Amer. Math. Soc.* **117** (1993), 343–354.
[75] C. Riedtmann, Degenerations for representations of quivers with relations, *Ann. Sci.École Norm. Sup.* **4** (1986), 275–301.
[76] C.M. Ringel, The spectrum of a finite dimensional algebra, Ring Theory, Lecture Notes in Pure Applied Math. 51, Marcel Dekker, 1979, pp. 535–597.
[77] C.M. Ringel, Finite dimensional hereditary algebras of wild representation type, *Math. Z.* **161** (1978), 236–255.
[78] C.M. Ringel, *Tame algebras and integral quadratic forms*, Lecture Notes in Math. 1099, Springer Verlag, 1984.
[79] C. M. Ringel, Representation theory of finite dimensional algebras, Representations of Algebras, London Mathematical Society Lecture Notes Series 116, Cambridge University Press, 1986, pp. 7–79.
[80] C.M. Ringel, The regular components of the Auslander-Reiten quiver of a tilted algebra, *Chinese Ann. Math.* **9B** (1988), 1–18.
[81] D. Simson, *Linear Representations of Partially Ordered Sets and Vector Space Categories*, Algebra, Logic and Applications Vol. 4, Gordon and Breach Science Publishers, 1992.
[82] A. Skowroński, Tame triangular matrix algebras over Nakayama algebras, *J. London Math. Soc.* **34** (1986), 245–264.
[83] A. Skowroński, Group algebras of polynomial growth, *Manuscr. Math.* **59** (1987), 499–516.

[84] A. Skowroński, Algebras of polynomial growth, Topics in Algebra, Banach Center Publications Vol. 26, Part 1, PWN, Warszawa, 1990, pp. 535–568.
[85] A. Skowroński, Simply connected algebras and Hochschild cohomologies, Representations of Algebras, CMS Conference Proceedings 14, AMS, 1993, pp. 431–447.
[86] A. Skowroński, Cycles in module categories, Finite Dimensional Algebras and Related Topics, NATO ASI Series, Series C: Mathematical and Physical Sciences 424, Kluwer Acad. Publ, 1994 309–345.
[87] A. Skowroński, Generalized standard Auslander-Reiten components, *J. Math. Soc. Japan* **46** (1994), 517–543.
[88] A. Skowroński, Cycle-finite algebras, *J. Pure Appl. Algebra* **103** (1995), 105–116.
[89] A. Skowroński, Simply connected algebras of polynomial growth, *Compositio Math.*, in press.
[90] A. Skowroński, Tame algebras with strongly simply connected Galois coverings, *Colloq. Math.*, in press.
[91] A. Skowroński, Tame quasitilted algebras, Preprint (Toruń, 1997).
[92] A. Skowroński and J. Waschbüsch, Representation-finite biserial algebras, *J. Reine Angew. Math.* **345** (1983), 172–181.
[93] A. Skowroński and G. Zwara, On degenerations of modules with nondirecting indecomposable summands, *Canad. J. Math.* **48** (1996), 1091–1120.
[94] A. Skowroński and G. Zwara, On the numbers of discrete modules over tame algebras, *Colloq. Math.*, in press.
[95] A. Skowroński and G. Zwara, On indecomposable modules without selfextensions, *J. Algebra*, in press.
[96] A. Skowroński and G. Zwara, Degenerations for indecomposable modules and tame algebras, Preprint (Toruń, 1996).
[97] L. Unger, The concealed algebras of minimal wild, hereditary algebras, *Bayreuther Math. Schriften* **31** (1990), 145–154.
[98] B. Wald and J. Waschbüsch, Tame biserial algebras, *J. Algebra* **95** (1985), 480–500.
[99] Y. Zhang, The structure of stable components, *Canad. J. Math.* **43** (1991), 652–672.
[100] G. Zwara, Degenerations for representations of extended Dynkin quivers, Preprint (Toruń, 1997).

A. Skowroński
Faculty of Mathematics and Informatics
Nicholas Copernicus University
Chopina 12/18, 87-100 Toruń
Poland
e-mail: skowron@mat.uni.torun.pl

Canadian Mathematical Society
Conference Proceedings
Volume **22**, 1998

Direct Limits of Finite Dimensional Algebras and Finite Groups*

Aleksandr E. Zalesskii

1. Introduction

This paper discusses certain new problems and ideas concerning the theory of simple locally finite associative and Lie algebras, as well as related problems about locally finite simple groups. Recent development provides evidence of closer relation between these topics than it earlier seemed. One can hope to use the theory of associative simple locally finite algebras in order to clarify structure and classification problems of so called 'diagonal' simple locally finite groups and Lie algebras. In group theory this class has been singled out by classification theory of simple locally finite groups. In Lie algebra theory the class has occurred in proving a version of Ado's theorem for locally finite Lie algebras.

Locally finite simple associative algebras were studied in the 1940's and 50's as an area of possible generalizations of classical theorems about finite dimensional algebras and artinian rings. The theory of $\mathbf{C}^*$-algebras and von Neumann regular rings made it clear that locally finite algebras are in fact asymptotic objects important for the theory of normed algebras and various applications. A prominent role was played by the work of Bratteli [Br] (1972) who reduced classification of

AMS Subject Classification (1991). 16K20, 16P10, 17B20.

* This paper was started during the author's visit to the University of Antwerp (February, 1994). Essential impulse to continue it was given by the workshop 'Locally finite simple groups' (Oberwolfach, 1996). The work was also partially supported by the INTAS grant 'Noncommutative algebra and geometry with focus on representation theory' during the author's visit to the University Paris VI (June-July, 1996). I am grateful to Professors F. Van Oystayen, M.-P. Malliavin and J. Alev for their hospitality, and to O. Kegel, K. Goodearl and A. Baranov for scientific discussions. I am indebted for the help of my colleague D. Evans and the referee in improving the final version of the manuscript.

so called approximatively finite $\mathbf{C}^*$-algebras to that of direct limits of finite dimensional semisimple associative algebras. For this reason locally finite algebras were studied from that time in the framework of the theory of $\mathbf{C}^*$-algebras. This study provides evidence of numerous links of the topic with other branches of mathematics. This can be seen from a remarkable survey of Vershik and Kerov [VK], the book [GHJ] etc.

Notation. $\mathbf{N}, \mathbf{Z}, \mathbf{Z}^+, \mathbf{Q}, \mathbf{R}, \mathbf{C}$ will denote, respectively, the set of natural, integer, non-negative integer, rational, real and complex numbers. $\mathbf{F}_p$ and $\bar{\mathbf{F}}_p$ stand for the field of p elements and its algebraic closure. For bravity we explore the term 'locally finite algebra' instead of 'locally finite dimensional algebra.'

To simplify the exposition, we always assume that *all algebras discussed are of countable dimension.*

2. Locally Finite Simple Associative Algebras

Locally finite simple associative algebras are our main topic of discussion, and our exposition here is fairly detailed. Our discussion of locally finite simple Lie algebras (mainly for a field of characteristic 0) in Section 4 and locally finite simple groups in Section 5 is sketchy. We only tried to show the links and possible applications of associative algebras to these topics.

2.1 Locally Simple Algebras

Let F be a field. The most obvious construction of a locally finite simple associative algebra is a direct limit of *simple* finite dimensional algebras. We call such algebras locally simple. A straightforward example is the *locally matrix algebra* $M(\infty, F)$ which can be defined as the union of matrix algebras $M(n, F)$ with embeddings $M(n, F) \to \mathrm{diag}(M(n, F), 0) \subset M(n+1, F)$, $n = 1, 2, \dots$. One arrives at an essentially more general construction by taking multiple embeddings as follows. Let $n = km$ and let $M(m, F) \to M(n, F)$ be defined to be $x \to \mathrm{diag}(x, \dots, x)$ (k times). More precisely, every sequence $\mathcal{S} = \{s_1, s_2, \dots\}$ of natural numbers $s_i > 1$ determines the embeddings

$$M(1, F) \to M(s_1, F) \to M(s_1 s_2, F) \to \dots \to M(s_1 \dots s_i, F) \to \dots$$

Denote the direct limit by $M(\mathcal{S}, F)$. Observe that $M(s_1 \dots s_i, F)$ can be expressed as the tensor product $\otimes_i M(s_i, F)$ so $M(\mathcal{S}, F)$ can be considered as the infinite tensor product $\otimes_{i=1}^{\infty} M(s_i, F)$. One can easily describe conditions of isomorphisms of $M(\mathcal{S}, F)$ for various $\mathcal{S}$. Let $\mathcal{S}' = \{s_1', s_2', \dots\}$. Set $t_i = s_1 \dots s_i$ and $t_i' = s_1' \dots s_i'$. Then $M(\mathcal{S}, F) \cong M(\mathcal{S}', F)$ if and only if for some subsequences of $\{t_i\}$ and $\{t_i'\}$ one has

$$M(t_{i_1}, F) \subseteq M(t'_{j_1}, F) \subseteq M(t_{i_2}, F) \subseteq M(t'_{j_2}, F) \subseteq \dots$$

This is possible if and only if the following condition holds:

(a) for every i there is j such that t_i divides t_j', and for every j there is k such that t_j' divides t_k.

One can observe that (a) defines an equivalence relation on the set of sequences $\mathcal{S}$. Express $t_i = \prod_l p_l^{\alpha_{il}}$ and $t'_i = \prod_l p_l^{\alpha'_{il}}$ where p_l are distinct primes. Then (a) means that for each fixed l for every i there exists j such that $\alpha'_{jl} \geq \alpha_{il}$, and for every j there exists k such that $\alpha'_{jl} \leq \alpha_{kl}$. It follows that (a) is equivalent to

(b) For a prime p let $\mathcal{S}(p)$ denote the maximum of the multiplicities of p in t_i, $i = 1, 2, \ldots$, so $\mathcal{S}(p) \in \mathbf{N} \cup \infty$. So $M(\mathcal{S}, F) \cong M(\mathcal{S}', F)$ if and only if $\mathcal{S}(p) = \mathcal{S}'(p)$ for all primes p.

Condition (a) and (b) can be expressed by several alternative forms.

(c) *Steinitz numbers*. A Steinitz (or supernatural) number is a formal product $\Pi_i p_i^{k_i}$ taken under all natural primes p where $k_i \in \mathbf{N} \cup \infty$. There is an obvious way to multiply a Steinitz number by a natural number. Set $\operatorname{Stz}(\mathcal{S}) = \Pi_l p_l^{\mathcal{S}(p_l)}$. (One can think about $\operatorname{Stz}(\mathcal{S})$ as an infinite formal product $\Pi_i s_i$.) Then $M(\mathcal{S}, F) \cong M(\mathcal{S}', F)$ if and only if $\operatorname{Stz}(\mathcal{S}) = \operatorname{Stz}(\mathcal{S}')$.
(d) *Subgroups of the additive group of the group of rational numbers*. Let $\mathbf{Q}_{\mathcal{S}}$ denote the additive group consisting of all $q \in \mathbf{Q}$ such that $qt_i \in \mathbf{Z}$ for some i. It is fairly clear that $\mathbf{Q}_{\mathcal{S}} = \mathbf{Q}_{\mathcal{S}'}$ if and only if $\operatorname{Stz}(\mathcal{S}) = \operatorname{Stz}(\mathcal{S}')$.
(e) *Subgroup systems*. Let $t_i\mathbf{Z}$ be the additive group of integer multiples of t_i. Let $\operatorname{Sgr}_{\mathcal{S}}(\mathbf{Z})$ stand for the set of the subgroups $\{t_i\mathbf{Z}\}$ of $\mathbf{Z}$. The condition above means that every group of $\operatorname{Sgr}_{\mathcal{S}}(\mathbf{Z})$ contains a group of $\operatorname{Sgr}_{\mathcal{S}'}(\mathbf{Z})$, and conversely. This condition induces an equivalence relations on the subgroup systems, $M(\mathcal{S}, F) \cong M(\mathcal{S}', F)$ if and only if $\operatorname{Sgr}_{\mathcal{S}}(\mathbf{Z})$ is equivalent to $\operatorname{Sgr}_{\mathcal{S}'}(\mathbf{Z})$.
(f) *Topology*. As $t_{i+1} > t_i$, we have $\cap_i t_i\mathbf{Z} = 0$. Therefore, taking $\operatorname{Sgr}_{\mathcal{S}}(\mathbf{Z})$ $\operatorname{Sgr}_{\mathcal{S}'}(\mathbf{Z})$ as a system of neighbourhoods of zero determines a Hausdorff topology $\operatorname{Top}_{\mathcal{S}}(\mathbf{Z})$ on $\mathbf{Z}$. Condition (e) exactly means that $\operatorname{Top}_{\mathcal{S}}(\mathbf{Z}) = \operatorname{Top}_{\mathcal{S}'}(\mathbf{Z})$. Thus, $M(\mathcal{S}, F) \cong M(\mathcal{S}', F)$ if and only if $\operatorname{Top}_{\mathcal{S}}(\mathbf{Z}) = \operatorname{Top}_{\mathcal{S}'}(\mathbf{Z})$.
(g) *Profinite groups*. Clearly, $\{t_i\mathbf{Z}\}$ is a descending series of subgroups of $\mathbf{Z}$. Observe that $\cap t_i\mathbf{Z} = 0$ as $t_{i+1} > t_i$ for all i. Let $\hat{\mathbf{Z}}_{\mathcal{S}}$ denote the completion of $\mathbf{Z}$ under the topology $\operatorname{Top}_{\mathcal{S}}(\mathbf{Z})$. Again, $M(\mathcal{S}, F) \cong M(\mathcal{S}', F)$ if and only if $\hat{\mathbf{Z}}_{\mathcal{S}} \cong \hat{\mathbf{Z}}_{\mathcal{S}'}$.
(h) *The dual group*. Instead of $\hat{\mathbf{Z}}_{\mathcal{S}}$ one can take the dual group $\operatorname{Hom}(\hat{\mathbf{Z}}_{\mathcal{S}}, \mathbf{C})$.
(i) *K_0-functor*, see the next section.

If one considers $M(n, \mathbf{C})$ as a $\mathbf{C}^*$-algebra under the involution given by transpose, then the multiple embeddings are $\mathbf{C}^*$-algebra homomorphisms. The algebras $\overline{M(\mathcal{S}, F)}$ where the overline stands for closure under the standard norm are known as Glimm algebras.

There are more complicated embeddings of matrix algebras than the multiple ones. Let $n = km + l$, and let $M(m, F) \to M(n, F)$ be the embedding defined by $x \to \operatorname{diag}(x, \ldots, x, 0, \ldots, 0)$ (l zeros, $x \in M(m, F)$). To have a direct limit, we choose a double sequence $\mathcal{D} = \{s_i, l_i\}$, set $n_{i+1} = n_i s_i + l_i$, and define $M(\mathcal{D}, F)$ to be the direct limit of the embeddings $M(n_i, F) \to M(n_{i+1}, F)$ defined above. By omitting certain members of the sequence $M(n_i, F)$ one arrives either at the case of multiple embeddings or that of the embeddings with $l_i > 0$ for all i. Then $M(\mathcal{D}, F)$ has no identity in contrast with the case of multiple embeddings. Thus, assuming $l_i > 0$ we indeed have a new class of locally finite simple F-algebras. If $\mathcal{D}' = \{s'_i, l'_i\}$ then $M(\mathcal{D}, F) \cong M(\mathcal{D}', F)$ implies the existence of subsequences of $\{n_{i_r}\}$ and $\{n'_{j_s}\}$

such that

$$M(n_{i_1}, F) \subset M(n'_{j_2}, F) \subset M(n_{i_2}, F) \subset M(n'_{j_2}, F) \subset \dots$$

with commutative diagram of embeddings for $k < l$

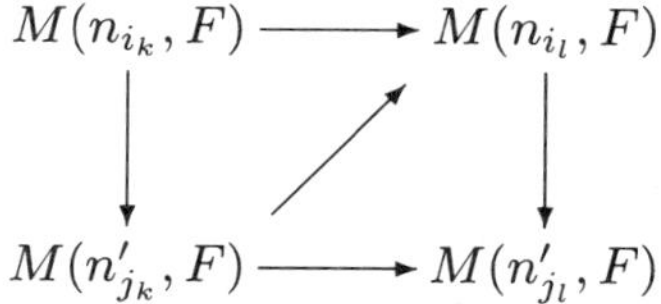

It is more difficult than above to express this condition in combinatorial terms. This was done by Dixmier [Di].

By the representation theory of matrix algebras, every embedding $\varepsilon : M(m, F) \to M(n, F)$ is equivalent to one described above, i.e. it is determined by a pair k, l such that $n = km + l$. If the ground field is algebraically closed, any embedding of simple finite dimensional algebras is equivalent to one of this shape.

Normally, one considers $M(n, F)$ as an F-algebra. However, for the theory of direct limits of $M(n_i, F)$ one can extend this point of view to obtain even more sophisticated types of embeddings. Let $A \subseteq \mathrm{Aut}\,(F)$ be a group of field automorphisms of F and F^A the subfield of A-fixed elements. The action of A on F extends to $M(n, F)$ in the natural way, i.e. by forcing $a \in A$ acts on every matrix entry of a matrix $M \in M(n, F)$.

Let $\varepsilon : M(m, F) \to M(n, F)$ be a ring representation. Then ε is completely reducible, and the irreducible constituents are either trivial or of shape $\rho_a : x \to a(x)$ ($a \in A$, $x \in M(m, F)$). Thus, ε is determined up to equivalence by the set l, k_a where k_a is the multiplicity of ρ_a in ε and $l = n - \sum_{a \in A} k_a$.

These embeddings are not F-algebra embeddings, however, they are F^A-algebra embedding. So one can form the direct limits of $M(n_j, F)$ with embeddings of this type. The limit algebra is of course an F^A-algebra. For F finite or locally finite such algebras are interesting for applications to locally finite group theory.

The next level of generality provides the problem of classifying the direct limits of simple finite dimensional F-algebras. The case where these algebras have a common identity was studied by Kurochkin [K]. In the next section we shall discuss a machinery that leads to classification of more general class of algebras in terms of K_0-functor. Therefore, the problems of classifying locally simple algebras has sense only as finding more simple system of invariants than the K_0-functor, in spirit of that of [Di].

2.2 Locally Semisimple Algebras

Definition 2.1. Let L be a locally finite algebra over a field F. We say that L is locally semisimple if every finite subset of L is contained in a *semisimple* subalgebra of finite dimension.

We shall see in the next section that certain simple locally finite algebras are not locally semisimple. The natural question whether every simple locally semisimple algebra is locally simple has a negative answer, see Bratteli [Br]. It is now well understood that this phenomenon is not exceptional.

Of course, every semisimple finite dimensional F-subalgebra of L is a finite direct sum of matrix algebras over finite skew field extensions of F. There is an important particular class of locally semisimple algebras which has attracted more attention. In [Go1] it is called a class of ultramatricial algebras; following [Go1] we give the following definition:

Definition 2.2. Let A be an algebra over a field F. We say that A is *matricial* if A is a finite direct sum of matrix algebras over F. We say that A is *ultramatricial* if A is a direct limit of matricial algebras over F.

The classification of ultramatricial algebras was obtained by Elliott [El], see Theorem 2.3 below. More precisely, Elliott classified ultramatricial algebras L in terms of systems of idempotents in L; later Handelman [Ha] and Goodearl [Go] showed that Elliott's invariant can be interpreted in terms of the K_0-functor. An exposition of this result can be found in [Go1] and [Go2].

The Grothendieck group $K_0(R)$ of a ring R is defined as follows. Let $\Pr(R)$ stand for the set of isomorphism classes of finitely generated projective R-modules. Then $K_0(R)$ is the quotient group F/N where F is the free abelian group with generators g_P ($P \in \Pr(R)$) and N is the subgroup generated by the elements $g_P + g_Q - g_{P+Q}$ ($P, Q \in \Pr(R)$). (Additive notation is standard for the group operation in $K_0(R)$.) Let ρ denote the image of $Pr(R)$ in $K_0(R)$. If $\varepsilon : R \to R'$ is an injective homomorphism then $P' = P \otimes_R R'$ is a projective R'-module. This gives rise to a group homomorphism $\kappa : K_0(R) \to K_0(R')$ such that $\kappa(\rho) \subseteq \rho'$. Furthermore, if $R = \lim_i R_i$, a direct limit of subrings, then $K_0(R) = \lim_i K_0(R_i)$. If R is artinian then $K_0(R)$ is a free abelian group of finite rank. It follows that for a locally finite algebra R the group $K_0(R)$ is torsion free. Let $K_0^+(R)$ stand for the semigroup generated by g_P ($P \in \Pr(R)$).

A multiple embedding $M(m, F) \to M(n, F)$ induces an embedding $\mathbf{Z} \cong K_0(M(m, F)) \to K_0(M(n, F)) \cong \mathbf{Z}$, and the index of the image subgroup is exactly n/m. Thus, one can form the direct limit of $K_0(M(t_i, F))$ with respect to the embeddings above. It follows that $\lim_i K_0(M(t_i, F)) = \lim_i \mathbf{Z}[1/t_i] = K_0(M(\mathcal{S}, F))$. (Here $\mathbf{Z}[1/t]$ denotes the additive subgroup of $\mathbf{Q}$ generated by $1/t$.) Observe that $K_0(M(\mathcal{S}, F))$ does not determine $M(\mathcal{S}, F)$; indeed, if $t_i = 2^i$, $t_i' = 3 \cdot 2^i$ then $K_0(M(\mathcal{S}, F)) = K_0(M(\mathcal{S}', F))$. (In fact, $K_0(M(\mathcal{S}, F))$ determines $M(\mathcal{S}, F)$ up to Morita equivalence, see [Go1, Corollary 15.27].) An additional invariant is $E_R \in K_0(R)$, the image in $K_0(R)$ of the regular R-module. More precisely, if there exists an isomorphism $a : K_0(M(\mathcal{S}, F)) \to K_0(M(\mathcal{S}', F))$ such that $a(E_{M(\mathcal{S},F)}) = E_{M(\mathcal{S}',F)}$ then $M(\mathcal{S}, F) \cong M(\mathcal{S}', F)$.

The following result provides a classification of ultramatricial algebras in terms of K_0-functor.

Theorem 2.3. ([El], see also [Ha, Go1, Go2]) *Let L, L' be ultramatricial algebras. Then $L \cong L'$ if and only if $(K_0^+(L), E_L)$ is isomorphic to $(K_0^+(L'), E_{L'})$ (i.e. there exists an ordered group isomorphism $\alpha : K_0(L) \to K_0(L')$ such that $\alpha(E_L) = E_{L'}$).*

The ideas used in classifying ultramatricial algebras have been further developed in [GH] and [HD]. Goodearl–Handelman [GH] classified direct limits of semisimple finite dimensional algebras over the real number field (see, also, Giordano [Gi]). Let A be such an algebra, we set $A^{\mathbf{C}} = A \otimes_{\mathbf{R}} \mathbf{C}$ and $A^{\mathbf{H}} = A \otimes_{\mathbf{R}} \mathbf{H}$ where $\mathbf{H}$ stands for the real quaternion algebra. One has embeddings $K_0(A) \to K_0(A^{\mathbf{C}}) \to K_0(A^{\mathbf{H}})$. In [GH] the classification is obtained in terms of the diagram isomorphisms

$$(K_0(A)^+, E_A) \to (K_0(A^{\mathbf{C}})^+, E_{A^{\mathbf{C}}}) \to (K_0(A^{\mathbf{H}})^+, E_{A^{\mathbf{H}}})$$

Giordano's classification is given in terms of higher K-functors. Hoan Duong [HD] extends the method to direct limits of direct sums of matrix algebras over skew fields provided these skew fields form a finite family.

Let G be a group and F a field. The group algebra FG of G over F is called *modular* if G contains no element of order $p = \text{char } F$.

Corollary 2.4. *Let G and G' be countable locally finite groups. Let F be a field such that the group algebras FG and FG' are not modular. Then $FG \cong FG'$ if and only if $(K_0^+(FG), E_{FG})$ is isomorphic to $(K_0^+(FG'), E_{FG'})$.*

Question 2.5. *Let G and G' be countable infinite simple locally finite groups. Is it true that G is isomorphic to G' if and only if $K_0(\mathbf{C}G)$ is isomorphic to $K_0(\mathbf{C}G')$?*

Problem 2.6. *Express the isomorphism condition of infinite complex group algebras of simple locally finite groups in group terms.*

There is a general result describing ordered abelian groups isomorphic to $K_0(L)$ for a ultramatricial algebra L, see [Go1, Go2, EFS].

Every ordered abelian group is isomorphic to a subgroup of a direct product of copies of the additive group of rational numbers. For an abelian group A, if every equation $mx = a$ $(a \in A, m \in \mathbf{N})$ has a solution in A, then A is called divisible. In this case A is a direct sum of copies of the additive group of rational numbers. The homomorphism $G \to \{1\}$ induces a surjective homomorphism $\varepsilon : K_0(\mathbf{C}G) \to K_0(\mathbf{C}) \cong \mathbf{Z}$. For an analysis of 2.5 it seems to be useful to consider the particular case:

Problem 2.7. *For which simple locally finite groups G is* $\ker(\varepsilon)$ *a divisible abelian group?*

Explicit calculation of $K_0(\mathbb{C}G)$ for a locally finite group G is in general a difficult problem. Nevertheless, there are examples of doing this for specified G (see [VK1], [VKo]).

2.3 Simple Non-locally Semisimple Algebras

The natural question on the existence of a (semi) simple locally finite algebra that are not locally semisimple is answered negatively. The first example of such an algebra was constructed by Farkas and Snider [FS] in order to show that a simple locally finite algebra is not necessary von Neumann regular (a locally semisimple algebra is von Neumann regular). Two more example, however, for imperfect ground field can be found in [Ha1] and [MR]. Later May [M] discovered a example of a von Neumann regular simple locally finite algebra that is not locally semisimple. A fairly simple example (essentially of the same nature as in [FS]) was recently given by Bahturin and Strade [BS-95] in the framework of their proof that there exists a simple locally finite Lie algebra that is not locally semisimple. We present the example of [BS] in Proposition 2.8 with a very elementary proof. Another construction can be deduced from group algebra theory, however, only for fields of prime characteristics, see Proposition 2.9.

Proposition 2.8. *Let K be a perfect field. There exists a simple associative locally finite K-algebra that is not a union of semisimple finite dimensional K-subalgebras.*

Proof. Let K be an arbitrary field, M_n the ring of $(n \times n)$-matrices over K, and L_n the subring of M_n formed by the matrices with zero n-th row. Observe that L_n has exactly one nonzero proper ideal J_n, which consists of the matrices of L_n with zeros outside the last column. Let $\varepsilon : M_n \to M_{2n}$, $m \to \mathrm{diag}(m, m)$, $m \in M_n$, be the embedding of M_n into M_{2n}. It is clear that $\varepsilon(L_n) \subset L_{2n}$, while $\varepsilon(J_n) \cap J_{2n} = \{0\}$. Let L denote the direct limit of L_{2^i} under the embeddings above ($i = 1, 2, \dots$).

Let I be a two-sided proper ideal of L. Set $I_{2^n} = I \cap L_{2^n}$. Then $L_{2^n} \neq I_{2^n} \neq \{0\}$ for some n. It follows that $L_{2^m} \neq I_{2^m} \neq \{0\}$ for $m > n$. As L_{2^m} has the only nonzero proper ideal, $I_{2^m} = J_{2^m}$. Therefore, $J_{2^m} = I_{2^m} \subset I_{2^{m+1}} = J_{2^{m+1}}$. However, J_{2^m} is not contained in $J_{2^{m+1}}$ by the above. Hence L is simple.

Suppose now that L is a union of semisimple subalgebras S_i, $i = 1, 2 \dots$ Then for any n there exist i and m such that $L_{2^n} \subset S_i \subset L_{2^m}$. By the Wedderburn-Malcev theorem S_i is conjugate by an inner automorphism of L_{2^m} to the subalgebra of M_{2^m-1} which is a complement of J_{2^m} in L_{2^m}. It follows that L_{2^n} is conjugate by an inner automorphism of L_{2^m} to a subalgebra of M_{2^m-1}. In fact, this is not the case even for $n = 1$. Indeed, if $x = \begin{pmatrix} 0 & 1 \\ 0 & 0 \end{pmatrix} \in L_2$ then the image of x in M_{2^m} is the matrix $y = \mathrm{diag}\left(\begin{pmatrix} 0 & 1 \\ 0 & 0 \end{pmatrix}, \dots, \begin{pmatrix} 0 & 1 \\ 0 & 0 \end{pmatrix}\right)$ (2^{m-1} blocks). However, M_{2^m-1} contains no conjugate of y. Indeed, y is a matrix of rank 2^{m-1} and $y^2 = 0$. It is clear that M_{2^m-1} contains no such a matrix.

Thus, there exist simple algebras that can not be expressed as a union of semisimple subalgebras of finite dimensions. In the other words, there exist simple locally finite algebras that are not locally semisimple. Analysis of [FS] shows that the class of such algebras is very large.

For fields of prime characteristic there is another construction. Recall that the group algebra FG of a group G over a field F is called *modular* if G contains no element of order $p = \mathrm{char}\ F$.

Proposition 2.9. ([Za]) *A modular group algebra FG of a locally finite group G is never locally semisimple.*

Observe that for G infinite simple the Jacobson radical $J(FG)$ of FG is trivial (see [PZ] and [Pa2]). Moreover, for the majority of simple locally finite groups G the augmentation ideal $\mathrm{Aug}\,(FG)$ of FG is a simple algebra (see [Za] for comments). Possibly, the simplest example of simplicity of $\mathrm{Aug}\,(FG)$ is provided by $G = PSL(2, \bar{\mathbf{F}}_p)$ for F of characteristic $\neq p$. Observe that Proposition 2.9 can be interpreted as a natural generalization of the classical Maschke theorem for finite group algebras:

Proposition 2.10. *The group algebra of a locally finite group is locally semisimple if and only if it is non-modular.*

This fact can be deduced from a theorem of McLaughlin–Auslander (see [McL], [Au]) saying that a locally finite group algebra FG is a von Neumann regular ring if and only if FG is not modular. On the other hand, 2.10 can be proved by using the following fact on finite group algebras:

Lemma 2.11. ([Za, Lemma 8.13]) *Let $G \subset H$ be finite groups. Suppose that FG is modular, i.e. $(|G|, \mathrm{char}\,(F)) > 1$. Then FG is contained in no semisimple subalgebra of FH.*

Our next goal is to associate with every locally finite algebra A a locally semisimple algebra $S(A)$, the ideal lattice of which is isomorphic to the semiprimitive ideal lattice of A. Recall that an ideal I of a ring A is called semiprimitive if $\mathrm{Rad}\,(A/I) = 0$ where Rad stands for the Jacobson radical.

Definition 2.12. Let A be a locally finite algebra over a field F. Let $A_1 \subset A_2 \subset \dots$ be a tower of finite dimensional F-subalgebras of A such that $\cup_i A_i = A$. Set $R_i = \mathrm{Rad}\,(A_i)$, $S_i = A_i/R_i$. For $i < j \in \mathbf{N}$ we have the homomorphism $A_i \to S_j$ induced by the projection $\pi : A_j \to S_j$. For every simple constituent S of S_j let V_S denote the natural S-module. Set $D = \mathrm{End}\,(V_S)$. A choice of a D-basis B of V_S compatible with a composition series of A_i-module $V_S|A_i$ defines a representation $\eta : S_i \to S$. This produces an embedding $\eta_{ij} : S_i \to S_j$. Another choice of a composition series and of B yields a representation η' equivalent to η. In particular, for $j = i+1$ we fix an embedding $\eta_{i,i+1}$. The direct limit $\lim_i S_i = S(A)$ with respect to the embeddings $\eta_{i,i+1}$ is called the *external Levi algebra of A*.

Observe that for $i < j$ the embeddings $S_i \to S_j$ and $A_i \to A_j$ are consonant in the sense that there exists some embedding η_{ij} such that $\eta_{ij}(S_i)$ coincides with the image of S_i in the sequence of the embeddings $S_i \to S_{i+1} \to \dots \to S_j$.

For transparency we describe η in a matrix form. Set $r = \dim_D(V_S)$. A choice of a composition series and of B in 2.12 yields a matrix representation $\theta : A_i \to M(r, D)$ of shape

$$\begin{pmatrix} \theta_1 & * & * \\ 0 & \ddots & * \\ 0 & 0 & \theta_q \end{pmatrix}$$

where θ_s $(1 \le s \le q)$ is an irreducible representation of A_i. It follows that θ_s can be viewed as a representation of S_i. We associate with θ the representation η of S_i given by $\mathrm{diag}(\theta_1, \dots, \theta_q)$.

Proposition 2.13. *If $A \cong B$ are locally finite F-algebras then $S(A) \cong S(B)$. In particular, $S(A)$ is independent of the choice of the tower $\{A_i\}$, so $S(A)$ is an invariant of A.*

Proof. By identifying A and B we obtain two towers of finite dimensional subalgebras in A, say $\{A_i\}$ and $\{B_j\}$. Set $T_j = B_j/R_j$. For every i there exist j, k, l such that $A_i \subset B_j \subset A_k \subset B_l$ with commutative diagram

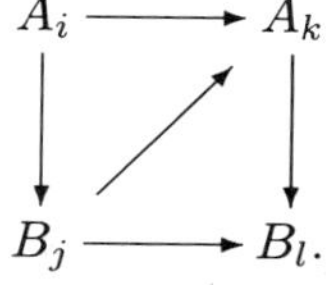

By omitting certain members of $\{A_i\}$ and $\{B_j\}$ and by reordering the remaining ones we obtain a sequence $A_1 \subset B_1 \subset A_2 \subset B_2 \subset \dots$ The procedure of 2.12 gives

rise the embeddings $S_1 \subset T_1 \subset S_2 \subset T_2 \subset \dots$ with commutative diagram

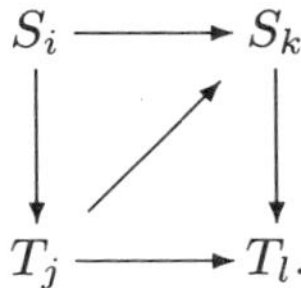

It follows that direct limits $\lim_i S_i$ and $\lim_i T_j$ coincide.

Theorem 2.14. *Let A be a locally finite algebra, and $S(A)$ the external Levi algebra of A. There exists an isomorphism between the lattice of semiprimitive ideals of A and that of ideals of $S(A)$.*

Corollary 2.15. *Let A be a simple locally finite algebra. Then $S(A)$ is simple.*

The assertion 2.15 seems to be fairly important. To some extent it reduces Problem 2.18 below about the classification of simple locally finite algebras to that of algebras with the same $S(A)$. One can observe that $K_0(A) \cong K_0(S(A))$.

To prove Theorem 2.14 we need some machinery. Let $\{A_i\}$ be as above. For every i let $\mathrm{Irr}_{\ i}$ be the set of all pairwise non-isomorphic irreducible A_i-modules.

Definition 2.16. Let $\Phi_i \subset \mathrm{Irr}_{\ i}$ be a subset. We say that $\{\Phi_i\}$ is an inductive system for $\{A_i\}$ if $\mathrm{Irr}\,(\Phi_j|A_i) = \Phi_i$ for every pair $i < j$. *(The symbol $\Phi_j|A_i$ stands for $\cup_{\phi\in\Phi_j} \mathrm{Irr}\,(\phi|A_i)$, and $\mathrm{Irr}\,(\phi|A_i)$ means the set of all composition factors of the restriction $\phi|A_i$ not regarding the multiplicities.)*

Example. Let M be an A-module. Let $\Phi_i(M)$ denote the set of all pairwise non-isomorphic composition factors of $M|A_i$. Then $\Phi(M) = \{\Phi_i(M)\}$ is an inductive system.

Inductive systems form a lattice under obvious inclusion.

Proposition 2.17. *The map $I \to \Phi(A/I)$ is an inverse isomorphism between the lattice of semiprimitive ideals of A and the lattice of inductive systems for any fixed tower $\{A_i\}$.*

Proof. Injectivity. Let $I \neq J$ be semiprimitive ideals of A. We can assume $J \not\subset I$. By replacing A by A/I we reduce to the case $I = 0$ so $\mathrm{Rad}\,(A) = 0$ and $\Phi_i(A/I) = \mathrm{Irr}_{\ i}$ as I is semiprimitive. We have to show that $\Phi_i(A/J) \neq \mathrm{Irr}_{\ i}$ for some i. Suppose the contrary. Then every irreducible A_i-module occurs as a constituent of A/I. Set $I_i = I \cap A_i$. Then

$$I_i \subseteq \bigcap_{M\in\Phi_i(A/I)} \mathrm{Ann}_{\ A_i}(M) = \mathrm{Rad}\,(A_i)$$

by a theorem of Jacobson. Hence I_i is nilpotent. Then I is locally nilpotent so $I \subseteq \mathrm{Rad}\,(A) = 0$.

Surjectivity. Let $\{\Phi_i\}$ be an inductive system for A_i. Let I_i denote the intersection of all left ideals X of A_i such that $\mathrm{Irr}\,(A_i/X) = \Phi_i$. Obviously, I_i is a two sided ideal of A_i. As $\mathrm{Irr}\,(\Phi_j|A_i) = \Phi_i$ for $i < j$, we have $I_i \subset I_j$. Hence $\cup I_i = K$ is an ideal of A, and $\mathrm{Irr}\,((A/K)|A_i) = \Phi_i$. Define the ideal I by the conditions $K \subseteq I$ and $I/K = \mathrm{Rad}\,(A/K)$. Then I is semiprimitive, and $\mathrm{Irr}\,((A/I)|A_i) = \Phi_i$.

Proof of Theorem 2.14. Obviously, every ideal of a locally semisimple algebra is semiprimitive. By 2.17 it suffices to establish an inverse isomorphism between the lattices of inductive systems of A and $S(A)$ for a tower $\{A_i\}$. We set $S(A_i) = A_i/\operatorname{Rad}(A_i)$. Then an inductive system Φ_i for A can be regarded as an inductive system for $S(A)$ by viewing an irreducible A_i-module as an $S(A_i)$-module, and conversely.

After the comments above it seems natural to state the following

Problem 2.18. *Classify the simple locally finite associative algebras (at least, over an algebraically closed field).*

Observe that a similar problem for Lie algebras was stated in 1974 by Amayo and Stewart [AS, page 396, Problem 35]; no progress has been achieved from that time.

Let A be an arbitrary finite dimensional algebra over an algebraically closed fieldand R its radical. Then $A = R + M$ where M is matricial. Let M^- and M^+ be the subalgebras generated by lower and upper triangular matrices in every simple component of M, M^0 the subalgebra generated by diagonal matrices. Set $A^- = M^-$, $A^0 = M^0$ and $A^+ = R + M^+$. The triple (A^-, A^0, A^+) will be called a *triangular structure* on A. Let A' be a finite dimensional algebra containing A, and R', M' its radical and a semisimple component. By the Wedderburn–Malcev theorem M is conjugate to a subalgebra of M'. The natural homomorphism $A' \to A'/R' \to M'$ maps A into a direct sum of matrix algebras. Obviously, every subalgebra of $M(n, F)$ can be transformed to a block-triangular form. It follows that a triangular structure of a subalgebra of A can be extended to a triangular structure of A'. This implies the following fact.

Proposition 2.19. *Let A be a countable locally finite associative algebra over an algebraically closed field. Then there exist locally nilpotent subalgebras A^- and A^+, a commutative subalgebra A^0 and an ultramatricial subalgebra M such that $A = A^- + A^0 + A^+$, $A^0, A^- \subset A^-$, $A^0, A^+ \subset A^+$, $A_0 = M_0$ and $A^- + = M^-$.*

The triple (A^-, A^0, A^+) is called a *triangular structure* on A.

3. Diagonal Simple Lie Algebras

One may expect that simple locally finite Lie algebras attract more attention than associative ones, because of numerous links with Lie group theory. However, this is not the case. In fact, only locally matrix Lie algebras of classical type were considered. By analogy with the results of §2.1 one may hope to classify locally simple Lie algebras. However, in contrast with the associative case, this problem turns out to be almost hopeless. We shall discuss below a nice class of simple Lie algebras for which that method may be efficient. This is the class of diagonal Lie algebras.

The idea of diagonality appeared for the first time in a group ring theory context, see [Za1]. It was understood later that the same notion plays a role in the Lie algebra theory. In fact, diagonal simple Lie algebras are the closest relatives of associative ones. For the particular case of locally simple Lie algebras the notion was used by Zhilinski [Zh]; in the form below it was introduced by Baranov [Ba2] in his investigation of locally finite quotients of universal enveloping algebras of locally finite Lie algebras.

Definition 3.1. Let $K \subset L$ be finite dimensional Lie algebras over an algebraically closed field F. The embedding $K \to L$ is called diagonal if the following conditions holds:

(i) simple quotients of K and of L are Lie algebras of classical type;
(ii) If V is the natural module for any of the simple components of $L/\operatorname{Rad}(L)$ then every nontrivial irreducible constituent X of $V|K$ is nontrivial for a unique simple component S of $K/\operatorname{Rad}(K)$, and $\dim(X)$ is equal to the dimension of the natural S-module.

Definition 3.2. Let L be a locally finite Lie algebra over an algebraically closed field F. L is called diagonal if L is a direct limit of finite dimensional Lie subalgebras L_i $(i = 1, 2, \dots)$ with diagonal embeddings $L_i \to L_{i+1}$.

Problem 3.3. *Describe Lie algebras that can be embedded into a locally finite associative algebras.*

This problem is an analogue of the problem on finite dimensional Lie algebras solved by the Ado Theorem. However, in contrast with the finite dimensional case, the class of Lie algebras that can be embedded into locally finite associative algebras is quite small. If the ground field is algebraic closed of characteristic 0, a description of this class is given by Baranov [B1, B2] under some restrictions (he assumes the algebras to be locally perfect). We confine ourself here to simple algebras:

Theorem 3.4. ([B1, B2]) *Let L be a simple Lie algebra over an algebraically closed field of characteristic 0. Then L can be embedded into a locally finite associative algebra if and only if L is diagonal.*

For locally simple Lie algebras Theorem 3.4 was established by Zhilinski [Zh].

Problem 3.5. *Extend* 3.4 *to simple Lie algebras over an algebraically closed field of prime characteristic.*

The notion of diagonality is trivial for simple associative algebras, i.e. every locally finite simple associative algebra A is diagonal. It follows that the Lie algebras $L(A)$ and $K^t(A)$ associated with A (see 3.6 and 3.7) are diagonal. Below $Z(L)$ stands for the center of L.

Theorem 3.6. ([He1]) *Let A be a simple associative ring, and $L(A)$ the Lie ring generated by Lie commutators $ab - ba$ $(a, b \in A)$. Then $L(A)/Z(L(A))$ is a simple Lie ring unless* $\operatorname{char}(A) = 2$ *and A is finite dimensional algebra of dimension at most* 4 *over its centroid.*

If A is a ring with involution t, one can consider the Lie ring $L^t(A) = \{a \in A : a^t = -a\}$. Let $K^t(A)$ stand for the commutator subring of $L^t(A)$, and $K_1^t(A)$ for the commutator subring of $K^t(A)$.

Theorem 3.7. ([He2, Ba, Mo]) $K^t(A)/Z(K^t(A))$ *is simple unless* $\operatorname{char}(A) = 2$ *or A is of dimension at most* 16 *over its centroid. If* $\operatorname{char}(A) = 2$ *then* $K_1^t(A)/Z(K_1^t(A))$ *is simple unless A is of dimension at most* 16 *over its centroid.*

By analogy to the associative algebra case it is natural to introduce a notion of ultraclassical Lie algebras:

Definition 3.8. A Lie algebra is called ultraclassical if it is a direct limit of direct sums of classical Lie algebras with diagonal embeddings.

Problem 3.9. *Classify the ultraclassical Lie algebras.*

The essence of the problem is in defining an equivalence relation of the embedding diagrams that reflects isomorphy of the limit algebras. Also, the combinatorial picture of the embedding diagrams for Lie algebras is similar to that for associative algebras. Possibly, Problem 3.8 is related with that of classification of associative locally semisimple algebras with involution.

There is a classification of diagonal locally simple Lie algebras due to Baranov and Zhilinski [BZ]. To state the result, we need some notation.

Let L be a tower of subalgebras $sl(n_i)$ with diagonal embeddings $\varepsilon_i : sl(n_i) \to sl(n_{i+1})$. Every ε_i is determined by a triple l_i, r_i, z_i of non-negative integers such that $n_i(l_i + r_i) + z_i = n_{i+1}$. This means that $\varepsilon_i(sl(n_i))$ contains the natural $sl(n_i)$-module with multiplicity l_i, its dual with multiplicity r_i, and the trivial $sl(n_i)$-module with multiplicity z_i. Thus we can write $L = sl(\mathcal{N})$ where $\mathcal{N} = \{(l_i, r_i, z_i)\}$ is the corresponding triple family. One can assume that $l_i \geq r_i$. Indeed, if $l_i \leq r_i$ for an infinite set of the indices i then one can suppose that all $l_i \leq r_i$ and replace l_i by r_i. Set $s_i = l_i + r_i$, $c_i = l_i - r_i$, $\mathcal{S} = \{s_i\}$, $\mathcal{C} = \{c_i\}$. Similar construction is valid for direct limits of algebras of type *so* (orthogonal) and *sp* (symplectic).For these cases $r_i = 0$ for all i.

It is convenient to add to the family an "algebra" of dimension 1. Formally we can and shall assume that $n_1 = 1$, $s_1 = c_1 = n_2$. Set $\delta_i = (s_1 \dots s_{i-1})/n_i$. Then

$$\delta_{i+1} = \frac{s_1 \dots s_i}{n_{i+1}} = \frac{s_1 \dots s_{i-1} s_i}{n_i s_i + z_i} = \frac{s_1 \dots s_{i-1}}{n_i + (z_i/s_i)} \leq \delta_i. \tag{1}$$

The limit $\delta = \lim_{i \to \infty} \delta_i$ is called the *density index* of $\mathcal{N}$ and is denoted by $\delta(\mathcal{N})$. Since $\delta_2 = s_1/n_2 = 1$, we have $0 \leq \delta \leq 1$. If $\delta = 0$ then the family is called *sparse*. If there exists i such that for any $j > i$ one has $\delta_j = \delta_i \neq 0$ then the family is called *pure*. In view of (1) this is equivalent to the following. There exists i such that for any $j \geq i$ we have $z_j = 0$. We say that the family is *dense* if and only if $0 < \delta < \delta_i$ for any i.

If there exists i such that $c_j = s_j$ (equivalently, $r_j = 0$) for any $j \geq i$, then the family $\mathcal{N}$ is called *one-sided*. Otherwise, it is called *two-sided*. If for any i there exists $j > i$ such that $c_j = 0$ (equivalently, $l_j = r_j$), then $\mathcal{N}$ is called (two-sided) *symmetric*. Otherwise it is called *non-symmetric*. In the latter case we may and shall assume that $c_i > 0$ for any $i \in \mathbf{N}$. Set $\sigma_i = \frac{c_1 \dots c_i}{s_1 \dots s_i}$. The limit $\sigma = \lim_{i \to \infty} \sigma_i$ is called the *symmetry index* of $\mathcal{N}$ and is denoted by $\sigma(\mathcal{N})$. Observe that $0 \leq \sigma \leq 1$. Two-sided non-symmetric families with $\sigma = 0$ are called *weakly non-symmetric*, and those with $\sigma \neq 0$ are called *strongly non-symmetric*.

We say that Steinitz numbers Π and Π' are $\mathbf{Q}$-equivalent and write $\Pi \overset{\mathbf{Q}}{\approx} \Pi'$ if there exist $a, b \in \mathbf{N}$ such that $a\Pi = b\Pi'$. We say that $\alpha \in \frac{\Pi}{\Pi'}$ if there exists $b \in \mathbf{N}$ such that $b\alpha \in \mathbf{N}$ and $b\alpha\Pi' = b\Pi$.

Theorem 3.10. ([BZ]) *Let* $\mathcal{N} = \{(l_i, r_i, z_i)\}$ *and* $\mathcal{N}' = \{(l'_i, r'_i, z'_i)\}$. *Set* $\delta = \delta(\mathcal{N})$, $\sigma = \sigma(\mathcal{N})$, $\delta' = \delta(\mathcal{N}')$, $\sigma' = \sigma(\mathcal{N}')$. *Then* $sl(\mathcal{N}) \cong sl(\mathcal{N}')$ *if and only if conditions* (1)–(6) *below hold;* $so(\mathcal{N}) \cong so(\mathcal{N}')$ *and* $sp(\mathcal{N}) \cong sp(\mathcal{N}')$ *if and only if conditions* (1)–(3) *hold.*

(1) *The families* $\mathcal{N}$ *and* $\mathcal{N}'$ *have the same density type (sparse, pure or dense).*

(2) $\mathrm{Stz}(\mathcal{S}) \overset{\mathbf{Q}}{\sim} \mathrm{Stz}(\mathcal{S}')$.
(3) $\frac{\delta}{\delta'} \in \frac{\mathrm{Stz}(\mathcal{S})}{\mathrm{Stz}(\mathcal{S}')}$ *for dense and pure families.*
(4) *The families* $\mathcal{N}$ *and* $\mathcal{N}'$ *have the same symmetry type (i.e. both they are one-sided, or two-sided symmetric, or two-sided weakly non-symmetric, or two-sided strongly non-symmetric).*
(5) $\mathrm{Stz}(\mathcal{C}) \overset{\mathbf{Q}}{\sim} \mathrm{Stz}(\mathcal{C}')$ *for two-sided non-symmetric families.*
(6) *There exists* $\alpha \in \frac{\mathrm{Stz}(\mathcal{S})}{\mathrm{Stz}(\mathcal{S}')}$ *such that* $\alpha\frac{\sigma}{\sigma'} \in \frac{\mathrm{Stz}(\mathcal{C})}{\mathrm{Stz}(\mathcal{C}')}$ *for two-sided strongly non-symmetric families. Moreover,* $\alpha = \frac{\delta}{\delta'}$ *if in addition the families are dense or pure.*

If $l_i = r_i$ for infinitely many i (i.e. $\mathcal{N}$ is symmetric) then $\lim sl_i \cong \lim so_{2i} \cong \lim sp_{2i}$. Precise description of such isomorphisms is given in [BZ].

Simple Lie algebra limits with $z_i = 0$ were studied by Yanson and Zhdanovich [YZ]; their results are less complete and are stated in an alternative form.

Theorem 3.10 may be regarded as a Lie algebra analog of a result of Dixmier [Di] on classification of locally simple associative algebras. We emphasize that 3.10 does not exhaust Problem 3.9, as 3.10 only deals with locally simple Lie algebras.

In view of 2.8 one should distinguish locally semisimple Lie algebras. This leads to the notion of an external Levi algebra as in 2.12, and an argument similar to 2.13 shows that every countable locally finite Lie algebra possesses a unique external Levi algebra $S(L)$. The argument is now valid only for the case of characteristic zero, as one needs the Weyl theorem on complete reducibility of representations of semisimple Lie algebras.

Theorem 3.11. ([Ba2]) *Let* L *be a countable locally finite Lie algebra over an algebraically closed field of characteristic 0. If* L *is simple then so is* $S(L)$.

Remark. In [Ba2] Baranov uses an alternative definition of $S(A)$ as a union of Levi subalgebras of a tower of finite dimensional subalgebras L_i of L. He does not prove that $S(A)$ does not depend on the choice of $\{L_i\}$.

We introduce a useful notion of a triangular structure on a Lie algebra.

Definition 3.12. Let L be a locally finite Lie algebra, and L^-, L^0, L^+ subalgebras of L. We say that the triple (L^-, L^0, L^+) forms a triangular structure of L if $L = L^- + L^0 + L^+$, L^0 is a semisimple commutative subalgebra, L^- and L^+ are locally nilpotent, $L^- \cap L^+ = 0$, and $[L^0, L^-] \subseteq L^-$, $[L^0, L^+] \subseteq L^+$ and $L^- + L_0$ is contained in a locally semisimple subalgebra of L.

Theorem 3.13. ([KZ]) *Let* L *be a locally finite Lie algebra of countable dimension (not necessary diagonal). Then* L *has a triangular structure.*

Remark. For L locally semisimple the result of 3.13 is contained in Bahturin and Benkart [BB].

Lemma 3.14. *Let* $M \subset L$ *be Lie algebras of finite dimension, and let the triple* (M^-, M^0, M^+) *forms a triangular structure of* M. *Then there exists a triangular structure* (L^-, L^0, L^+) *of* L *such that* $M^- \subset L^-$, $M^0 \subset L^0$ *and* $M^+ \subset L^+$. *Moreover, there exist semisimple subalgebras* S, T *of* M, L, *respectively, such that* $M^- + M_0 \subset S \subset T$ *and* $L^- + L_0 \subset T$.

Proof of Theorem 3.13. As L is of countable dimension, L can be expressed as a union of a tower of subalgebras L_i of finite dimension. By Lemma 3.14 for every i

there exists a triangular structure (L_i^-, L_i^0, L_i^+) of L_i and semisimple algebras S_i with $L_i^- + L_i^0 \subset S_i \subset L_i$ such that $L_i^- \subseteq L_{i+1}^-$, $L_i^+ \subseteq L_{i+1}^+$, $L_i^0 \subseteq L_{i+1}^0$ and $S_i \subset S_{i+1}$. Set $L^+ = \cup_i L_i^+$, $L^- = \cup_i L_i^-$, $L^0 = \cup_i L_i^0$. Obviously, (L^-, L^0, L^+) is a triangular structure of L, and $S = \cup_i S_i$ is a locally semisimple subalgebra of L containing $L^- + L^0$.

Observe that the notion of a triangular decomposition studied by Moody and Pianzola [MP] is much stronger. One of the additional conditions in [MP] requires L^+ to be a direct sum of one dimensional L_0-submodules. Probably, a simple locally finite Lie algebra with this condition is locally matrix.

4. Diagonal Locally Finite Simple Groups

Normally, a classical group C is the group of automorphisms of a non-degenerate bilinear, sesquilinear or quadratic form on a vector space V of finite dimension over a field F. We slightly extend the content of this term by including subgroups of C containing C', the commutator subgroup of C. Also, we add groups between $SL(V)$ and $GL(V)$. *We assume in this section that F is a finite field of characteristic p.* We call the $\mathbf{F}_pC$-module V (by viewing V as a space over $\mathbf{F}_p$) the *natural module for C*. A *nearly natural* $\mathbf{F}_pC$-module is defined to be either natural module or its dual module.

Let X be a finite group. We say that X is *almost classical of characteristic p* if X is perfect and $X/O_p(X)$ is a central product of perfect classical groups over finite subfields of $\bar{\mathbf{F}}_p$. (Here $O_p(X)$ stands for maximal normal p-subgroup of X.) Thus, $X/O_p(X) = X_1 \cdots X_l$ where X_i $(i = 1, \dots, l)$ are $SL(n_i, F_i)$, $Sp(n_i, F_i)$, $SU(n_i, F_i)$ or $\Omega(n_i, F_i) = O'(n_i, F_i)$. Define V_X to be $\sum V_C$ where C runs over simple quotients of X and V_C is the sum of the natural $\mathbf{F}_pC$-module, its dual module and the trivial module 1_C. In fact, we consider V_X as an $\mathbf{F}_pX$-module in the obvious way. Below $\ker(M) =: \{g \in G : gm = m \text{ for all } m \in M\}$.

Definition 4.1. Let G, H be two almost classical groups of characteristic p. Let $\varepsilon : G \to H$ be an embedding, so V_H also is an $\mathbf{F}_pG$-module. We say that ε is diagonal if for every nontrivial composition factor M of $V_H|G$ the group $G/\ker(M)$ is classical and M is a nearly natural module for $G/\ker(M)$.

In other words, G projects into the classical quotients of H in such a way that every nontrivial composition factor of G in $V_H|G$, the restriction of $V|H$ to G, is nearly natural for a unique classical quotient of G.

Definition 4.2. A locally finite group G is called diagonal if $G = H/Z$ with $Z \subset Z(H)$, and H can be expressed as a direct limit of finite almost classical subgroups H_i of characteristic p with diagonal embeddings $H_i \to H_{i+1}$. The prime p is called the defining characteristic of G.

There is a kind of a classification theorem for simple locally finite groups due to Meierfrankenfeld [Me1, Me2] saying that every such group either is diagonal or can be described to some extent in terms of alternating groups. This forces one to pay attention to diagonal groups. In view of the discussion in the previous sections it is natural to link diagonal groups with the adjoint groups of simple locally finite algebras (or rings). Observe that a simple ring is an algebra over a field, see [J, Ch.V, §3, Theorem 1].

Let A be an associative ring, $U(A)$ be the commutator subgroup of the adjoint group of A, and $\bar{U}(A)$ its quotient modulo the center.

Theorem 4.3. ([Am]) *Let A be a simple associative ring containing at least three pairwise orthogonal idempotents. Then $\bar{U}(A)$ is a simple group.*

If A has an involution t, one can consider the group $A^t = \{a \in U(A) : a^t = a^{-1}\}$. Let $C^t(A)$ stand for the commutator subgroup of A^t and $\bar{C}^t(A)$ for its quotient modulo the center. The following problem goes back to Herstein [He1] (without any restriction on A).

Problem 4.4. *Let A be a simple locally finite ring (i.e. every finite set of elements belong to a finite subring). Under what conditions is $\bar{C}^t(A)$ a simple group?*

In the case A is not artinian there is probably no publication concerning this problem. I expect that the group $\bar{C}^t(A)$ is simple provided the centroid F is not of characteristic 2. If $\operatorname{char}(F) = 2$, A^t may have a noncentral normal abelian subgroup. Possibly, for an infinite ring A the group $\bar{C}^t(A)$ is simple if and only if the F-span of A^t coincides with A. Observe that $\bar{U}(A)$ can be interpreted as $\bar{C}^t(A_1)$ where A_1 is the direct sum of A and the opposite ring and t permutes the summands.

Lemma 4.5. *Let A be a simple associative locally finite ring. Then $\bar{U}(A)$ is a simple diagonal locally finite group.*

Proof. The simplicity of $\bar{U}(A)$ follows from 4.3 if A is not artinian. For A artinian this is well known. Next, by [J, Ch.V, §3, Theorem 1] A is an algebra over a field. It follows that A is a locally finite $\mathbf{F}_p$-algebra. Then $A = \lim_i A_i$ where A_i are finite $\mathbf{F}_p$-algebras. The embeddings $A_i \to A_j$ for $i < j$ induce diagonal embeddings $U(A_i) \to U(A_j)$. Indeed, let S be a simple quotient of A_j. As an $\mathbf{F}_p$-algebra S has a unique irreducible module W (this is exactly the standard irreducible S-module viewed as a vector space over $\mathbf{F}_p$), every non-trivial composition factor M of $W|A_i$ is non-trivial for a unique simple quotient T of A_i. Then M viewed as $\mathbf{F}_pU(T)$-module is nearly natural.

It is not hard to show that $\bar{C}^t$ is diagonal as well. May one hope that diagonal groups can be described in terms of the groups $\bar{U}(A)$ or $\bar{C}^t$ for a suitable A and t? There is at least one counterexample: the commutator subgroup of the finitary orthogonal group over a field of characteristic 2. More complicated examples can be constructed by direct limits of orthogonal groups in characteristic 2. However, it is probable that there does exist a regular way to derive diagonal groups from simple associative rings.

It may be useful to introduce an external Levi group for a diagonal group by imitating Definition 2.12.

Definition 4.6. Let G be a countable locally finite diagonal group in characteristic $p > 0$. Let $G_1 \subset G_2 \subset \dots$ be a tower of almost classical subgroups of G with diagonal embeddings $G_i \to G_j$ for $i < j$, and $G = \cup_i G_i$. Set $S_i = G_i/O_p(G_i)$. For $i < j \in \mathbf{N}$ we have the homomorphism $G_i \to S_j$ induced by the projection $G_j \to S_j$. Then a choice of a basis B of V_{S_j} compatible with a composition series of $\mathbf{F}_pG_i$-module V_{S_j} defines a representation $\eta : S_i \to S_j$; another choice of a composition series and of B yields a representation η' equivalent to η. For $j = i + 1$ we fix an embedding $\eta_i : S_i \to S_{i+1}$ arising from η's. The direct limit $\lim_i S_i = S(G)$ is called the *external Levi group* of G.

Proposition 4.7. *If $G \cong G'$ are two locally finite diagonal groups of characteristic p, then $S(G) \cong S(G')$. In particular, $S(G)$ is independent of the choice of a tower $\{G_i\}$, so $S(G)$ is an invariant of G.*

Proof. By identifying G and G' we obtain two towers of finite subgroups of G, say $\{G_i\}$ and $\{G'_j\}$. Let $T_j = G'_j/O_p(G_j)$. Hence for every i there exist j, k, l such that $G_i \subset G'_j \subset G_k \subset G'_l$ with commutative diagram

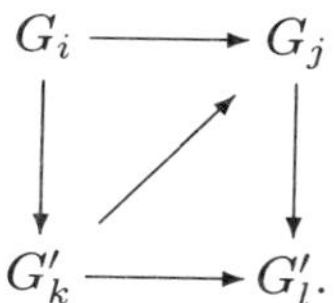

From this we obtain embeddings $S_i \subset T_j \subset S_k \subset T_l$ with commutative diagram

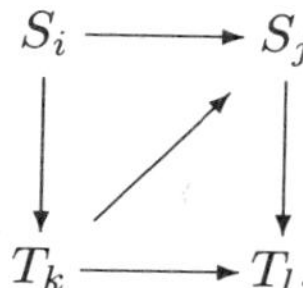

It follows that the direct limits $\lim_i S_i$ and $\lim_i T_j$ produce isomorphic groups.

Theorem 4.8. *Let G be a perfect diagonal locally finite group of characteristic p and $S(G)$ the external Levi group. Suppose that $O_p(G) = 1$. Then $S(G)/Z(S(G))$ is simple if and only if $G/Z(G)$ is simple.*

Proof. "Only if". Let $G_1 \subset G_2 \subset$ be a tower of finite almost classical subgroups of G such that $G = \cup_i G_i$. By a theorem of Kegel one can assume that all G_i are perfect. Set $S_i = G_i/O_p(G_i)$. Let N be a non-central proper normal subgroup of G. Set $N_i = G_i \cap N$ and $\bar{N}_i = N_i/(N_i \cap O_p(G_i)) \subset S_i$. Suppose first that $\bar{N}_i \subseteq Z(S_i)$ for infinitely many i. Then $X_i =: [N_i, G_i] \subset O_p(G_i)$ and $X_i \subset X_j$ for $i < j$. Hence $X =: \cup X_i$ is normal in G, and $O_p(X) = X$ implies $X = 1$. Then $X_i = 1$ and $N_i \subset Z(G)$. This implies $\bar{N} \subseteq Z(G)$, a contradiction. Thus, there are finitely many i with $\bar{N}_i \subseteq Z(S_i)$. By omitting G_i with this property we have $\bar{N}_i \not\subseteq Z(S_i)$ for all i. In particular, all N_i are not solvable.

For a group K let K^c denote the minimal normal subgroup of K such that K/K^c is solvable. Obviously, K^c is characteristic in K. Clearly, $N_i^c \subset N_j^c$ for $i < j$, and $\cup_i N_n^c$ is a normal subgroup of G. So we can assume that all N_i are perfect. Then all $\bar{N}_i$ are perfect.

It follows that $\bar{N}_i = Y_{i1} \dots Y_{il_i}$ where Y_{ij} are perfect classical groups. We can order the quasi-simple multipliers of S_i so that $S_i = \bar{N}_i Y_{i,l_{i+1}} \dots Y_{ir_i}$. Next show that $\bar{N}_i \subset \bar{N}_j$ for $i < j$. Indeed, if not, then the projection of $\bar{N}_i$ into Y_{j,l_j+1} is non-central; by construction of S_i the image of N_i in Y_{j,l_j+1} under the projection $G_j \to S_j$ is noncentral. This contradicts the fact that $N_i \subset N_j$.

Thus, $\bar{N}_i \subset \bar{N}_j$, so $\cup_i \bar{N}_i$ is a normal subgroup of $S(G)$.

Prove the "if" part. Let $\bar{N}$ be a normal subgroup of $S(G)$, and set $\bar{N}_i = S_i \cap \bar{N}$. Express $S_j = Y_{j1} \dots Y_{jr_j}$ where Y_{jl} are perfect classical groups. By omitting certain G_i we can assume that $\bar{N}_i \neq S_i$.

Set $N_i = X_i^c$ where X_i is the pullback of $\bar{N}_i$ in G_i. Then N_i is normal in G_i. Let $j > i$ and $\varepsilon : G_j \to S_j$ be the natural projection. Show that $\varepsilon(N_i) \subset \bar{N}_j$. By reordering Y_{jl} one can assume that $\bar{N}_j = Z(\bar{N}_j)Y_{j1} \ldots Y_{jk}$ with $k < r_j$. Suppose that for some $l > k$ the projection L of $\varepsilon(N_i)$ into Y_{jl} is non-trivial. As $\varepsilon(N_i)$ is perfect, $L \not\subset Z(Y_{jl})$. It follows that the composition factors of L on the natural module V for Y_{jl} are not all trivial. However, the projection of L onto the sum of the composition factors is equivalent to the projection of $\bar{N}_i$ by construction of S_i. This means that the projection of $\bar{N}_i$ into Y_{jl} is non-trivial, a contradiction.

Thus, $\varepsilon(N_i) \subset \bar{N}_j$. It follows that $N_i O_p(G_j) \subset N_j O_p(G_j)$. Then $N_i = N_i^c \subseteq (N_i O_p(G_j))^c \subseteq (N_j O_p(G_j))^c = N_j$. Set $N = \cup N_i$. Then N is a normal subgroup of G as N_i is normal in G_i for every i. Obviously, $1 \neq N \neq G$.

5. Direct Limits of Algebraic Groups

The fact that a diagonal direct limit of the orthogonal group of characteristic 2 is not always described in terms of the adjoint groups of associative rings forces one to look for another apporoach that can reflect the 'linear' nature of these groups. It is natural to pay attention to direct limits of algebraic groups. Of course, the embeddings should be algebraic groups morphisms.

Definition 5.1. We say that a group is locally algebraic if G is a direct limit of ordinary algebraic groups with embeddings being algebraic group morphisms.

For characteristic 0 there is no essential difference in understanding locally finite Lie algebras and locally algebraic groups. For the prime characteristic case there are a number of anomalies in comparison with the behaviour of algebraic groups. To illustrate this, we state the following result:

Theorem 5.2. ([Ku]) *Let U stand for P. Hall's universal locally finite countable group, p a prime. Then U can be expressed as a direct limit of the groups $SL(n_i, \bar{F}_p)$ for suitable n_i such that the subsequent embeddings are morphisms of algebraic groups.*

Thus, the characteristic of the ground field is *not* an invariant of a direct limit of algebraic groups. Even worse, in the construction of [Ku] for any $q = p^k$ the groups $SL(n_i, q)$ map into $SL(n_{i+1}, q)$, and the direct limit of $SL(n_i, q)$ coincides with U. This means that one can *not* in general distinguish between direct limits of algebraic groups and their finite field subgroups.

Problem 5.3. *Let G be a simple diagonal group. Is it true that there exists a locally algebraic group H and an automorphism α of H such that G is isomorphic to H^α, the subgroup of α-fixed elements of H?*

References

[Am] S.A. Amitsur, Invariant submodules of simple rings, *Proc. Amer. Math. Soc.* **7** (1956), 987–989.

[AS] R.K. Amayo and I. Stewart, *Infinite dimensional Lie algebras*, Leiden, 1974, pp. 425.

[Au] M. Auslander, On regular group rings, *Proc. Amer. Math. Soc.* **8** (1957), 658–664.

[BB] Yu. Bahturin and G. Benkart, Highest weight modules for locally finite Lie algebras, *J. Algebra* (to appear).

[BS] Yu. A. Bahturin and H. Strade, Locally finite-dimensional simple Lie algebras, *Mat. sb.* **185** (1994), 3–32. (Russian)

[BS-95] Yu. A. Bahturin and H. Strade, Some examples of locally finite simple Lie algebras, *Arch. Math.* **65** (1995), 23–26.

[B1] A.A. Baranov, Diagonal locally finite Lie algebras and a version of Ado theorem, submitted to J. Algebra.

[B2] A.A. Baranov, Simple diagonal locally finite Lie algebras, (in preparation).

[BZ] A.A. Baranov and A.G. Zhilinski, Classification of diagonal direct limits of Lie algebras and classical groups, (in preparation).

[Ba] W.E. Baxter, Lie simplicity of a special class of associative rings, II, *Trans. Amer. Math. Soc.* **87** (1958), 63–75.

[Br] O. Bratteli, Inductive limits of finite dimensional $\mathbf{C}^*$-algebras, *Trans. Amer. Math. Soc.* **171** (1972), 195–234.

[Di] J. Dixmier, On some $\mathbf{C}^*$-algebras considered by Glimm, *J. Functional Analysis* **1** (1967), 182–203.

[Ef] E.G. Effros, Dimensions and $\mathbf{C}^*$-algebras, *Amer. Math. Soc.*; Providence (1981).

[EHS] E.G. Effros, D.E. Handelman and C.L. Shen, Dimension groups and their affine representations, *Amer. J. Math.* **102** (1980), 385–407.

[El] G.A. Elliott, On the classification of inductive limits of sequences of semisimple finite dimensional algebras, *J. Algebra* **38** (1976), 29–44.

[FS] D.R. Farkas and R.L. Snider paper Locally finite dimensional algebras, *Proc. Amer. Math. Soc.* **81** (1981), 369–372.

[Ga] S. Gauger, Lokal endlich-dimensionale einfache Lie-Algebren, Diplomarbeit, Albert-Ludwigs-Universitat Freiburg, 1992.

[Gl] J. Glimm, On a certain class of operator algebras, *Trans. Amer. Math. Soc.* **95** (1960), 381–340.

[Gi] T. Giordano, Classification of real approximatively finite $\mathbf{C}^*$-algebras, *J. reine und angew. Math.* **385** (1988), 161–194.

[Go1] K.R. Goodearl, *Von Neumann regular rings*, Pitman, London, 1979.

[Go2] K.R. Goodearl, *Notes on real and complex* $\mathbf{C}^*$*-algebras*, Shiva Publ., Nantwich, 1982.

[GH] K.R. Goodearl and D.E. Handelman, Classification of ring and $\mathbf{C}^*$-algebra direct limits of finite dimensional semisimple real algebras, *Memoirs Amer. Math. Soc.* **69** (1987), no.372, 147pp.

[GHJ] F.M. Goodman, P. de la Harpe and V.F.R. Jones, *Coxeter graphs and towers of algebras*, Springer, Berlin, 1989.

[H] J. Hall, Locally finite simple groups of finitary linear transformations, In: 'Finite and locally finite groups', Kluwer Academic publisher, Dordrecht, 1995, pp. 147–188.

[Ha] D. Handelman, K_0 of von Neumann and AF $\mathbb{C}^*$-algebras, *Quart. J. Math.* **29** (1978), 427–441.

[Ha1] D. Handelman, Centers of rank-metric completions, *Can. J. Math.* **37** (1985), 1134–1148.

[He1] I. Herstein, The Lie ring of a simple associative ring, *Duke Math. J.* **22** (1955), 471–476..

[He2] I. Herstein, Lie and Jordan systems in simple rings with involution, *Amer. J. Math.* **78** (1956), 629–649.

[He3] I. Herstein, Lie and Jordan structures in simple associative rings, *Bull. Amer. Math. Soc.* **67** (1961), 517–531.

[HD] Hoan Duong, An invariant for locally finite dimensional semisimple algebras, Ph. D. Thesis, University of Ottawa, 1995.

[J] N. Jacobson, *Structure of rings*, Amer. Math. Soc., Providence, 1956.

[KZ] O. Kegel and A. Zalesski, Triangular decomposition in locally finite Lie algebras, (unpublished).

[KS] N.V. Kroshko and V.I. Sushchansky, Direct limits of symmetric and alternating groups with strictly diagonal embeddings, Arch. Math., (submitted).

[K] V.M. Kurochkin, On the theory of locally simple and locally normal algebras, *Mat. Sb.* **22** (1948), 443–454. (Russian)

[Ku] M. Kuzucuoglu and A. Zalesski, The Hall universal group as a direct limit of algebraic groups, *J. Algebra* (to appear).

[M] M.K. May, A von Neumann regular ring that is locally finite but not locally semisimple, *Comm. Algebra* **16** (1988), 103–113.

[Ml] J.E. McLaughlin, A note on regular group rings, *Michigan Math. J.* **5** (1958), 127–128.

[M1] U. Meierfrankenfeld, Non-finitary locally finite simple groups, In: 'Finite and locally finite groups', Kluwer Academic publisher, Dordrecht, 1995, pp. 189–212.

[M2] U. Meierfrankenfeld, Structure of simple locally finite groups of p-type, (in preparation).

[MR] P. Menal and R. Raphael, On epimorphism-final rings, *Comm. Algebra* **12** (1984), 1871–1876.

[Mo] M.S. Montgomery, Lie structure of simple rings of characteristic 2, *J. Algebra* **15** (1970), 387–407.

[MP] R. Moody and A. Pianzola, *Lie algebras with triangle decompositions*, Wiley, New York, 1995, pp. 685.

[Ol] G.I. Ol'shanski, Infinite dimensional classical groups of finite R-rank: description of representations and asymptotic theory, *Func. Anal. and Appl.* **18** (1984), 28–42.

[Pa1] D.S. Passman, *The algebraic structure of group rings*, Wiley, New York, 1977.

[Pa2] D.S. Passman, Semiprimitivity of the group algebras of infinite simple groups of Lie type, *Proc. Amer. Math. Soc.* **121** (1994), 399–403.

[PZ] D.S. Passman and A.E. Zalesskii, On the semiprimitivity of modular group algebras of locally finite simple groups, *Proc. London Math. Soc.* **67** (1993), 243–276.

[Po] S.C. Power, *Limit algebras: an introduction to subalgebras of* $\mathbf{C}^*$*-algebras*, Longman, Essex, 1992.

[VK1] A.M. Vershik and S.V. Kerov, K_0-functor (the Grothendieck group) of the infinite symmetric group, *J. Sov. Math.* **28** (1985), 549–568.

[VK2] A.M. Vershik and S.V. Kerov, Locally semi-simple algebras: Combinatorial theory and K_0-functor, *J. Sov. Math.* **38** (1987), 1701–1734.

[VKo] A.M. Vershik and K.P. Kokhas, Computation of the Grothendieck group of the algebra $\mathbb{C}[\mathrm{PSL}(2,k)]$ where k is a countable algebraically closed field, *Algebra i analyz* (Sankt-Peterburg) **2** (1990), 98–106; English translation: *Leningrad Math. J.* **4** (1991), 1251–1259.

[Wa] A. Wagner, On the classification of the classical groups, *Math. Zeitschr.* **97** (1967), 66–76.

[YZ] I.A. Yanson and D.V. Zhdanovich, The set of the direct limits of Lie algebras of the type A, *Comm. Alg.* **24**(3) (1996), 1125–1156.

[Za1] A.E. Zalesskii, Group rings of inductive limits of alternating groups, *Leningrad Math. J.* **2** (1991), 1287–1903.

[Za] A.E. Zalesskii, Group rings of simple locally finite groups, In: "Finite and locally finite groups", Kluwer Academic Publ., Dordrecht, 1995, pp. 219–246.

[Zh] A.G. Zhilinski, Coherent systems of representations of inductive families of simple complex Lie algebras, *Doklady Acad. Sci. Belarus* **36** (1992), 9–13. (Russian)

A. E. Zalesskii
School of Mathematics
University of East Anglia
Norwich NR4 7TJ
UK
and
Institute of Mathematics
Academy of Sciences of Belarus
Minsk, 220072
Belarus